Chemical Engineering Handbook

Volume I

Chemical Engineering Handbook
Volume I

Edited by **Serena Gibson**

NY RESEARCH PRESS
New York

Published by NY Research Press,
23 West, 55th Street, Suite 816,
New York, NY 10019, USA
www.nyresearchpress.com

Chemical Engineering Handbook: Volume I
Edited by Serena Gibson

International Standard Book Number: 978-1-63238-074-6 (Hardback)

Printed in China.

Contents

Preface

Chemical engineering involves processing chemicals, in order to improve just about anything and everything that you see around you. It deals with chemicals and the processes that create or convert them. This discipline requires comprehensive knowledge of Mathematics, Physics, and Economics, in order to solve technical problems. In fact, experts of this branch are sometimes referred to as 'universal engineers' because of their rich scientific and technical mastery.

Chemicals are involved in almost all production processes, which is why this branch of engineering is an extremely significant and relevant one in present times. Chemical engineering can be broadly classified into Chemical Processing and Chemical Producing.

The fundamental principle of Chemical engineering is to convert raw-materials into more valuable forms and facilitating the process by using chemicals. The applications of Chemical engineering can be found in a lot of areas. Chemical engineers improve food processing techniques, thereby optimizing the quantity and quality of food. Its principles are used to construct models for reactor analysis. In industries, chemicals are used to ensure optimum production. From mass production of drugs to refining petroleum products, making synthetic fibers to pollution control, all these fields utilize principles of Chemical Engineering in one way or the other.

I would like to thank all the contributors of this book for sharing their knowledge with me and maintaining the time parameters. I would also like to thank my family for their support and trust in me.

Editor

Advancement of Hydro-Desulfurization Catalyst and Discussion of Its Application in Coal Tar

Zongkuan Liu, Lei Zhang*, Jian Jiang, Cheng Bian, Zichao Zhang, Zifeng Gao
Chemical Engineering Department, School of Chemical Engineering, Xi'an Jiaotong University, Xi'an, China

ABSTRACT

This paper describes the influences of active metal, promoter and chelating agent on the properties of hydro-desulfurization catalyst. The use of chelating agent, especially its combination with common promoters e.g., EDTA-P, has an important meaning to develop highly active catalyst, specifically to unify the active metal dispersion degree and sulfurization degree in some extent, however, they are contradictory in conventional cognition. In the aspect of carriers, composition and nanometer carriers have more excellent performances in acidity, pores structure and metal-carrier interaction than common carriers, and are the developing trend in the future and should be a breakthrough mainly in preparation methods. We also pointed out the decisive factors to improve the activity of the catalyst: higher sulfurization degree of active metal oxide and higher aspect ratios of active phase crystal morphology, and the proper acidity and pores structure can be considered the key factors for deep desulfurization whose mainly obstacle is the desulfurization of large rigid molecules, e.g., dibenzothiophene and 4, 6-dimethyl substituted dibenzothiophene. Based on above that, We discussed the suitable hydrodesulfurization (HDS) catalyst for coal tar, aiming at providing some theoretical guidance for the "design" of coal tar HDS catalyst.

Keywords: Carrier; Active Metal; Promoter and Chelating Agent; HDS Catalyst; Metal-Carrier Interaction; Coal Tar

1. Introduction

China's current energy situation is rich in coal, lack of natural gas and less of oil. In recent years, with the development of the national economy, the contradiction between supply and demand of oil and gas resources is more and more prominent. Using coal tar to prepare fuel oil will alleviate the energy crisis in a certain extent. Coal tar is the main byproduct of coal gasification and coal coking process, its output was about 15 million tons with the annual growth rate about 13.68% in 2009 in China [1, 2].

Hydrotreating is the main technology to prepare fuel oil from coal tar, and hydrodesulfurization (HDS) is an important link in the process of which core is HDS catalyst. Traditional HDS catalyst does increasingly not meet the requirements and the following two reasons could account for it: on one hand, coal tar is heavy and the unsaturated compounds which are abundant in coal tar have evil influences (resinification reaction reacts between unsaturated compounds and chemical reagent, and thus creating much precipitation and slagging) on coal tar hydro-refining [2]; on the other hand, the environmental protection laws are more and more strict with sulfur content of fuel oil. Pawelec *et al.* [3] pointed out that the activity of the new catalyst should be 3.2 times the conventional catalyst when requiring diesel sulfur content reduces from 500 ppm to 15 ppm with not changing the technological parameters. It is press for designing and developing new HDS catalyst. HDS catalyst includes loaded type and un-loaded type, this paper mainly studied on the former.

The catalyst design process is very complex. In the past, preparation of catalyst depends on the rich "formula" experiences; in recent years, along with the development of surface physics, surface chemistry and organic catalytic mechanism, as well as the use of precision instrument in catalyst characterization, making catalyst "design" be possible. The components that can be distinguished in the catalyst include active metal, promoter or chelating agent and carrier.

Active metal is the main source of hydrogenation activity, it should be chosen according to the properties of raw oils or model molecules and product quality standards [4-12]. Promoter is an important part of the catalyst and a lot of confidential materials are concentrated in it. It is essential to optimize its kinds and content in the process of catalyst preparation [13-22]. The available carriers in HDS catalyst are as the following types ac-

*Corresponding author.

cording to the chemical compositions: silicon-aluminum molecular sieves [23,24], metallic oxides [25,26], mesoporous silica molecular sieves [10,11,27,28] and composites made up by these materials [7,8,23,26]. The most significant characteristics of carriers are its ability to disperse metal and possession of B, L type acids. These two kinds of acids play different roles in hydrogenation reaction, but synergy exists between them [29].

The three parts of catalyst are interrelation and interaction. Hence, a comprehensive consideration should be given in the process of catalyst preparation. In the actual process of catalyst "design", "Attending to one thing and losing sight of another" is a frequently encountered problem. For one instance, weak metal-carrier interaction is beneficial to the improvement of sulfurization degree of the active metal oxide, but the dispersion degree of the active phase may be reduced. For another instance, although large pores can weaken the diffusion resistance and reduce secondary cracking of the reactant molecules, that would make the initial activity of the catalyst and the metal capacity of carrier reduced. Consequently, the problem of the catalyst "design" is actually an "optimization" problem. The objective function should be the maximization of enterprise's interests, the constraint conditions should be the nature of raw oils, quality standards of products and laws of environmental protection, and the optimization variables are main parameters of catalyst design: metal types, metal content and crystal morphology of active phase, promoter or chelating agent types and content, and carrier's surface properties. The main purpose of this paper is to provide certain theoretical guidance for the preparation of coal tar HDS catalyst on the basis of system analysis and introduction of the main parameters of the HDS catalyst.

2. Active Components

2.1. Conventional Active Components

In general, conventional hydrogenation active sites include the simple substance of noble metal and non-noble metal sulfide. But the use of noble metal is limited to some extent because of its expensive price. So, keep the activity of noble metals catalyst not decline while making the consumption of noble metals reduced is the goal that researchers always pursue. In this respect, Nakamura, *et al.* [30] proposed a nanometer-catalyst technology to reduce the dosage of noble metals. Its specific process was schematically shown in **Figure 1**. Its basic concept is that the catalyst is made up of the substrate to restrain the sintering of noble metal and the separation material to inhibit the agglutination of the substrate. The structure can greatly reduce the dosage of noble metal because it can keep noble metal be nanometer particle. To make an appropriate bonding force between metal and carrier is its key point, because it is not conducive to improving the catalytic activity whether the bonding force is too strong or too weak, and thus selecting a proper carrier is important. Weak resistance capability to sulfur is the other reason resulting in its limited application. Some studies [8,9] showed that the resistance capability to sulfur can be enhanced when catalyst possesses smaller metal particles, proper acidity (mainly Lewis acid sites with high electronegativity) and formation of alloy. This may relate to the inhibition of generation of low activity Pd_4S in these circumstances.

Co-Mo-S, Ni-Mo-S and Mo-Ni-W-S are commonly used active phases of non-noble metal HDS catalysts. In general, their hydro-refining activity increases as the sequence [4,5,31,32]. This may because different promoter metals (Co, Ni) have different modulation ability to S-metal bonding energy. Literature [3] confirmed this view and pointed out that the change process of the activity of Co(Ni)Mo(W)S is a volcano curve with the enhancement of S-metal bonding energy. Therefore, moderate S-metal bonding strength is one of the crucial factors to improve the catalytic activity.

2.2. Non-Noble Metal Phosphide and Noble Metal Phosphide

Metal phosphides are commonly obtained through the reduction of their corresponding oxides or chlorides by temperature-programmed method in H_2 atmosphere [33, 34]. The reduction temperature affects the particle size of active phase and hence has significant implications to

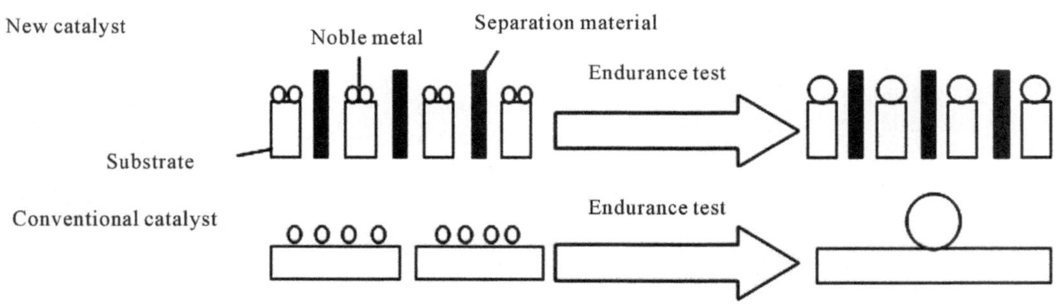

Figure 1. The new catalyst design concept about restraining sintering of the noble metal.

HDS activity; therefore, selecting an appropriate reduction atmosphere to lower the reduction temperature is momentous.

Non-noble metal phosphides have highly HDS activity which is higher than normal Co(Ni)Mo(W)S. Particularly, the activity of crystalline Ni_2P is the highest among them because of its higher intrinsic activity and dispersion degree compared to the other phosphides [10]; Another advantage of Ni_2P is its HDS activity will not be inhibited with the competition of HDN (hydrodenitrification) [4,6,10]. Inversely, the HDS activity of Co(Ni)Mo(W)S would be restricted due to the competition of nitrogen compounds in H and the edges of active phase [35]. The disadvantage of Ni_2P is its very low dispersion degree, only 1/3 times the active phases of industrial hydrotreating catalyst. Consequently, to improve its dispersion degree is the key to its practical application. JI research group [13-16] have studied the influences of W, Mo, Li, Na, K, Mg, Ca, Sr, Ba and B on the dispersion degree of Ni_2P based on SBA-15, manifesting that W, Mo, Ca and B promoted its dispersion degree, and thus the HDS activity of the catalyst was enhanced, but that property is closely related to the promoters content (as the discussion in 5.1.3 in this paper); In addition, Ni/P molar ratio is also a major parameter in process of the catalyst preparation because of its affection on the types of active phases. If the ratio is too high, there will be more formation of low active $Ni_{12}P_5$. In general, this ratio is between 1.2 - 1.4 [6,13,14,36], and is lower than the ideal ratio 2.

Recently, it is worth mentioning that Kanda *et al.* [33] synthesized a series of noble metal phosphides: Rh-P, Pd-P, Ru-P, Pt-P. They made the corresponding noble metal chloride precursors primarily decomposed in N_2 atmosphere, and then reduced by temperature-programmed method in H_2 atmosphere. Their HDS activity for thiophene follows the order: Rh-P > Pd-P > Ru-P > Pt-P, activity and stability of Rh-P are the best and are better than Ni-P.

2.3. Non-Noble Metal Carbides and Nitrides

Chemical bond in carbides and nitrides has part of the properties of metal bond. Their preparation methods mainly differ in the reaction atmosphere and temperature [12]. Nitrides are commonly generated in NH_3 and N_2 atmosphere, the former's advantage over the latter's is to inhibit the formation of impurities and lower the reduction temperature. The formation process of carbides is the oxide precursors reduced primarily by temperature-programmed method in CH_4-H_2 atmosphere, and then utilize N_2 containing low quantity of O_2 of which volume fraction is less than 1% to passivate the carbides. One of the remarkable properties of metal carbides is to embellish the acidity of silica-aluminum molecular sieves [20, 21].

2.4. Others

Lan *et al.* [4] made Ni_2P and Mo-Ni-W combined catalyst Ni_2P-NiMoW/γ-Al_2O_3 which showed more excellent performances than conventional catalysts, that is, it can make diesel oil achieve deep desulfurization, simultaneously, the increase of cetane number.

ReS_2 also can be used for hydrogenation active site of HDS catalyst. Its HDS activity is positive correlation to the carrier's ratio of silica to aluminium, *i.e.*, the acidity of the catalyst plays an inhibitory role for HDS activity [7], but oppositely for traditional HDS catalysts [28,37].

3. The Content of Active Components

Metal content including total metals content and the proportion of different metals in one catalyst is an important parameter in catalyst design process. Metal content usually affects the dispersive state of active metal, and the dispersive state would also impact the acids distribution [38], metal-support interaction [39] and the number of hydrogenation active sites etc. Overall, a highly active catalyst should be higher total metals content and appropriate metal dispersion degree, *i.e.*, in the case of the excellent dispersion state of active metals, the higher total metals content is conducive to improving the catalytic activity.

3.1. Influences on Acids

In normal circumstances, the appropriate acids distribution and moderate acidity strength are beneficial to HDS activity of catalyst. That is because, on one hand, B, L acids are in favor of the adsorption of reactant molecules [28] and hydrogen overflow, and thus are useful to cleave C-S [37,40]; on the other hand, the acids can make the alkyl-substituted dibenzothiophene (DBTs) isomerize to molecules owning weaker steric hindrance for -S-.

There are three regulation mechanisms of metal active on acid amount and acids type: to generate new acid, ion exchange and covering effect. Acidity is changed as the change of metal content. When the active metal content increases within a certain range, the amount of L acid (weak acid) is increased but inversely for B acid [4,41]. The increase of L acid amount is due to the empty orbits in active metals which can generate Lewis acid that is different from the Lewis acid site centering in Al_3^+; the reduction amount of B acid is because of ion exchange between metal cation and proton, along with covering effect of metal components on part of Si-OH-Al (strong B acid center) and OH-Si (weak B acid site); Study [28] on the acids distribution of NiW/Al-SBA-15 and Al-

SBA-15 proved this point, the former's L, B acid amount are 1.6, 0.5 times the latter, respectively. Study in literature [38] also supported this view. But literature [42] showed that the active metal also made B acid amount increase, this may relate to the conversion of tetra-coordinate metal species on the support surface [43].

3.2. The Consideration for Metals Content Ratio

There is synergistic effect in multi-components active metal catalyst, and the effect can make catalyst be better performance. e.g., HDS activity of PdAu/HS-HMS for thiophene was three times Pd/HS-HMS due to the synergistic effect between Pd and Au [9], and the deactivation rate of the former was also significantly lower than the latter. The main reason is that the former has the formation of the alloy made up of Pd and Au, and thus inhibits formation of PdS_4. The synergy closely relates to the metals content ratio. Consequently, both the total content and ratios of the metals content should be taken into account in the process of catalyst design. Literatures [44, 45] discussed the impact of molar ratios of Pt/(Pt + Pd) and Ni/(Ni + W) on the catalytic activity, demonstrating that the former or the latter possessed the highest hydrogenation activity for naphthalene or HDS activity for thiophene when the ratios are 0.7 or 0.28, respectively. Study on Mo-Ni_2P/SBA-15/cordierite monolithic catalyst [16] indicated that its HDS activity for DBT was optimal when the molar ratio of Mo/Ni + Mo was 0.18, however, when the ratio was 0.26 in NiMo/Al_2O_3-nmY, the optimal desulfurization degree of diesel was up to 99.6% [23]. Through the latter two examples, we can find that the ratios of Mo/Ni + Mo are different, that may because the active phases are different in the two catalysts. So, the ratio is not invariable, and should be optimized in accordance with the active phase、carrier and preparation method.

4. The Active Phase Crystal Morphology

The active phase crystal morphology has vital influence on the activity of the catalyst. It is closely relevant to the types and amount of the active phases, as well as the number of the corner and edge active sites and lattice defects. In addition, it can also determine whether the hydrogenation active sites can be well exposed to the reactant molecules or not. The main parameters used to describe the active phase crystal morphology include stacking degree, length and curvature of the lamellae.

Stacking degree has major impact on the types of active phase and the amount of corner and edge active sites. Generally speaking, multilayer stacking shows better performance than the single-layer. This is mainly manifested in two aspects: First, multilayer stacking can generate more type II active sites than the single-layer due to its weaker interaction with carrier except the basal layer [46]. The top layer possesses the highest activity because of the weakest steric hindrance and higher unsaturation degree of the brim and corners active sites [8], in addition, the top layer and middle layers also differ in the desulfurization mechanism, the brim of the top layer owns capability of DDS (direct desulfurization) and HY D (hydrogenation desulfurization) routes while the middle layers only have DDS activity [47]. Second, multilayer stacking can produce more lattice defects compared to monolayer stacking and thus is beneficial to chemisorption [26,28]. However, the number of stacking layers is not the more the better; too many stacking layers will lead to a ratio decrease of vertical adsorption, as well as the corners and edges active sites [8,39,47]. Therefore, it is major to optimize stacking layers in order to balance the amount of corners and edges active sites and II type active sites.

However, the multi-layer stacking is not the necessary condition to produce type II active sites. Sometimes, monolayer stacking may be the II active sites. Parola *et al.* [48] pointed out the prerequisite of generating type II active sites is weaker metal-carrier interaction. Moreover, in case of inherently weak metal-support interaction, the increase of active phase stacking layers, sometimes, leads to a detriment effect on its HDS activity [28], Study [8] about NiMo-NTA/Al_2O_3 (NTA = nitrilo triacetic acid that can significantly weaken the metal-carrier interaction) showed that its intrinsic HDS activity for DBT significantly reduced with the increase of active phase stacking layers because of activity of the top layer higher than others, and more stacking layers may mean lower proportion of the top layer [26]. Nevertheless, when the stacking layers are few, it is not conducive to the plane adsorption of reactant of catalyst [25,28]. Therefore, even in the case of weak intrinsic metal-support interaction, and hence generating a large number of type II active sites, an optimal point of the stacking layers should also exists, rather than the fewer the better.

In short, stacking layers of the active phase has important influences on HDS activity. However, the relation between active phase crystal morphology and HDS activity can't be determined only depending on the stacking degree of the crystal morphology. Shimada *et al.* [46] summarized the relation, indicating that the higher aspect ratios (approximately equal to the thickness divided by length) of the crystal morphology, the higher intrinsic activity of the catalyst. Many studies [5,23,26,28,32,49] have supported the view. That may because this kind of crystal morphology can maximize the amount of corner and edge active sites, and weaken steric hindrance of the active sites so that it could be well exposed to the reactant molecules, as well as generate more type II active sites. Finally, we can also safely get such a conclusion according to those literatures: moderate metal-support interaction is the decisive factor for achieving of the highest aspect ratios of crystal morphology.

The bending of the active phase lamellae also has an important impact on HDS activity and HDS mechanism [47]. This is because, on one hand, the curved lamellae would make the edge of metal layer expose more active sites, on the other hand, it make the S-Mo (W) bond tight ,and thereby increasing the unsaturation degree of the active metal and hence creating new active sites.

5. Promoters and Chelating Agents

5.1. Promoters

The modulation capability of promoters for catalyst performances is not only relevant to the nature of the promoter itself and quantity of the promoter, but also connected with the impregnation order of active metal and the promoter [18,19,22,50,51]

5.1.1. Influences on Metal-Support Interaction

Metal-support interaction is the decisive factor to the sulfurization degree and crystal morphology of active metal. Therefore, utilizing promoter to modulate metal-support interaction is important. In general, the modulation of the metal-support interaction actually refers to "weaken" this interaction.

Metal-support interaction is modulated by promoter mainly in three ways: First, it can be weakened by modulating the coordination state of active metal. Different coordination states of the active metal bring about different intensity of metal-support interaction, more specifically, tetrahedron coordinated active metal species commonly have stronger binding force with support than octahedron species [26,28]. P modulates the metal-carrier interaction is just through that mechanism [19]. But the improvement degree of metal-support interaction is mainly associated with the impregnation sequence of P [18], to be specific, co-impregnation improvement capacity is higher than the surface impregnation; second, metal-support interaction is modulated through synergy effect between the active metal and promoter, the promoters mainly should be transition metals, e.g., Fe [50], Co [51]. To these promoters, The content must be suitable because excessive content of promoter may inversely strengthen the metal-support interaction due to the improvement of the dispersion degree of active phase in this case [50]; third, promoters could achieve the modulation ability through "isolation" of the active metal and support in physical space. For B modified catalyst B-MoNi/Al_2O_3 and MoNi-B/Al_2O_3 [22], the former which impregnated B and MoNi in sequence has better modulation capability than the latter. This is because B (w0.6%) could be monolayer distribution on the carrier surface in B-MoNi/Al_2O_3, which "separates" active metal and carrier in some extent and thereby weaken the metal-carrier interaction.

5.1.2. Influences on the Acids Distribution and Acid Amount

The major impacts of promoter on the acid amount and acids distribution are fulfilled through the following five kinds of mechanisms: first, it is modulated by ion exchange, mainly some metal promoters; second, the modulation is achieved in the way of promoting the dispersion degree of active metal; third, "covering" effect; fourth, the acidity also can be modulated by some promoters that are inherently acidic, such as transition metal ions; fifth, the modulation can be obtained by substitution reaction between skeleton aluminum and promoter. Sometimes, one promoter may have, simultaneously, several roles to modulate, therefore, to optimize its content and impregnation orders to get desired acidity is important.

Mg, K, P and W all own the ability to modulate the acidity of catalyst. Mg can promote the dispersion degree of metal component and hence make the metal surface present more electron-deficient sites; as a result, the number of L acid is increased. e.g., the density of L acid of (w) 2% Mg-(w) 7% Mo_2C/HY is nearly doubled compared to (w) 7% Mo_2C/HY [20]. K usually makes the amount of B acid significantly reduced due to proton exchange between hydroxyl hydrogen on carrier surface and K [20,21]. However, study [17] on CoMo/γ-Al_2O_3 showed K mainly made L acid amount reduced. Study [18] about P-NiW/Y-SiO_2-TiO_2 showed that both surface impregnation and co-impregnation methods resulted in significantly reduction of the total amount of acid, but in the case of co-impregnation, although P covered part of acid sites on the carrier's surface, simultaneously, the dispersion degree of active metals was improved, so that the amount of L acid was increased. The reason for the increase of L acid amount is the same to Mg. In addition, P may substitute for the carrier's framework aluminum, more specifically, a framework aluminum hydroxyl of the zeolite would be substituted by two phosphorus hydroxyl, making B, L acid amounts modulated [24]. Transition metal ions such as W have empty orbits; their addition can make a well increase of the amount of acid [15, 45]

5.1.3. Influences on the Pores Distribution and Specific Surface Area

The realization of impact of promoters on pores distribution and specific surface area is mainly based on its influences on metal dispersion state.

Generally, with the increase of promoter content, the specific surface area and pore volume increase primarily and then decrease, *i.e.*, there is an optimum point. For B-MoNi/Al_2O_3 [22], W-Ni_2P/SBA-15 [15], Mo-Ni_2P/SBA-15/cordierite [16], when B (w) of 0.8%, W (w) 3%, Mo (w) 4.2%, their specific surface area and pore volume are maximal, respectively. Such effect of B, W and Mo can be attributed to the impact of promoter content on the dispersion state of active metal. When the promoter content is low, the dispersion degree of the active metal is improved and thus the surface properties are ameliorated; on the contrary, if the content is too high, the active phase (or precursor) may aggregate together and block portion of the pores, resulting in the decrease of surface area and pore volume. But there are exceptions, e.g., Mg-Mo_2C/HY [20,21], the pore volume and specific surface area primarily decrease and then increase with the increasing content of Mg.

5.2. Chelating Agents

Chelating agents are typically organic compounds with donor atom having two or more available electrons. They are capable of binding metal ions to form chelates, such as NTA [8,52], EDTA [53], and CyDTA etc. Their common feature is that they can delay the sulfurization of promoter metal (Co) until the active metals (Mo, W) are fully sulfurized, avoiding the production of Co_9S_8 phase that is thermodynamically unstable, which is conducive to the formation of the active phase of Co-W(Mo)-S. Another remarkably notable feature of the chelating agents is that they can dramatically weaken the metal-support interaction, but scarcely any impact on the dispersion degree of active phase, or even improve the dispersion degree of active phase [49,52]. The nature of chelating agents breaks through the traditional cognition which the improvement of metal dispersion can be achieved only by enhancing the metal-support interaction, and makes sulfurization and dispersion degree of the active phase unified in a certain extent, providing more ways for people to prepare highly active catalyst. For instance, Yu *et al.* [49] prepared NiW/Al_2O_3, NiW-P/Al_2O_3, NiWP/EDTA/Al_2O_3 three catalysts, their HDN activity for quinoline were improved as the order. The addition of EDTA in NiWP/EDTA/Al_2O_3 made the average number of the stacking layers and the length of active phase crystal morphology reduced 0.3 layers and 1.2 nm, respectively, with respect to NiWP/Al_2O_3, making the aspect ratios of NiWP/EDTA/Al_2O_3 improved; moreover, the sulfurization degree of active metal is also pretty increased because of weak metal-carrier interacttion due to the addition of EDTA. If NiW/Al_2O_3 is modified directly by EDTA, it may cause too high dispersion degree of the active phase, which is not conducive to the molecules adsorption and is harmful to the increase of corner and edge active sites. Thus, P actually played a role like bridge in this experiment.

6. Carriers

In the preparation process of carrier, pore structure should be primarily design and topology analysis in order to balance the specific surface area and diffusion resistance (mainly determined by pores diameter) [11,40,54, 55]. Generally speaking, the relation between pore diameter and specific surface area is mutually restrictive. But, the principal restricted condition for HDS activity is different for different reactants. In terms of macromolecules, the relatively larger pore diameter is more important than larger specific surface area [55,56]. Secondly, choosing a suitable carrier having moderate binding force with active metal is crucial. When the binding force is too strong, It will lead to the formation of the Mo(W)-O-Al and thus the lower type I active site during sulfurization, whereas the formation of the type II active site requires Mo(W)-O-Al bond is broken (at least partially broken). Oppositely, when the interaction is too weak, there is no doubt that it is not conducive to improving the dispersion degree of the active site and hence does harm to the HDS activity [26].

The most commonly used carrier of HDS catalyst is γ-Al_2O_3 in industry. But it chiefly has two disadvantages, one is the dispersed pores distribution and short pores diameter concentrating in shorter than 5 nm resulting in large diffusion limit for reactant and intermediate products [11]; the other is that its strong interaction with active metal oxide. These shortcomings make it increasingly can't meet the requirements of desulfurization. The typically mesoporous silica molecular sieve materials SBA-15, HMS and MCM-41 as the carrier of hydrofining catalyst have potential value in use. Their advantages are large specific surface area (about 1000 m^2/g), uniform controlled mesoporous (5 - 30 nm) and stable skeleton structure. Their disadvantages are weaker acidity which should be strengthen on the original basis in order to isomerize DBTs to molecules owning weaker steric hindrance for -S- [23,37], and weaker hydrothermal stability and mechanical strength, as well as too weak interaction with active metal. Those several factors limit its industrial application so that the modification study should be carried on which is commonly based on their surface Si-OH.

At present, the commonly used modification method of those disadvantages is to introduce Al [28,37,55], Ti [54], Zr [56], W [45], etc. into the carrier's frame. The methods to add heteroatom are usually co-impregnation (grafted directly) and post-grafting. the former's weaknesses are that the difficulty to introduce heteroatom in the case of low PH (lower than 2) [28], and to obtain larger and regular pores of catalyst [55,56]; its merit is it makes the catalyst possess higher specific surface area than post-grafting; The advantages of the latter are pore expanding and isomorphous substitution [55]; its shortcoming is pore plugging and thus making the specific surface area significantly lower than the catalyst obtained by co-impregnation method. The principle of post-grafting is firstly hydrolyzing the aluminum source (e.g. ammonium hexafluoroaluminate) to be $[Al(OH)_4]^-$, and then isomorphously substituting. Inspired by the princeple of post-grafting, Li *et al.* [28] developed a new method named high-temperature hydrothermal treatment (HTHT) to solve the pore blockage problem of post-grafting. $[Al(OH)_4]^-$ can be easily formed by Al^{3+} and Al-OH in condition of high-temperature hydrothermal; the other key of the method is that PH should be isoelectric point nearby to generate lots of Si-OH, and finally through the following reaction to complete graft.

$$\text{Si-OH} + \text{HO-Al} \rightarrow \text{Si-O-Al} + H_2O$$

Besides the preparation method, a crucial parameter need to be optimized is the content of heteroatom, which is closely related to the dispersion degree of active phase and acids distribution of catalyst etc. e.g., the HDS performance of NiMo/TiSBA-15 on coke light gas oil (KLGO) was the best when Si/Ti mole ratio was 20. Research of NiMo/Zr-SBA-15 (synthesized directly) also had similar result [56].

Along with more and more heavy oil and increasingly strict laws of environmental protection, single carrier based catalyst already cannot meet the demand. Composite carrier commonly shows more excellent HDS activity than single carrier, such as Al_2O_3-ZrO_2 [57], Al_2O_3-SiO_2 [40], TiO_2-SiO_2 [31], Beta-SBA-15 [37], TiO_2-SBA-15 [25] and ZrO_2-SBA-15 [25,26,32,56]. In those carriers preparation process, the most important design parameter is the proportion of carrier contents in one composite carrier, because the proportion has an important relation with pores diameter, specific surface area, acids distribution and intensity, as well as metal-carrier interaction [31, 40]. e.g., ZrO_2-SBA-15, the optimal quality ratio of ZrO_2/SBA-15 is generally 1:3 [25,26,32]; as to SiO_2-Al_2O_3, when SiO_2/Al_2O_3 quality ratio was 1:10, the HDS activity and inactivation rate can reach a satisfactory balance [40].

It is greatly worth to mention that carrier on nanometer level has dramatically better performance on acidity, pore structure and carrier-metal interaction than carrier on micrometer level [23,58]. e.g., HDS degree of NiMo/Al_2O_3-nmY was increased by 0.8% compared to NiMo/Al_2O_3-umY, and the rate constant of the former is 3.46 times the latter [23]. In addition, the carriers on nanometer level can protect molecular sieve from poisoning and shorten diffusion path for raw material [58]. In total, its advantages are highly obvious. But, its preparation faces many problems, such as the need of expensive structure-directing-agent, and demanding strict reaction temperature in the synthesis process etc. In addition, the cordierite honeycomb ceramics monolithic catalyst [16] has the advantages of low bed pressure drop and high mass transfer efficiency, etc. over traditional hydrogenation catalysts, and hence also attracts researchers' attention From the above analysis, we can see that carrier's development trend is multicomponent and smaller particle size.

7. Discussion on Desulfurization Catalyst That Is Suitable for Low Temperature Coal Tar

Coal tars can be divided into three categories according to the coal pyrolysis temperature: high temperature coal tar, medium temperature coal tar and low temperature coal tar. The low temperature coal tar is the most suitable one to produce fuel oil because of its lower aromatics and asphalt content and higher alkane content. Typically, sulfur compounds in coal tar include thiol, sulfoether, disulfides, thiophene series and dibenzothiophene series. The commonly grading sequence of coal tar hydrotreating catalysts is: protection catalyst, demetallization catalyst, HDS catalyst, HDN catalyst, mild hydrocracking catalyst, which calls for hydrofining catalysts have some cracking ability. A pilot research [59] showed that catalysts used in heavy oil hydrotreating process are feasibly used for hydrotreating of low temperature coal tar. However, coal tar has its own characteristics; the following will theoretically discuss the suitable HDS catalyst for low temperature coal tar in the case of this kind of grading sequence.

First of all, its acid density and strength should be controlled, more specifically, the acidity should be stronger than demetallization catalyst and weaker than hydrodenitrification catalyst. That is because, on one hand, nitrogen content (0.48% - 1.13%) is high in coal tar [2], and the nitrogen with alkaline will poison the catalyst and inhibit HDS because of the competition effect of HDN; on the other hand, the moderate acidity can make HC and HDS well occur, moreover, can inhibit the formation of coke. Secondly, coal tar is a complex mixture, the sizes of sulfur compound molecules in coal tar are greatly different, that means these molecules need different degree of cracking and hydrogenation; moreover, aromatics content is very high in coal tar, and mainly tricyclic or above anthracene series. Those kinds of characteristics of coal tar require catalyst possesses different pores distribution and active sites distribution, namely, the activity in larger pores should be higher than in smaller pores in theory, which will match to the different reaction degree that is required by mixed molecular. That needs to explore more elaborate catalyst preparation methods. Thirdly, HDS catalyst is relatively near the front of grading sequence; hence, material flowing through HDS catalyst is very heavy, the catalyst should have powerful capacity of treating aromatics including aromatics adsorption and diffusion, as well as more effective hydrogenation and ring opening ability. This requires catalyst possesses more lattice defects, secondary pores and excellent synergy between hydrogenation and cracking sites, In addition, appropriate number of large pores (100 - 500 nm) is also important. This primarily need to develop a new metal loading concept to increase the quantity of lattice defects, and amount of secondary pores can be added through hydrothermal treating and synthesis of composite molecular sieve. Finally, coal tar contains much asphalt whose structure is very complex, the polycyclic aromatic hydrocarbons in asphalt are bonded by the sulfur bridge bond, aliphatic bond and metal porphyrin bond. The structures of nickel and vanadium porphyrin compounds are similar to polycyclic aromatic hydrocarbons in asphalt, leading to abundant poisonous metals exists in it, moreover, part of the asphalt would generate carbon deposit by condensation polymerization and graphitization reaction, these carbon deposit can jam the channel and cover the active sites of catalyst and thereby making HDS catalyst inactivation. To solve this problem, the hydrogenation and cracking sites should have well synergetic effect. For the problem of catalyst poisoning, it can be eased by adding some nanometer oxides in support [58].

8. Conclusions and Prospect

1) Transition phosphide, carbide and nitride catalyst possess potential practical value for oil hydrotreating. But there are also some drawbacks must be further improved. The main drawback of the phosphide is the lower dispersion degree; the use rate of unit mass catalyst related to the phosphide sites is low. Transition metal carbides and nitrides catalyst own higher initial activity for hydrorefining, but in the using process, the active phase will be gradually sulfurized, making the catalytic activity decline. To further overcome these shortcomings, it will make them become a new generation catalyst for hydrotreating. In addition, the combination of non-sul-fide metal catalyst with the sulfide catalysts may also be a research direction.

2) Promoter or chelating agent can improve the active phase dispersion degree, acids distribution, pores distribution and metal-carrier interaction. Metal-carrier interaction is closely relevant to the crystal morphology of active phase. The using of promoter-chelating has important significance to develop novel catalyst preparation methods. We need to further develop new promoter (chelating agent), on one hand, to expand the catalyst preparation methods and to improve its thermal stability, mechanical properties and activity, on the other hand, to weaken the contradictions related to the performances of catalyst.

3) Composite and nanometer carriers can provide more appropriate acidity, pore structure and metal-carrier interaction than conventional carriers, but its thermal stability and mechanical properties are need to be further improved. To develop new preparation method is the key for the use of the composite and nanometer carriers.

The catalyst "design" problem is an "optimization" problem, and a system analysis should be given in process of catalyst preparation. The crucial factors to improve the activity of HDS catalyst are higher sulfurization degree and higher aspect ratios of the active phase crystal morphology. In addition, moderate acidity and proper pores structure could be considered the key factors of deep desulfurization.

9. Acknowledgements

We are grateful to many of our colleagues for stimulating discussions on desulfurization and hydrogenation. We would like to acknowledge the financial supports from Science and Technology Innovation Project of Shanxi Province in China (2011KTZB03-03-01) and Major Scientific and Technological Innovation Project of Shaanxi in China (2008zkc03205).

REFERENCES

[1] Y. Zhang and L. F. Zhao, "Study on Hydro Catalysis of Middle/Low Temperature Coal Tar to Clean Fuel," *Coal Conversion*, Vol. 32, No. 3, 2009, pp. 49-51.

[2] B. Q. Ma, P. J. Ren, Z. B. Yang and S. K. Wang, "Preparation of Fuel Oil from Coal Tar," Chemical Industry Press, Beijing, 2011.

[3] B. Pawelec, R. M. Navarro, J. M. Campos-Martin and J. L. G. Fierro, "Towards near Zero-Sulfur Liquid Fuels: A Perspective Review," *Catalysis Science & Technology*, No. 1, 2011, pp. 23-42.

[4] L. Lan, S. H. Ge, K. H. Liu, Y. D. Hong and X. J. Bao, "Synthesis of Ni_2P Promoted Trimetallic Ni MoW/γ-Al_2O_3 Catalysts for Diesel Oil Hydrotreatment," *Journal of Natural Gas Chemistry*, Vol. 20, No. 2, pp. 117-122.

[5] L. Zhang, X. Y. Long, D. D. Li and X. D. Gao, "Study on High-Performance Unsupported Ni-Mo-W Hydrotreating Catalyst," *Catalysis Communications*, Vol. 12, No. 11, 2011, pp. 927-931.

[6] Y. Zhao, M. W. Xue, M. H. Cao and J. Y. Shen, "A Highly Loaded and Dispersed Ni_2P/SiO_2 Catalyst for the Hydrotreating Reactions," *Applied Catalysis B: Environmental*, Vol. 104, No. 3-4, 2011, pp. 229-233.

[7] C. Sepulveda, V. Bellière, D. Laurenti, N. Escalona and R. García, "Supported Rhenium Sulfide Catalysts in Thio-

phene and 4,6-Dimethyldibenzo Thiophene Hydrodesulfurization: Effect of Acidity of the Support over Activities," *Applied Catalysis A*: *General*, Vol. 393, No. 1-2, 2011, pp. 288-293.

[8] A. E. Coumans, D. G. Poduval, J. A. Rob van Veen and E. J. M. Hensen, "The Nature of the Sulfur Tolerance of aAmorphous Silica-Alumina Supported NiMo(W) Sulfide and Pt Hydrogenation Catalysts," *Applied Catalysis A*: *General*, Vol. 411-412, 2012, pp. 51-59.

[9] V. L. Parola, M. L. Testa and A. M. Venezia, "Pd and PdAu Catalysts Supported over 3-MPTES Grafted HMS Used in the HDS of Thiophene," *Applied Catalsis B*: *Environmental*, Vol. 119-120, 2012, pp. 248-255.

[10] A. Infantes-Molina, J. A. Cecilia, B. Pawelec, J. L. G. Fierro and E. Rodríguez-Castellón, "Ni_2P and CoP Catalysts Prepared from Phosphite-Type Precursors for HDS-HDN Competitive Reactions," *Applied Catalysis A*: *General*, Vol. 390, No. 1-2, 2010, pp. 253-263.

[11] P. E. Boahene, K. K. Soni, A. K. Dalai and J. Adjaye, "Application of Different Pore Diameter SBA-15 Supports for Heavy Gas Oil Hydrotreatment Using FeW Catalyst", *Applied Catalysis A*: *General*, Vol. 402, No. 1-2, 2011, pp. 31-40.

[12] S. Chouzier, T. Czeri, M. Roy-Auberger, C. Pichon and C. Geantet, "Decomposition of Molybdate-Hexamethylenetetramine Complex: One Single Source Route for Different Catalytic Materials," *Journal of Solid State Chemistry*, Vol. 184, No. 10, 2011, pp. 2668-2677.

[13] Q. Ma, P. H. Zeng, S. F. Ji, H. Liu and C. Y. Li, "Effect of Metal Promoters on the Structure and Performance of the Ni_2P/SBA-15 Catalyst for Hydrodesulfurization," *Acta Petrolei Sinica* (*Petroleum Processing Section*), Vol. 27, No. 2, 2011, pp. 175-180.

[14] P. F. Zhao, S. F. Ji, N. Wei, Q. Ma and H. Liu, "Effect of Boron Promoter on the Structure and Hydrodesulfurization Activity of Ni2P/SBA-15 Catalysts," *Acta Physico-Chimica Sinica*, Vol. 27, No. 7, 2011, pp. 1737-1742.

[15] N. Wei, S. F. Ji, Y. M. Guan, H. Liu and C. Y. Li, "Influence of W on Structure and Hydrodesulfurization Performance of W-Ni_2P/SBA-15 Catalysts," *Acta Petrolei Sinica* (*Petroleum Processing Section*), Vol. 27, No. 6, 2011, pp. 852-858.

[16] Y. N. Guo, P. H. Zeng, S. F. Ji and N. Wei, "Effect of Mo Promoter Contenton Performance of Mo-Ni_2P/SBA-15/ Cordierite Monolithic Catalyst for Hydrodesulfurization," *Chinese Journal of Catalysis*, Vol. 31, No. 3, 2010, pp. 329-334.

[17] M. X. Qin, B. Yu, J. Yang and H. J. Li, "Effects of Metal Modification on Selectivity of Co-Mo/γ-Al_2O_3 Hydrodesulifcation Catalyst," *Industrial Catalysis*, Vol. 17, No. 5, 2009, pp. 45-49.

[18] Y. S. Zhou, Q. Wei, T. Zhang and X. J. Tao, "Effect of Phosphoric Modification on the Heavy Oil Hydrotreating Performance of NiW/CTS Catalysts," *Journal of Fuel Chemistry and Technology*, Vol. 39, No. 10, 2011, pp. 766-770.

[19] D. W. Hui, Q. H. Yang, S. L. Sun and C. F. Niu, "Effect of Phosphorus on the Performance and Active Component Structure of MoCoNi/Al_2O_3 Catalyst," *Petroleum Processing and Petrochemicals*, Vol. 42, No. 5, 2011, pp. 1-4.

[20] S. J. Ardakani and K. J. Smith, "A Comparative Study of Ring Opening of Naphthalene, Tetralin and Decalin over Mo2C/HY and Pd/HY Catalysts," *Applied Catalysis A*: *General*, Vol. 403, No. 1-2, 2011, pp. 36-47.

[21] X. B. Liu, S. J. Ardakani and K. J. Smith, "The Effect of Mg and K Addition to a Mo2C/HY Catalyst for the Hydrogenation and Ring Opening of Naphthalene," *Catalysis Communications*, Vol. 12, No. 6, 2011, pp. 454-458.

[22] S. K. Maity, M. Lemus and J. Ancheyta, "Effect of Preparation Methods and Content of Boron on Hydrotreating Catalytic Activity," *Energy & Fuels*, Vol. 25, No. 7, 2011, pp. 3100-3107.

[23] H. L. Yin, T. N. Zhou, Y. Q. Liu, Y. M. Chai and C. G. Liu, "NiMo/Al_2O_3 Catalyst Containing Nanosized Zeolite Y for Deep Hydride Sulfurization and Hydrodenitrogenation of Diesel," *Journal of Natural Gas Chemistry*, Vol. 20, No. 4, 2011, pp. 441-448.

[24] Q. Wei, Y. S. Zhou, X. J. Tao, T. Zhang and Y. D. Liu, "Catalytic Activities of PY Zeolite Supported Hydrotreating Catalyst," *Acta Petrolei Sinica* (*Petroleum Processing Section*), Vol. 28, No. 1, 2012, pp. 15-20.

[25] T. Klimova, O. Gutiérrez, L. Lizama and J. Amezcua, "Advantages of ZrO_2- and TiO_2-SBA-15 Mesostructured Supports for Hydride Sulfurization Catalysts over Pure TiO_2, ZrO_2 and SBA-15," *Microporous and Mesoporous Materials*, Vol. 133, 2010, pp. 91-99.

[26] O. Y. Gutiérrez and T. Klimova, "Effect of the Support on the High Activity of the (Ni)Mo/ZrO_2-SBA-15 Catalyst in the Simultaneous Hydridesulfurization of DBT and 4,6-DMDBT," *Journal of Catalysis*, Vol. 281, 2011, pp. 50-62.

[27] R. Nava, A. Infantes-Molina, P. Castao, R. Guil-López and B. Pawelec, "Inhibition of CoMo/HMS Catalyst Deactivation in the HDS of 4,6-DMDBT by Support Modification with Phosphate," *Fuel*, Vol. 90, No. 8, 2011, pp. 2726-2737.

[28] Y. Li, D. H. Pan, C. Z. Yu, Y. Fan and X. J. Bao, "Synthesis and Hydrodesulfurization Properties of NiW Catalyst Supported on High-Aluminum-Content, Highly Ordered, and Hydrothermally Stable Al-SBA-15," *Journal of Catalysis*, Vol. 286, 2012, pp. 124-136.

[29] N. Malicki, G. Mali, A. A. Quoineaud, P. Bourges and L. J. Simon, "Aluminium Triplets in Dealuminated Zeolites Detected by ^{27}Al NMR Correlation Spectroscopy," Microporous and Mesoporous Materials, Vol. 129, No. 1-2, 2010, pp. 100-105.

[30] M. Nakamura, "Nanometer Catalyst Technology to Reduce the Dosage of Noble Metal," *Foreign Internal Combustion Engine*, Vol. 3, 2011, pp. 35-39.

[31] R. K. Sharma, B. S. Rana, D. Varma, R. Kumar and R. Tiwari, "3-D Mesoporous Titanosilicate Support for Highly Effective Hydrodesulfurization Catalysts," *Micro-*

porous and Mesoporous Materials, Vol. 155, 2012, pp. 177-185.

[32] D. Valencia and T. Klimova, "Effect of the Support Composition on the Characteristics of NiMo and CoMo/(Zr)SBA-15 Catalysts and Their Performance in Deep Hydrodesulfurization," *Catalysis Today*, Vol. 166, 2011, pp. 91-101.

[33] Y. Kanda, C. Temma, K. Nakata, T. Kobayashi and M. Sugioka, "Preparation and Performance of Noble Metal Phosphides Supported on Silica as New Hydrodesulfurization Catalysts," *Applied Catalysis A: General*, Vol. 386, No. 1-2, 2010, pp. 171-178.

[34] S. W. Gong, L. J. Liu, H. F. He and Q. X. Cui, "Dibenzothiophene Hydrodesulfurization over MoP/SiO_2 Catalyst Prepared with Sol-Gel Method," *Korean Journal of Chemical Engineering*, Vol. 27, No. 5, 2010, pp. 1419-1422.

[35] T. Kan, H. Y. Wang, H. X. He, C. S. Li and S. J. Zhang, "Experimental Study on Two-Stage Catalytic Hydroprocessing of Middle-Temperature Coal Tar to Clean Liquid Fuels," *Fuel*, Vol. 90, No. 11, 2011, pp. 3404-3409.

[36] Z. H. Yang, L. C. Li, Y. F. Wang, J. L. Liu and X. Feng, "Preparation of Nickel Phosphide/Mesoprous-TiO_2 Catalyst and Its Hydrodesulfurization Performance," *Chinese Journal of Catalysis*, Vol. 33, No. 3, 2012, pp. 508-517.

[37] D. Q. Zhang, A. J. Duan, Z. Zhao, X. Q. Wang and G. Y. Jiang, "Synthesis, Characterization and Catalytic Performance of Meso-Microporous Material Beta-SBA-15-Supported NiMo Catalysts for Hydrodesulfurization of Dibenzothiophene," *Catalysis Today*, Vol. 175, No. 1, 2011, pp. 477-484.

[38] J. Francis, E. Guillon, N. Bats, C. Pichon and A. Corma, "Design of Improved Hydrocracking Catalysts by Increasing the Proximity between Acid and Metallic Sites," *Applied Catalysis A: General*, Vol. 409-410, 2011, pp. 140-147.

[39] G. Q. Cui, J. F. Wang, H. F. Fan, X. Y. Sun and Y. Jiang, "Towards Understanding the Microstructures and Hydrocracking Performance of Sulfided Ni-W Catalysts: Effect of Metal Loading," *Fuel Processing Technology*, Vol. 92, No. 12, 2011, pp. 2320-2327.

[40] C. Leyva, J. Ancheyta, A. Travert, F. Maugé and L. Mariey, "Activity and Surface Properties of $NiMo/SiO_2$-Al_2O_3 Catalysts for Hydroprocessing of Heavy Oils," *Applied Catalysis A: General*, Vol. 425-426, 2012, pp. 1-12.

[41] L. B. Pierella, C. Saux, S. C. Caglieri, H. R. Bertorello and P. G. Bercoff, "Catalytic Activity and Magnetic Properties of Co-ZSM-5 Zeolites Prepared by Different Methods," *Applied Catalysis A: General*, Vol. 347, No. 1, 2008, pp. 55-61.

[42] S. G. A. Ferraz, F. M. Z. Zotin, L. R. R. Araujo and J. L. Zotin, "Influence of Support Acidity of NiMoS Catalysts in the Activity for Hydrogenation and Hydrocracking of Tetralin," *Applied Catalysis A: General*, Vol. 384, No. 1-2, 2010, pp. 51-57.

[43] Y. J. Wang, H. J. Zhang, B. Sun and X. G. Tian, "Research Progress of WO_3/ZrO_2 in Alkane Isomerization," *Acta Petrolei Sinica (Petroleum Processing Section)*, Vol. 25, No. 2, 2009, pp. 283-290.

[44] M. X. Du, Zh. F. Qin, H. Ge, X. K. Li and Z. J. Lü, "Enhancement of Pd-Pt/Al_2O_3 Catalyst Performance in Naphthalene Hydrogenation by Mixing Different Molecular sSieves in the Support," *Fuel Processing Technology*, Vol. 91, No. 11, 2010, pp. 1655-1661.

[45] L. Dimitrov, R. Palcheva, A. Spojakina and K. Jiratova, "Synthesis and Characterization of W-SBA-15 and W-HMS as Supports for HDS," *Journal of Porous Materials*, Vol. 18, No. 4, 2011, pp. 425-434.

[46] H. Shimada, K. Sato, K. Honna, T. Enomoto and N. Ohshio, "Design and Development of Ti-Modified Zeolite-Based Catalyst for Hydrocracking Heavy Petroleum," *Catalysis Today*, Vol. 141, No. 1-2, 2009, pp. 43-51.

[47] A. Nogueira, R. Znaiguia, D. Uzio, P. Afanasiev and G. Berhault, "Curved Nanostructures of Unsupported and Al_2O_3-Supported MoS_2 Catalysts: Synthesis and HDS Catalytic Properties," *Applied Catalysis A: General*, Vol. 429-430, 2012, pp. 92-105.

[48] V. L. Parola, B. Dragoi, A. Ungureanu, E. Dumitriu and A. M. Venezia, "New HDS Catalysts Based on Thiol Functionalized Mesoporous Silica Supports," *Applied Catalysis A: General*, Vol. 386, No. 1-2, 2010, pp. 43-50.

[49] G. L. Yu, Y. S. Zhou, Q. Wei, X. J. Tao and Q.Y. Cui, "A Novel Method for Preparing Well Dispersed and Highly Sulfided NiW Hydridenitrogenation Catalyst," *Catalysis Communications*, Vol. 23, 2012, pp. 48-53.

[50] M. Ke, X. Y. Wang, Z. Q. Zhang, Z. Z. Song and Q. Z. Jiang, "Preparations and Properties of Selective Hydrodesulfurization Catalysts for FCC Gasoline," *Petrochemical Technology*, Vol. 40, No. 2, 2011, pp. 133-139.

[51] C. Q. Li, J. W. Li, X. K. Shang, H. Wang and G. D. Sun, "Dibenzothiophene Hydrodesulfurization Performance of WP/MCM-41 Catalysts Containing Cobalt," *Acta Petrolei Sinica (Petroleum Processing Section)*, Vol. 27, No. 6, 2011, pp. 859-865.

[52] M. A. Lélias, P. J. Kooyman, L. Mariey, L. Oliviero and A. Travert, "Effect of NTA Addition on the Structure and Activity of the Active Phase of Cobalt-Molybdenum Sulfide Hydrotreating Catalysts," *Journal of Catalysis*, Vol. 267, No. 1, 2009, pp. 14-23.

[53] M. A. Lélias, E. Le. Guludec, L. Mariey, J. van Gestel and A. Travert, "Effect of EDTA Addition on the Structure and Activity of the Active Phase of Cobalt–Molybdenum Sulfide Hydrotreatment Catalysts," *Catalysis Today*, Vol. 150, No. 3-4, 2010, pp. 179-185.

[54] K. K. Soni, K. C. Mouli, A.K. Dalai and J. Adjaye, "Effect of Ti Loading on the HDS and HDN Activity of KLGO on NiMo/TiSBA-15 Catalysts," *Microporous and Mesoporous Materials*, Vol. 152, 2012, pp. 224-234.

[55] K. C. Mouli, K. Soni, A. Dalai and J. Adjaye, "Effect of Pore Diameter of Ni-Mo/Al-SBA-15 Catalysts on Thehy-

drotreating of Heavy Gas Oil," *Applied Catalysis A: General*, Vol. 404, No. 1-2, 2011, pp. 21-29.

[56] P. Biswas, P. Narayanasarma, C. M. Kotikalapudi, A. K. Dalai and J. Adjaye, "Characterization and Activity of ZrO_2 Doped SBA-15 Supported NiMo Catalysts for HDS and HDN of Bitumen Derived Heavy Gas Oil," *Industrial & Engineering Chemistry Research*, Vol. 50, No. 13, 2011, pp. 7882-7895.

[57] M. A. Al-Daous and S. A. Ali, "Deep Desulfurization of Gas Oil over NiMo Catalysts Supported on Alumina-Zirconia Composites," *Fuel*, Vol. 97, 2012, pp. 662-669.

[58] T. Kimura, K. Sakashita, X. H. Li and S. Asaoka, "Catalytic Roles of Nano-Sized Oxides Composed with Zeolite for Hydrocracking, Catalytic Cracking and Reforming," *Catalysis Surveys from Asia*, Vol. 15, 2011, pp. 259-266.

[59] C. Li, W. A. Deng, X. W. Li, B. Q. Mu and S. F. Li, "Pilot Tests of Medium/Low Temperature Coal Tar Oil Hydro-Upgrading with Heavy Oil Hydro-Treating Catalysts," *Petroleum Refinery Engineering*, Vol. 41, No. 9, 2011, pp. 32-35.

2

Bifurcation Analysis of a Coupled Acetylcholinesterase/Choline Acetyltransferase Enzymes Neurocycle

Hamza M. Habib[1], G. Ibrahim[2], Osama Noman Saleh[2*]
[1]Department of Mathematics, College of Science, Salman Bin Abdualziz University, Al-Kharj, KSA
[2]Department of Basic Engineering Science, Faculty of Engineering, Menoufia University, Shebin El-Kom, Egypt

ABSTRACT

A diffusion-reaction, two-compartment model was used to explore the bifurcation and chaotic behavior of acetylcholinesterase (AChE) and cholineacetyltransferase (ChAT) coupled enzymes system. The effects of hydrogen ion feed concentrations, choline (Ch) and acetylcholine (ACh) feed concentrations, as bifurcation parameters on the system performance are studied. It is found that hydrogen ions play an important role in creating potential differences through the plasma membranes. Detailed bifurcation analysis over a wide range of parameters is carried out in order to uncover some of the qualitative changes of the system such as hysteresis, multiplicity, Hopf bifurcation, boundary crises bifurcation, periodic transient, and other complex dynamics. Some of the obtained results relate to the phenomena occurring in the physiological experiments like periodic stimulation of neural cells and irregular functioning of acetylcholine receptors. The model depends on real kinetics expressions and parameters obtained from the literature, so the results can be used to direct more systematic research on cholinergic disorder.

Keywords: Neurocycle Modelling; Acetylcholinesterase/Choline Acetyltransferase; Bifurcation; Enzyme; Nonlinear Dynamics

1. Introduction

Acetylcholine (ACh) serves as the transmitter of nerve impulses at cholinergic synapses. In humans and homoeothermic animals, it affects transmission from motor nerves to skeletal muscles, from preganglionic parasympathetic and sympathetic fibers to neurons in the autonomic ganglia. ACh is also the transmitter at some synapses in the ventral neural system, as shown in **Figure 1**.

The released ACh diffuses in the synaptic gap and reacts with the cholinergic receptor protein, located in the post-synaptic membrane. The reaction with the receptor starts the chain of events resulting in the excitation or inhibition of postganglionic cell [1,2]. The substrates for the bio-synthesis of ACh are acetyl coenzyme A (acetyl-CoA) and choline. The synthesis is catalyzed by cholineacetyltransferase (ChAT).

Acetylcholine (ACh) plays a well recognized role in the nerve excitation [1,3-5]. It is found in cholinergic synapses that provide a stimulatory transmission in the nervous system. Its complete neurocycle constitutes a coupled two-enzyme system with the following two simultaneous events:

Activation Event: ACh is synthesized from choline and acetyl coenzyme A (Acetyl-CoA) biocatalyzed by the enzyme choline acetyltransferase (ChAT) and immediately stored in small vesicular compartments closely attached to the cytoplasmic side of the presynaptic membranes [2,3].

Degradation Event: Once ACh has completed its activation duty; the synaptic cleft degradation begins for removing the remaining ACh. This occurs when the ACh is consumed biocatalyzed by the acetylcholinesterase (AchE) to form choline and acetic acid [2,4,6,7].

In enzyme membrane systems, the pH decreases due to the local production of hydrogen ions. Because of the amphoteric properties of the proteinaceous membrane; the lower the pH the lower the density of the negative fixed charges in the membrane [8]. Artificial AChE membrane exhibits a potential difference electrical response similar to excitable membranes when ACh is injected on one side [9]. The steady state potential due to the enzyme activity for increasing and decreasing substrate concentrations exhibits a hysteresis behavior. Due to the auto-catalytic effect resulting from the production

*Corresponding author.

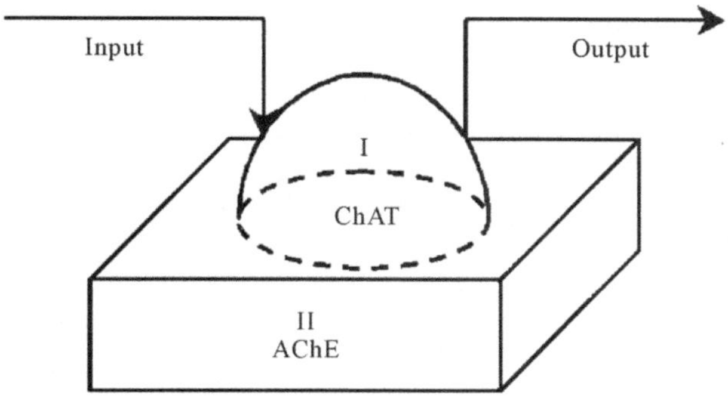

Figure 1. Two-enzymes/two-compartment model.

of hydrogen ions and the existence of diffusional resistances, hysteresis phenomenon develops in a definite range of parameters. The amphoteric properties of the membrane help to transfer the hysteresis of the internal pH into a hysteresis in membrane potential [9].

The two-enzymes/two-compartments model used in this investigation differs from other membrane excitation models such as of Rose-Hindmarsh model for action potential [10,11] which is a modification of Fitzhugh model [12]. This later model was developed to simulate the repetitive, patterned and irregular activity seen in molluscan neurons. Holden and Fan [13-15] and Fan and Holden [16] investigated the dynamic behavior of membrane excitation using a three-variable model of action potential which showed clearly the existence of different dynamic modes like simple periodic, bursting periodic and chaotic behavior.

Ibrahim and co-workers [17,18] investigated one half of the neurocycle in a two-compartment model using AChE as the only enzyme. The model unveiled bifurcation, instability, chaos and hyperchaos.

Moustafa *et al.* [19-21] investigated the complete neurocycle with special emphases on the choline recycling.

Parag *et al.* [22] investigated the bifurcation, stability and chaotic behavior of a two compartment model describing the complete neurocycle (AChE/ChAT) using general kinetic expressions. In this paper a more realistic kinetic expressions and parameters were implemented in order to further our insight into such important neurocycle.

2. The Simplified Diffusion-Reaction Two-Enzymes/Two-Compartment Model

A simplified model for the AChE/ChAT enzymes system inside the neural synaptic cleft is shown in **Figure 1**. The complete neurocycle of the acetylcholine, as a neurotransmitter, is simulated as a simplified two-enzymes/two-compartment model. Each compartment is modeled as a constant flow, constant volume, isothermal, continuously stirred tank reactor (CSTR). The two compartments are separated by a nonselective, permeable membrane as shown in **Figures 2** and **3**.

Assuming that all the events are homogeneous in all vesicles, and by using the appropriate dimensionless state variables and parameters, we consider the behavior for a single synaptic vesicle as described by this simple two compartment model, where I and II denote compartments 1 and 2 respectively.

It is assumed that acetylcholine is synthesized in the presynaptic cell by the enzyme choline acetyltransferase due to an activation reaction. The stimulatory neurotransmitter acetylcholine is synthesized through the reaction $R_{(1)}$ as follows:

$R_{(1)}$:

$$\text{Choline} + \text{Acetyl-CoA} \xrightarrow{\text{ChAT}} \text{Acetylcholine} + \text{CoA}$$

Acetylcholine is hydrolyzed (destroyed) in the postsynaptic cell by the acetylcholinesterase enzyme through a degradation reaction $R_{(2)}$ (where the stimulatory neurotransmitter acetylcholine is degraded) as follows:

$R_{(2)}$:

$$\text{Acetylcholine} + H_2O \xrightarrow{\text{AChE}} \text{Choline} + \text{Acetate} + H^+$$

- Both reactions are considered substrate inhibited and hydrogen ion rate dependent. This leads to a non-monotonic behavior of the reaction rates on both the substrate and pH.
- From an enzyme kinetics point of view, the kinetic expression evaluated by Louis B. Hersh *et al.* (1977) [23] was used. This expression describes the choline-acetyltransferase reaction. It relates the speed of reaction by the two reactants (Choline and acetate). The saturation constant k_s for this reaction is assumed to be dependent on H^+ ions concentration

The proposed rate expressions are as follows:

$$R_1(S_2, S_3, H_1) = \frac{V_{m1} S_{21} S_{31}}{K_s\left(\frac{K_{h1}}{H_1} + 1 + \frac{H_1}{K'_{h1}}\right) + K_{s2} S_{21} + K_{s3} S_{31} + S_{21} S_{31}} \quad (1)$$

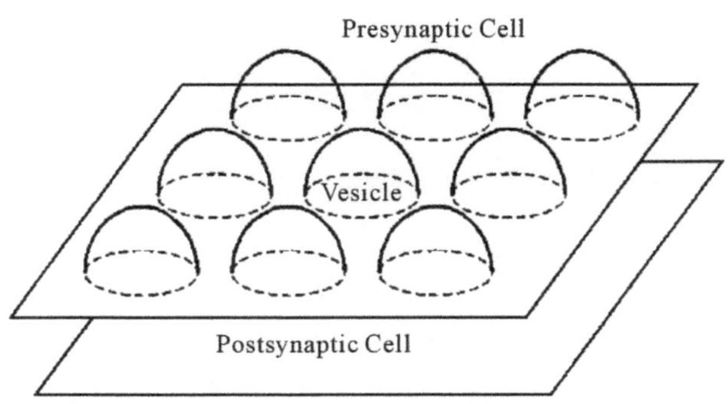

Figure 2. Schematic representation of synaptic cleft.

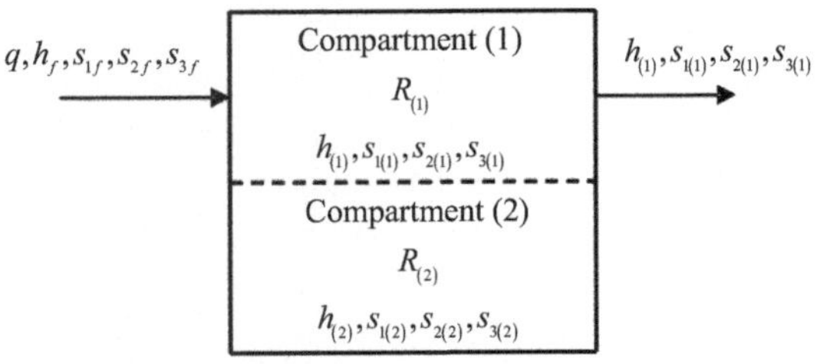

Figure 3. Two-enzymes/two-compartment model.

$$R_2\left(S_{12}, H_2\right) = \frac{V_{m2} S_{12}}{K_{s1}\left(\frac{K_{h2}}{H_2} + 1 + \frac{H_2}{K'_{h2}}\right) + S_{12} + S_{12}^2 / K_i} \quad (2)$$

The rate expressions R_1 and R_2 in a dimensionless form are

$$r_1 = \frac{R_1}{V_{m1}} = \frac{s_{21} s_{31}}{k_d\left(\frac{\delta_3}{h_1} + 1 + \delta_2 h_1\right) + \beta s_{21} + \frac{s_{31}}{\beta} + s_{21} s_{31}} \quad (3)$$

$$r_2 = \frac{R_2}{V_{m2}} = \frac{s_{12}}{\left(\frac{1}{h_2} + 1 + \delta_1 h_2\right) + s_{12} + \alpha s_{12}^2} \quad (4)$$

where

$$h_j = \frac{H_j}{K_{h2}}, \quad s_{1,j} = \frac{S_{1,j}}{K_{s1}}, \quad s_{2,j} = \frac{S_{2,j}}{K_{s2}}, \quad s_{3,j} = \frac{S_{3,j}}{K_{s3}},$$

$$\delta_1 = \frac{K_{h2}}{K_{h2}^{/}}, \delta_2 = \frac{K_{h2}}{K_{h1}^{/}}, \delta_3 = \frac{K_{h1}}{K_{h2}^{/}}$$

$$\alpha = \frac{K_{s1}}{K_i}, \quad \beta = \frac{K_{s2}}{K_{s3}}, \quad K_d = \frac{K_s}{K_{s2} K_{s3}}$$

And S_{1j}, S_{2j}, S_{3j} and H_j stand for acetylcholine, choline, acetate and hydrogen ions respectively.

The material balance equations for the two compartments are summarized in the dimensionless form as follows:

1) For Hydrogen Ions [H⁺]

a) For compartment (1)

$$\frac{dh_1}{dT} = \left(h_f - h_{(1)}\right) - \alpha_h\left(h_{(1)} - h_{(2)}\right) - \gamma\left(\frac{1}{h_f} - \frac{1}{h_{(1)}}\right) + \gamma\alpha_{oh}\left(\frac{1}{h_{(1)}} - \frac{1}{h_{(2)}}\right) \quad (5)$$

b) For compartment (2)

$$\frac{dh_2}{d\tau} = \alpha_h V_r\left(h_1 - h_2\right) - \alpha_{oh}\gamma V_r\left(\frac{1}{h_1} - \frac{1}{h_2}\right) + \beta_1 r_2 \quad (6)$$

where h_j is the dimensionless hydrogen ions concentration, $\tau = \frac{qt}{V_1}$ is the dimensionless time, $V_r = \frac{V_1}{V_2}$ is the dimensionless volume relating the two compartments; α_h is the dimensionless membrane permeability for hydrogen ions, α_{oh} is the dimensionless membrane permeability for hydroxyl ions;

$$\gamma = \frac{K_w}{K_{h2}^2}, \quad \beta_1 = \frac{V_1 V_{m2} E_2}{K_{h2} q}.$$

2) For Acetylcholine ACh [S₁]

a) For compartment (1)

$$\frac{ds_{11}}{d\tau} = s_{1f} - s_{11} - \alpha_{s1}\left(s_{11} - s_{12}\right) + \beta_2 r_1 \quad (7)$$

b) For compartment (2)

$$\frac{ds_{12}}{d\tau} = \alpha_{s1} V_r\left(s_{11} - s_{12}\right) - \beta_3 r_2 \quad (8)$$

where

$$\alpha_{s1} = \frac{\alpha'_{s1} A_m}{q}, \quad \beta_2 = \frac{V_1 V_{m1} E_1}{K_{s1} q}, \beta_3 = \frac{V_1 V_{m2} E_2}{K_{s1} q}$$

3) For Choline [S₂]

a) For compartment (1)

$$\frac{ds_{21}}{d\tau} = s_{2f} - s_{21} - \alpha_{s2}\left(s_{21} - s_{22}\right) - \beta_4 r_1 \quad (9)$$

b) For compartment (2)

$$\frac{ds_{22}}{d\tau} = V_r \alpha_{s2}\left(s_{21} - s_{22}\right) + \beta_5 r_2 \quad (10)$$

where

$$\beta_4 = \frac{V_1 V_{m1} E_1}{K_{s2} q}, \beta_5 = \frac{V_1 V_{m2} E_2}{K_{s2} q}, \quad \alpha_{s2} = \frac{\alpha'_{s2} A_m}{q}$$

4) For Acetate [S₃]

a) For compartment (1)

$$\frac{ds_{31}}{d\tau} = s_{3f} - s_{31} - \alpha_{s3}\left(s_{31} - s_{32}\right) - \beta_6 r_1 \quad (11)$$

b) For compartment (2)

$$\frac{ds_{32}}{d\tau} = \alpha_{s3} V_r\left(s_{31} - s_{32}\right) + \beta_7 r_2 \quad (12)$$

where

$$\alpha_{s3} = \frac{\alpha'_{s3} A_m}{q}, \quad \beta_6 = \frac{V_1 V_{m1} E_1}{K_{s3} q}, \beta_7 = \frac{V_1 V_{m2} E_2}{K_{s3} q}$$

The relatively simple two-enzyme/two-compartment model is thus represented by the set of Equations (3) through (12) which represents the nonlinear system of equations having eight state variables $\left[h_{(j)}, S_{1(j)}, S_{2(j)}, S_{3(j)}, j = 1, 2\right]$. The system parameters values are given in **Table 1**.

Table 1. System parameters evaluation.

Value	Parameter
1	Kd
1.2	V_r
0.032	γ
0.5	α_{s3}
0.5	α_{s2}
0.5	α_{s1}
0.5	α_{oh}
30	α_h
50	β_7
50	β_6
50	β_5
100	β_4
50	β_3
10	β_2
50	β_1
0.5	β
0.55	α
0.5	δ_3
0.1	δ_2
0.1	δ_1
2.0	S_{3f}
2.0	S_{2f}
0.003	H_f

3. Numerical Tools, Presentation Techniques and Computational Resources

The static and dynamic bifurcation analysis was performed. The feed concentrations of acetylcholine and hydrogen ions S_{1f} and pH_f was taken as the bifurcation parameter. The bifurcation diagrams are obtained using the software package AUTO 97 of Doedel *et al.* [24]. This package is able to perform both steady state and dynamic bifurcation analysis, including the determination of entire periodic branches. In some cases where multiplicity of steady-state was found in a very narrow range of the bifurcation parameter or where catastrophic changes in periodic branches occurred, the AUTO 97 package failed to capture the complete periodic branches. In these cases, simulation techniques were used to complete the whole picture of the bifurcation diagrams.

Poincare' presentation techniques were also used. The discrete points (return points) were obtained by intersecting the trajectories and a hyper surface (Poincare' surface). These discrete points of intersection were taken so that the trajectories intersected the hyper surface transversally and crossed it in the same direction.

The DGEAR subroutine available in the IMSL Libraries for FORTRAN with automatic step size was used to ensure accuracy for stiff differential equations for numerical simulation of periodic and chaotic attractors. The program employed by Elnashaie *et al.* [25] was also used to plot the Poincare' diagram. The routine enjoys high accuracy. In many cases, a bound on the allowable error as small as 10^{-15} was necessary to obtain accurate results.

4. Results and Discussion

Hydrogen (H^+) ion is a very simple element because it contains only one proton and its size is very small. H^+ ions play an extremely vital role in all metabolic processes, and transport of ions such as Na^+, K^+, and Ca^{2+} occurring in the living organisms. For example, the function of protein components can be altered from hydrophobicity to hydrophilicity because of its ability to release or bind H^+ ions [26]. It is observed that most pathological environments and ions diffusion are accompanied by observed pH changes [27,28].

During exocytosis the synaptic vesicle fuses with the surface membrane and undergoes a pH jump. When the synaptic vesicles is inside the presynaptic nerve terminal its internal pH ranges from 5.2 to 5.5 [29,30], and after fusion, the inside of the vesicle comes in contact with the extracellular median with a pH of about 7.25 this jump on pH affects the opening of non-specific ion channels [31].

Stanley M. Parsons [32] states that the value of transmembrane pH and electrical gradient (ΔpH and $\Delta\psi$ respectively) are about 1.4 pH units and 39 m volt respectively for filled cholinergic and monoaminorgic secretory vesicles.

The plasma membrane extrudes hydrogen ions from the cell to generate a proton (H ion) motive force with a membrane potential of −120 to −160 m volt (negative inside). This process takes place via H^+-ATPase mechanism of transport [33].

In enzyme membrane systems, the pH decreases due to the local production of hydrogen ions or to the active transport of protons. And because of the amphoteric properties of the plasma membrane the lower the pH value the lower the value of the negative fixed charges located at the membrane surface [34]. Artificial AChE membrane exhibits potential differences similar to excitable membrane when ACh is injected on one side [9]. The last article shows that the steady state potential due to the enzyme activity exhibits a hysteresis and oscillatory behavior. Due to the auto-catalytic effects resulting from the production of hydrogen ions and the existence of diffusional resistances, hysteresis phenomenon develops in finite range of parameters. The amphoteric properties of the membrane help to transfer the hysteresis of the internal pH into a hysteresis in membrane potential [9].

The bifurcation analysis is concerned with the way that steady state solutions of the system model vary with one of the system parameters. The feed concentration of acetylcholine (S_{1f}) and pH_f feed were chosen as the main bifurcation parameters through this investigation. Static and some dynamic characteristics of this biosystem are investigated using the software AUTO 97 [24] and dynamic simulation techniques. For each case, all the parameters (other than the bifurcation parameter S_{1f}, pH_f feed) were kept constant at the values in **Table 1**. The default values of **Table 1** are used, unless the contrary is mentioned. At each region of bifurcation analysis, a sample of the possible correspondence with real physiological values is given. Although some of the parameters were obtained from direct experimental results, but chosen in a heuristic way from plausible alternatives, it is interesting to notice that the state variables are similar to the physiological expected values.

An extensive literature review leads us to the values for different parameters in the model.

- pH in the range of 6.95 - 7.15 was measured in human brain [35].
- pH in the range of 6.95 - 7.35 was reported in a feline model [36].
- ACh in a rat brain was found to be in the range of 0.22×10^{-5} kmol/m^3. (Free ACh) to 1.77×10^{-5} (total ACh) [2]. While, in guinea pig cerebral cortex the range was 0.31×10^{-5} (free ACh) to 1.67×10^{-5} kmol/m^3 (total ACh) [2].
- ACh concentration in human placenta is reported to be in the range of 3.0×10^{-5} - 55.5×10^{-5} kmol/m^3 [37].
- ACh in the isolated rings of rat pulmonary artery was measured to be in the range of 0.001×10^{-5} - 3.0×10^{-5} kmol/m^3 [38].
- Choline concentration in mouse rat brain is about 1.15×10^{-4} kmol/m^3 [2].
- Choline concentration in human plasma is in the range of 0.01×10^{-4} - 0.7×10^{-4} kmol/m^3 [39].

In the following, the investigation will be emphasized on the autonomous periodic and aperiodic behavior of the biosystem under consideration, in addition to the behavior when considering the quantal nature of the neurotransmitter releases.

4.1. System Behavior in Response to Constant Fixed Input

Figures 4(a)-(c) show the bifurcation diagram at the set of system parameters listed in **Table 1**, the bifurcation diagram is shown for a wide range of the bifurcation parameter ($8.2 > pH_f > 8.8$).

With respect to **Figures 4(a)** and **(b)**, the *S* shape behavior exhibits a transfer of the system behavior from high conversion of ACh hydrolysis (in terms of ACh concentration in compartment 2 and membrane potential) to low conversion states with pH increase. The bifurcation diagram characterized by the existence of two Hopf bifurcation points (at $pH_f = 8.4754$ and 8.5232) and two static limit points. The bifurcation behavior is characterized by the following features:

- At $pH_f \in (8.2, 8.398)$: Unique steady state solutions on the high conversion steady state branch exist.
- At $pH_f \in (8.398, 8.4754)$: Bistability between steady state and periodic or aperiodic solutions exist.
- At $pH_f \in (8.4754, 8.5232)$: Unique periodic or aperiodic solutions exist.
- At $pH_f \in (8.5232, 8.8)$: Unique steady state solutions on the low conversion steady state branch exist.

The first Hopf bifurcation point (at $pH_f = 8.4754$) is a subcritical one where a set of unstable periodic solutions emerging in its neighborhood and this organization of the state space make it possible to exhibit a bistability.

Figure 4(c) shows the bifurcation diagram in terms of pH changes in the two compartments, the results show that the pH differences between the two compartments changed dramatically corresponding to the variation of the resultant hydrolysis process of ACh that is represented by **Figure 4(b)**. The pH difference changes from a value close to 0.8 at $pH_f = 8.2$ to a value close to 1.2 at the first Hopf bifurcation points ($pH_f = 8.4754$). The pH difference deceases dramatically at the solution sets of low conversion domain (close to 0.1 in pH units). These results are close to the experimental results reported by Stanley M. Parsons [32].

The membrane potential was estimated (assuming hydrogen ions were the only charged ions in the system) using the following formula [40]:

$$\text{membrane potential} = -58.0\log_{10}\frac{H_1}{H_2},$$

Thus the positive value (see **Figure 4(a)**) of the membrane potential means that the negative charges are localized on the membrane surface of compartment I. It is noticed that the membrane potential exhibit the same behavior as pH difference [the membrane potential increases with the increase of pH difference and vice versa]. The region of high conversion shows high membrane potentials [changes from a value close to 42 m volts up to a value of 76 m volt as a maximum value in the periodic region]. The average values of membrane potential at the periodic region are represented by the star symbols in **Figure 4(a)**, these values are estimated as follows:

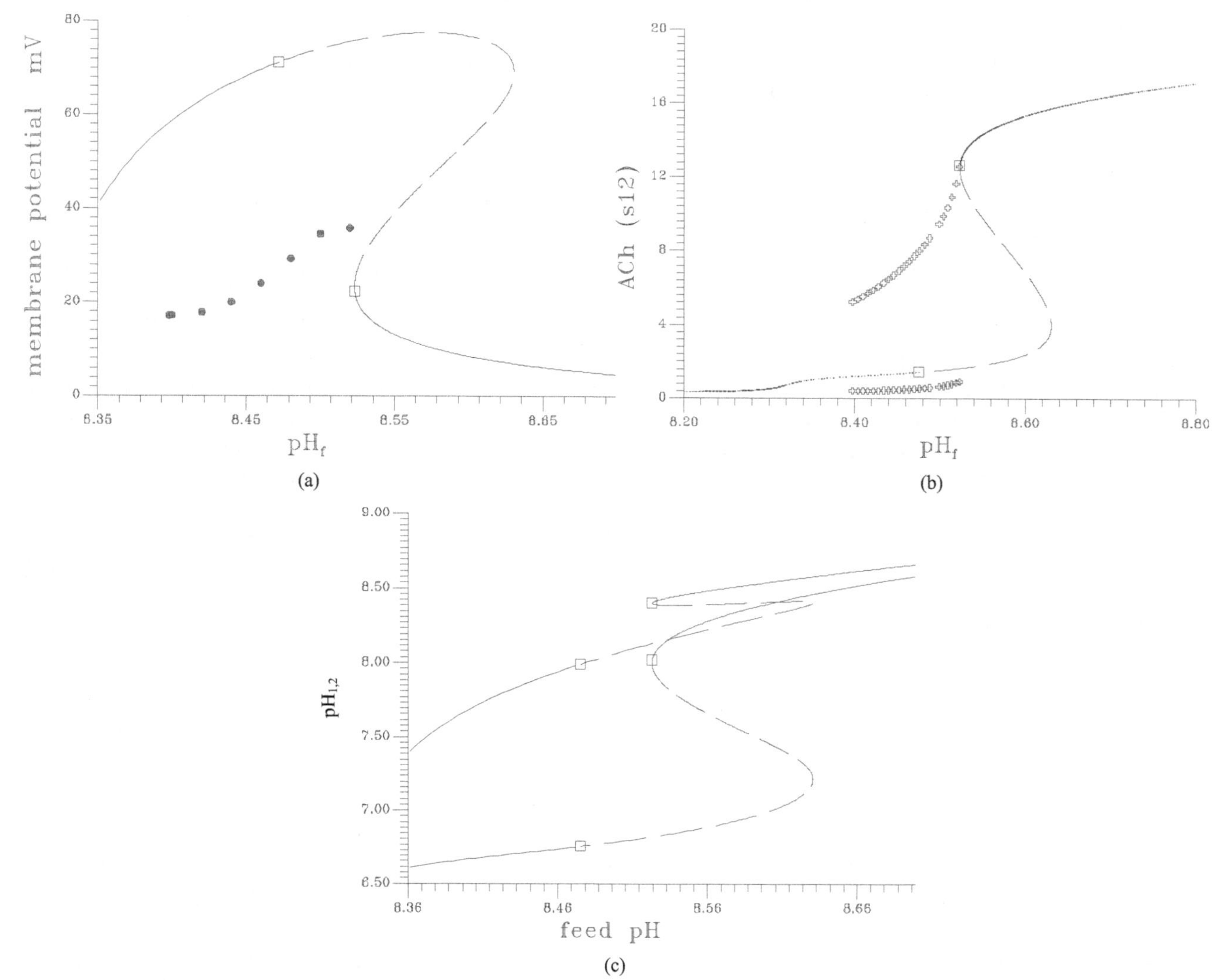

Figure 4. (a) Bifurcation diagram [pH_f vs membrane potential], produced at $s1_f = 20$ and the rest of system parameters as listed in Table 1], [solid circles represents the average value of the stable periodic branch, ··· unstable stationary branch, ____ stable stationary branch]; (b) Bifurcation diagram [pH_f vs ACh concentration in compartment 2; $s12$], produced at $s1_f = 20$ and the rest of system parameters as listed in Table 1. [The symbols indicates the stable periodic branches, solid line indicates the stable stationary branch, dashed line indicates the unstable branch]; (c) The lower curve [w.r.t $pH_{1,2}$] represents compartment 2 and the other curve for compartment 1.

$Ave = \frac{1}{\Delta\tau}\int_{\tau 1}^{\tau 2} f(x)\,\mathrm{d}\tau$ where, x represents the state variable of the system.

It is noticed that the average membrane potentials at the periodic and bistable (where two kind of attractors exist: the stationary and the periodic one) regions are less than that of the high conversion region. At the low conversion region the membrane potential decreases dramatically to values less than 20 m volts (near the second Hopf bifurcation point).

However a value of membrane potentials which occurs due to hydrogen ions concentration difference of the same range has been recorded by Heven Sze *et al.* [33].

Generally, ions in body fluids can be classified into two categories: the first one is buffer ions such as bicarbonates, and the second one is non buffer ions which are either strong ions or electrolytes [41]. These strong ions have no effect on buffering capacity as they are completely ionized at normal pH [41]. In this work, the effect of bicarbonate on pH is ignored. This assumption is supported and validated by Constable *et al.* [41]. Furthermore bicarbonate ions do not contribute to cholinergic ACh reactions that lead to the net synthesis or degradation of ACh [41,42]. In addition, bicarbonate can be effective through alterations in respiratory activity not in ACh cholinergic neurons [41,42]. Our model considers that ionic charges contributed by $[CO_3]^{2-}$ and $[HCO_3]^{-}$ are quantitatively unimportant.

It is noticed that the variation in pH and consequentially the membrane potential are a little bigger than those recorded in actual physiological situations, this is due to the fact that the model ignores any effects of other ions (generation and transport) but hydrogen ions, and the assumption of fully ionization of the acetic acid. These effects may play a basic role in controlling and regulating of charges on both membrane sides. On the other side, the use of an artificial membrane like the proposed one for this model or like the membrane which was used by Friboulet *et al.* [9] and eliminating any charged molecules but hydrogen ions may give closer results to that illustrated by **Figure 4(a)**. Friboulet *et al.* [9] used an artificial membrane which was injected by acetylcholine from one side and the hydrolysis reaction takes place in the other side found a hysteresis and oscillations in pH values and a pH variation close to the model results.

Figures 5(a)-(c) show the bifurcation diagrams where the ACh feed concentration is the bifurcation parameter. This diagrams are constructed using the same set of parameters as **Figure 4** and $pH_f = 8.5287874$. The bifurcation diagrams are characterized by the existence of four Hopf bifurcation points and four static limit points: the first HB_1 (from the left) occurs at $S1_f = 19.80467$, the second HB_2 at $S1_f = 21.55055$, the third one occurs at $s1_f = 31.5689$ and the fourth HB_4 at $S1_f = 31.86253$.

The AUTO 97 package failed to capture the periodic branch that emerging from these Hopf bifurcation points, so simulation techniques are used to estimate these branches (see **Figure 5(b)**, an elongation of **Figure 5(a)**). The Hopf bifurcation points are of the subcritical type where a set of unstable periodic solutions emerging in its neighborhood and this organization of the state space make it is possible to exhibit bistability.

The steady-state branches have two regions of multiplicity of steady-states separated by a high conversion steady-state branch. The first region of multiplicity, from the left-hand side of **Figure 5(a)**, lies between the two

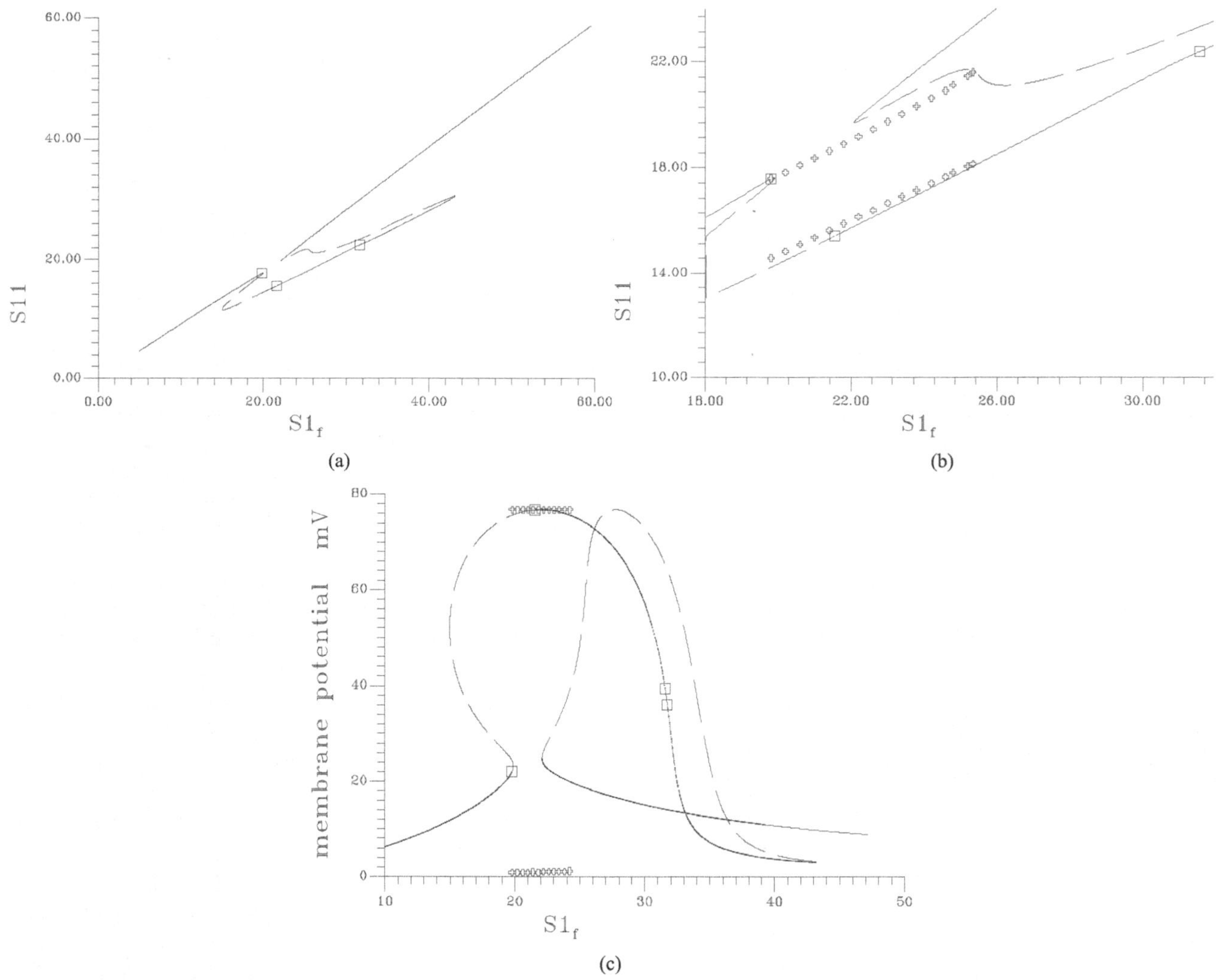

Figure 5. (a) Bifurcation diagram [$S1_f$ vs S11]-[This diagrams are constructed using the same set of parameters as Figure 4 and $pH_f = 8.5287874$]; (b) An enlargement of (a); (c) Bifurcation diagram [$S1_f$ vs membrane potential].

static limit points SLP_1 [$s1_f$= 19.90996] and SLP_2 [$s1_f$= 14.96821], the second region of multiplicity lies between the third SLP_3 [$s1_f$ = 43.18598] and SLP_4 [$s1_f$ = 22.08557]. On the right of the Hopf bifurcation point HB_1 at this point, the steady-state branch loses its stability to unstable steady-state of saddle-focus type, where six of the eigenvalues are of negative sign (the other two are complex conjugates of positive real values). The negative sign of the eigenvalues of these saddle-focus type steady-states correspond to a six-dimensional stable manifold W^s that enter the saddle point called the separatrix, while the positive eigenvalues corresponds to the two-dimensional unstable manifold W^u. The same type of saddle-focus unstable steady-states are born directly before gaining stability at HB2 ($s1_f$= 21.55055). There is a range between the two Hopf bifurcation points HB_1, HB_2 where $S1_f \in$ (19.80467, 21.55055) have three unstable static branches, this range occurs at $s1_f \in$ (19.80467, 19.90996). Two of these unstable steady states are of saddle-focus type and the third of saddle node type, where six out of the eight eigenvalues are real and negative and the rest are of real positive value.

The reinjection of W^s into W^u gives rise to homoclinic connections. The reinjection of unstable manifold W^u of one of the three unstable steady-states to another gives rise to heteroclinic connections also. On the other side, the transversal intersections between immersed manifolds (stable or unstable manifolds cannot intersect in the state space but immersed manifolds can) should create homoclinic points. Absence of blackholes in this region makes such connections easily formed in state space. The points on the two saddle-focus branches were found to satisfy the Sil'nikov's conditions [43]. These conditions with the expected homoclinic orbits existing in the state space generate a complex situation. Therefore two major expectations should be considered in this region between HB_1 and HB_2 within the range of multiple unstable steady states:

- The existence of three static points of the saddle type with their possible manifolds connections may create more than one basin of attraction and bistability of cyclic attractors should be expected.
- In the neighborhood of the homoclinical points chaos should be expected based on Sil'nikov's theorem and the generalization of Rossler *et al.* [44] as discussed in Ibrahim and Elnashaie [18]. The cyclic behavior in this region should be influenced by the spiralling-out effect of the saddle-focus which will play a recognized role in organizing the system dynamics.

However, the results show the following bifurcation characteristics as **Figures 5(a)-(c)** indicate:

- At $s1_f \in (5, 19.8046)$ unique stationary steady states characterize this region.
- At $s1_f \in (19.8046, 19.80467)$ bistability between stationary steady states and cyclic attractors exists.
- At $s1_f \in (19.80467, 21.55055)$ unique cyclic attractors dominate this region.
- At $s1_f \in (21.55055, 2208557)$ bistability between cyclic and stationary steady states attractors exists.
- At $s1_f \in (22.08557, 25.1)$ three stable solutions exist in the state space. One of them is cyclic attractor.
- At $s1_f \in (25.1, 31.56897)$ two stable stationary states exist.
- At $s1_f \in (31.56897, 31.86253)$ bistability between cyclic and stationary steady states attractors exists.
- At $s1_f \in (31.86253, 43.18598)$ two stable stationary states exist.
- At $s1_f$> 43.18598 a unique stationary state exists.

Figure 6 shows a periodic burst in a time trace representation at $s1_f$= 20. [the region of unique cyclic attractors], each cycle of this burst involved one smooth big cycle and one spike, this kind of complex oscillations is of course a casual event in the dangerous boundaries discussed before.

The branch of periodic bursts continues with increasing $s1_f$ [$s1_f$> 20] without remarkable changes in its shape until it dies out homoclinically at the second multiplicity region as illustrated by **Figure 5(b)**. The range where $s1_f \in (19.8046, 20)$ shows a burst bifurcation to chaos, a one-dimensional Poincare' map is constructed for this range, see **Figure 7**, where the Poincare' hyperplane is carefully chosen. According to Javier *et al.* [45], the Poincare' plane must be transverse to the bi-dimensional stable manifold of the corresponding saddle-focus. Therefore the Poincare' plane used crosses through the central hole of the attractors. On this basis, the Poincare' plane in this case is taken at a value of $s12$ = 12.4974075. Because of the expected bi-stability phenomenon and to avoid the attraction of any other attracting sets that may exist in the same state space, the initial conditions necessary to obtain the trajectories intersections with the Poincare' plane are changed automatically every step of the $s1_f$ state space to insure that the trajectories are retained in the same domain of attraction in the state space. It is clear from the Poincare' map of **Figure 7** that the burst oscillations go through an inverse period adding sequence with the increase of $s1_f$ until it bifurcates to chaos which terminates suddenly as the causality of crises bifurcations. **Figure 6(b)** shows a time trace representation to a complex periodic attractor selected near the point of chaos appearance at $s1_f$= 19.80462. Comparing to **Figure 6(a)**, the selected attractor have a period close to ten times of the period characterizing the attractor of **Figure 6(a)**. This means that the selected attractors are very close to dangerous boundaries creating crises conditions (*i.e.* homoclinicity conditions and/or heteroclinicty conditions).

Figures 8(a)-(c) show time trace representation of

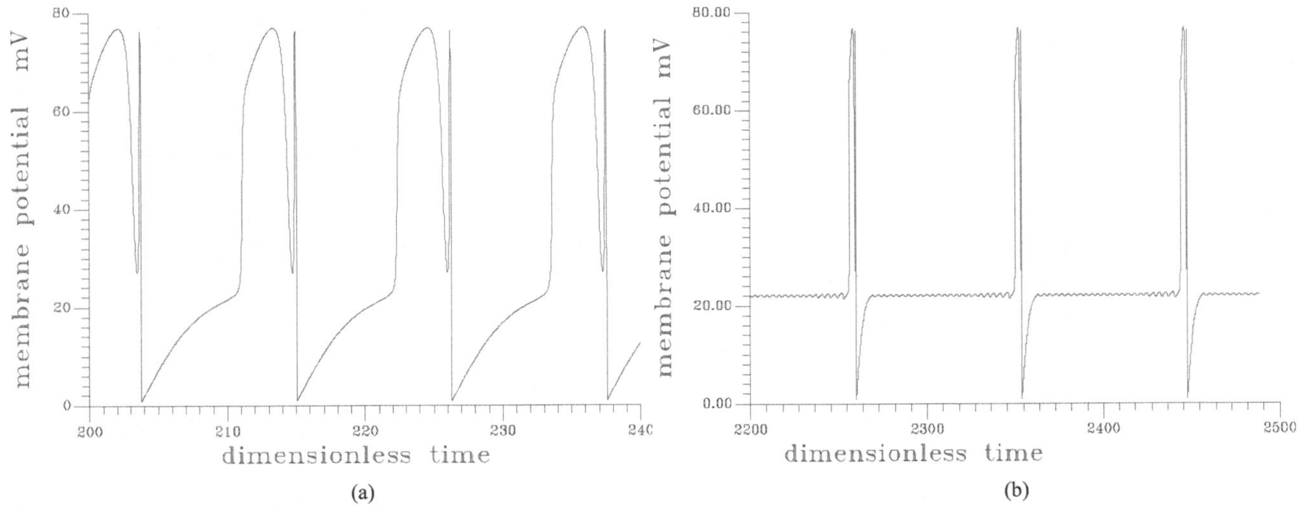

Figure 6. (a) Periodic bursts of one spike at $s1_f$ = 20 and the rest of the system parameters as Figure 5; (b) Complex periodic attractor at $s1_f$ = 19.84062.

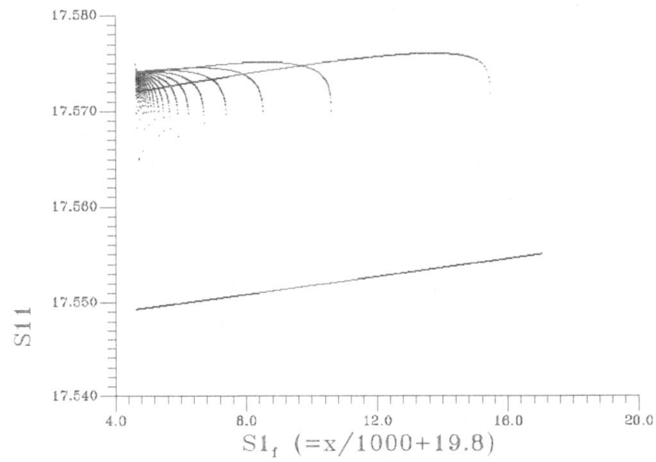

Figure 7. One dimension pioncare map: the hyper plan surface taken at $S12$ = 12.4974075.

three sets of attractors:

- **Figure 8(a)** shows the bistability evolved at $s1_f$ = 21.8 where a periodic attractor coexist with a stationary state attractor at the same state space. The biosystem alternate between the two attractors according to initial conditions.
- **Figure 8(b)** shows the time trace at $s1_f$ = 22.8, where three stable attractors coexist at the same state space dividing it into three domains of attraction: one for the bursts type of attractors and the other two attractors are of stationary type.
- **Figure 8(c)** shows the time trace at $s1_f$ = 31.7, where two attractors coexist at the same state space: one of smooth and low amplitude oscillations type and the other is of stationary state type.

It is known from bifurcation theory [46,47] that interactions between a SLP and HB point may lead to a variety of behavior ranging from simple periodicity to complex periodic behavior. In the beginning of this analysis it is helpful to build an overall picture of the possible bifurcation mechanisms that the system may exhibit. This task is best achieved by showing both the loci of the static limit points and the Hopf points in a two-parameter continuation diagram.

Figures 9(a)-(c) show the SLP curve (dashed line) and the HB curve (solid line) for instance in the parameter space ($s1_f$, pH_f), ($s1_f$, gama) and ($s1_f$, $s2_f$) respectively. It can be seen from **Figure 9(a)**, that the loci of the HB points consist of two curves that form two minima at pH_f = 7.8591 and 8.5217. An oscillatory behavior should be expected then in the model for any value of the pH_f larger than these values up to a value of pH_f = 8.524 for the first minimum and a value of 8.5936 for the second minimum. It can be also seen from the shape of the HB curves that no oscillatory behavior occurs below pH_f = 7.8591. Similarly to the HB curve the loci of the static limit points (SLP curve) shown by dashed lines consist of two maxima, two minima and a small cusp, the two

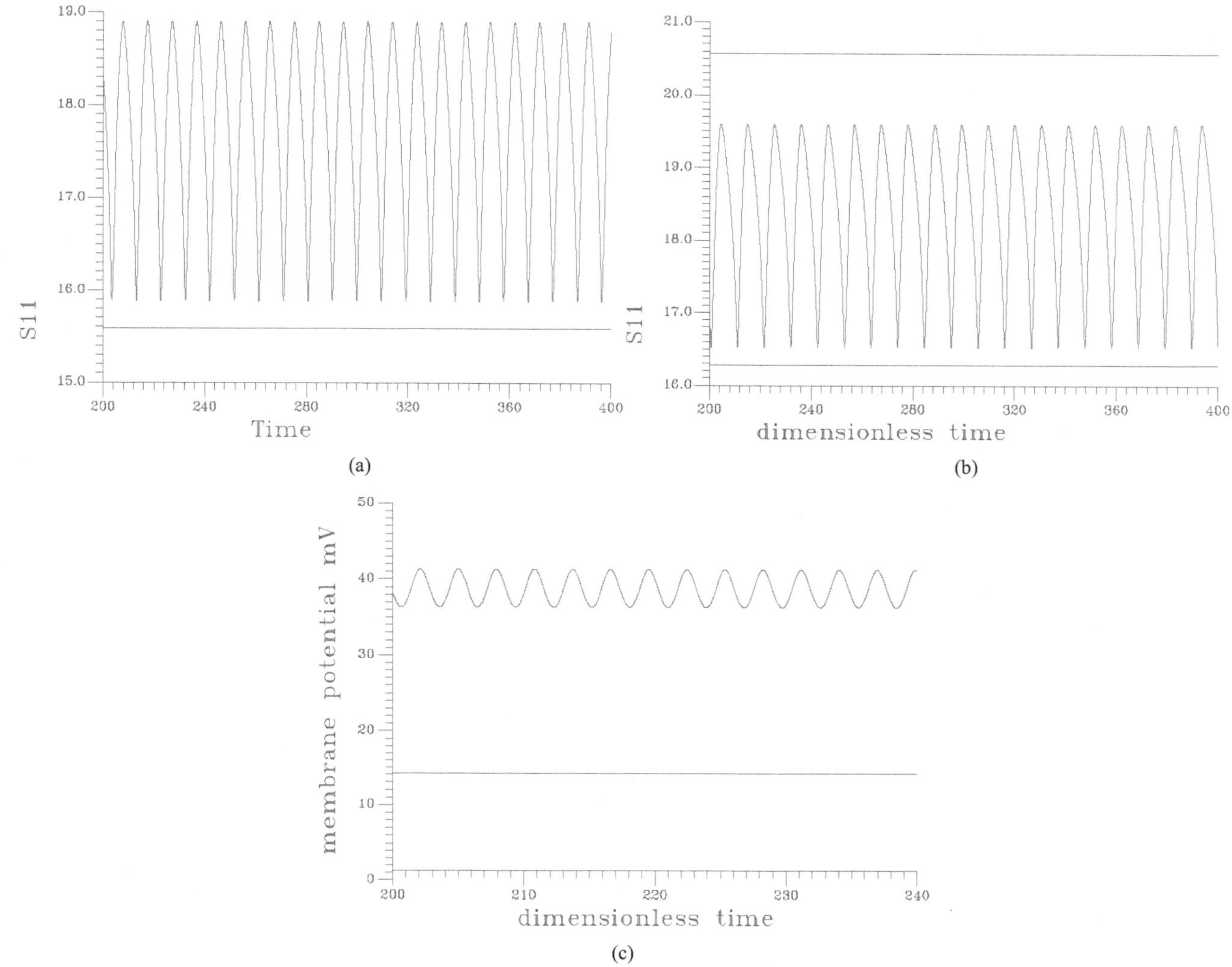

Figure 8. (a) Shows the time trace at $s1_f = 21.8$ [initial conditions for periodic attractor: y1 = 0.008, y2 = 0.005, y3 = 2, y4 = 1.5, y5 = 0.4, y6 = 2, y7 = 2, y8 = 1] [initial conditions for stationary solution: y1 = 0.0087, y2 = 0.184, y3 = 15.6, y4 = 1.58, y5 = 1.15, y6 = 15.17, y7 = 1.15, y8 = 15.2]; (b) Shows the time trace at $s1_f = 22.8$ [initial conditions for periodic attractor: y1 = 0.008, y2 = 0.005, y3 = 2, y4 = 1.5, y5 = 0.4, y6 = 2, y7 = 2. y8 = 1] [initial conditions for low conversion stationary solution: y1 = 0.0087, y2 = 0.184, y3 = 15.6, y4 = 1.58, y5 = 1.15, y6 = 15.17, y7 = 1.15, y8 = 15.2], [initial conditions for high conversion stationary solution: y1 = 0.00394, y2 = 0.00947, y3 = 20.58, y4 = 15.5, y5 = 1.44, y6 = 6.522, y7 = 15.5, y8 = 6.5]; (c) Shows the bistability evolved at $s1_f = 31.7$ initial conditions of periodic solution: y1 = 0.0938, y2 = 0.4365, y3 = 22.387, y4 = 1.554, y7 = 2, y8 = 1, y5 = 0.412, y6 = 21.244, initial conditions for stationary solution: y1 = 0.008, y2 = 0.005, y3 = 2, y4 = 1.5, y5 = 0.4, y6 = 2, y7 = 2, y8 = 1.

minima occur at pH_f = 8.4287 and 8.5217, the two maxima occur at pH_f = 8.6052 and 8.6973. Static limit points are then expected in the model for any values of feed pH lower than 8.6973 and the rest of the system parameters are kept constant as in **Table 1**.

It should be noted that points of maximum and minimum of SLP and HB curves are special points since they are considered as degenerate points. Interesting dynamic behavior can be found at the vicinity of these degenerate points in addition to qualitative changes in bifurcation behavior are excepted when passing these points.

The diagram of **Figure 9(a)** also shows that the HB curve crosses the SLP curve, respectively, at some point. These points are also degenerate points since they create the so-called type ***Fl*** degeneracy [46,47]. The two curves can also be seen to collapse in one line along the right branch of the diagram. This gives birth to another kind of degeneracy, *i.e.*, the so-called ***Fz*** degeneracy [46,47].

From practical point of view, one can expect close to these degenerate points some interesting dynamic behavior ranging from periodic solutions, quasiperiodic solutions to other global bifurcation phenomena. The two-parameter continuation diagrams of **Figures 9(a)-(c)** were, therefore, helpful in providing a "big picture" of the potential bifurcation mechanisms in the system. In this regard, the continuity diagram of **Figure 4(a)** ob-

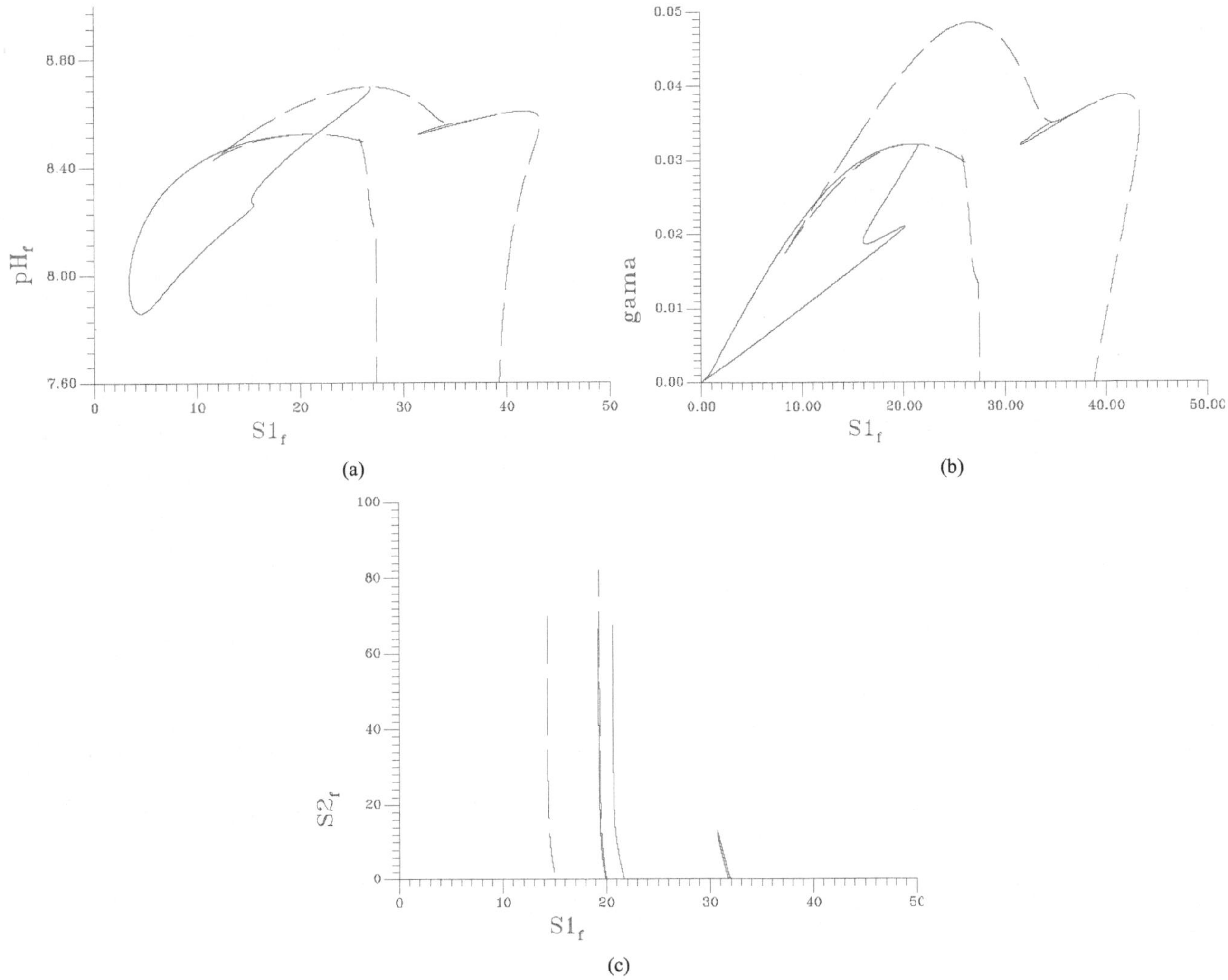

Figure 9. (a) Two parameter continuation [$S1_f$ vs pH_f]; (b) Two parameter continuation [$S1_f$ vs γ(gama)]; (c) Two parameter continuation [$S1_f$ vs $S2_f$].

tained for $s1_f = 20$ should be considered as only the tip of the iceberg as far as the richness of the system is concerned. To have a better understanding of the different bifurcation mechanisms in the system, the diagram of **Figure 9(a)**, is divided into different regions and the static and dynamic bifurcation for each region is investigated:

- $pH_f < 7.8591$, this region characterized by the existence of two limit points creating a range of multiplicity of stationary states as represented by **Figure 10(a)**.
- $pH_f \in (7.8591, 8.4287)$, this region characterized by the existence of two HB bifurcation points and two SLP, a sample result representing this region is shown by **Figure 10(b)** at $pH_f = 8.2$. The bifurcation diagram shows the emerging of a range of unique periodic or non-periodic solutions in addition to the multiplicity range.
- $pH_f \in (8.4287, 8.5217)$, this region is characterized by the existence of two HB bifurcation points and four SLP. A sample result representing this region is shown by **Figure 10(c)** at $pH_f = 8.452$. The bifurcation diagram shows the emerging of a second region of multiple stationary states in addition to the bifurcation behavior noticed at $pH_f = 8.2$ of **Figure 10(b)**.
- $pH_f \in (8.5217, 8.524)$, this region is characterized by the existence of four HB bifurcation points in addition to four SLP. A sample result representing this region was illustrated early by **Figure 10(d)**.
- $pH_f \in (8.524, 8.5936)$ this region is characterized by the existence of three HB bifurcation points and four SLP. A sample result representing this region is illustrated by **Figure 10(e)**, at $pH_f = 8.55$. The bifurcation diagram shows the emerging of isolated stationary solution (ISOLA).
- $pH_f \in (8.5936, 8.6973)$, this region is characterized by the disappearance of the second region of multiplicity and a shrinkage isolated solution as repre-

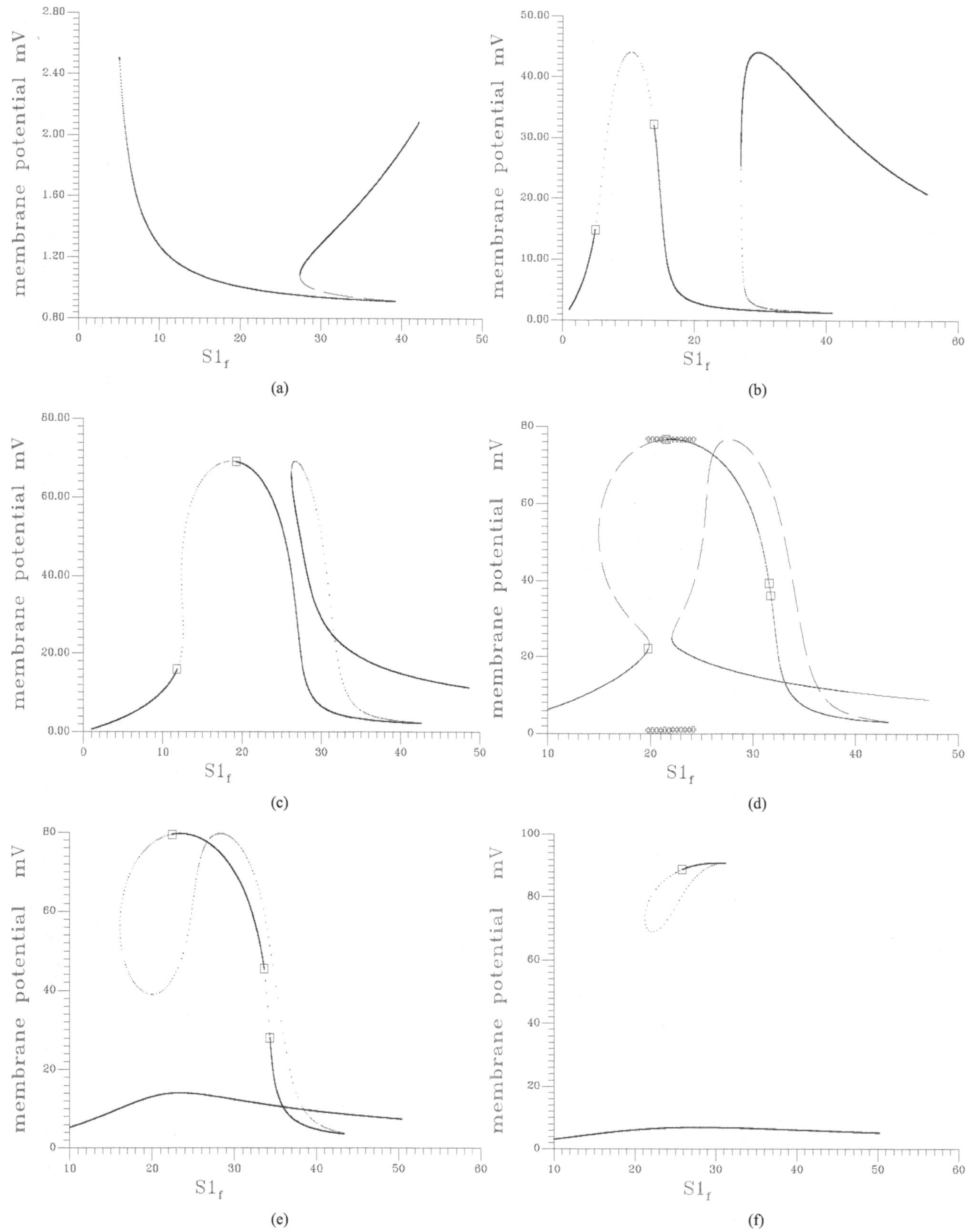

Figure 10. (a) Bifurcation diagram at pH_f = 7.6 [one region of multiple steady states]; (b) Bifurcation diagram at pH_f = 8.2; (c) Bifurcation diagram at pH_f = 8.452; (d) Bifurcation diagram at pH_f = 8.5287874; (e) Bifurcation diagram at pH_f = 8.55; (f) Bifurcation diagram at pH_f = 8.65.

sented by **Figure 10(f)**.

Figures 9(b) and **(c)** show the two-parameter continuation diagrams: ($s1_f$ vs gama) and ($s1_f$ vs $s2_f$) respectively. Variety of different one-parameter bifurcation may be obtained. **Figure 11** shows a special one-parameter bifurcation diagram, where periodic or non-periodic attractors dominate the system behavior for any value of $s2_f$ (feed concentration of choline) greater than the value of the bifurcation parameter at the HB point.

One of the interesting dynamic bifurcation phenomenon obtained is the so-called periodic transient: An orbit in a dissipative system can show transient periodic behavior, *i.e.* before the orbit falls down onto the final attractor it seems periodic for a long, but finite, time. **Figures 12(a)-(c)** show this interesting bifurcation phenolmenon. However in dangerous boundaries of the bifurcation parameter, just before the saddle-node bifurcation, the situation is as follows: an exponentially fast attraction along the contracting direction and slow linear drift along the other. These different two time scales cause this phenomenon. Trajectories starting from a given neighborhood are initially attracted to this point which will become a normal attractor as the bifurcation parameter passes through a critical point, then they spend a certain time in the close vicinity of this point, and finally they all leave this region and never return to it [47].

4.2. Response to Quantal Releases of Acetylcholine

Acetylcholine in rat brain was found to be in the range of 2.2 micromole/lit up to 17.7 micromole/lit. while in the guinea pig cerebral cortex the range is 3.1 micromole/lit (free ACh) to 16.7 micromole/lit (total ACh) [37]. ACh in human plancenta is reported to be in the range of 3 micromole/lit to 555 micromole/lit [37].

The quantal nature of acetylcholine release was discovered at the neuromuscular junction and the demonstrated in a variety of central and peripheral synapses [49, 50]. At the neuromuscular and nerve-electro-plaque synapses, the size of evoked ACh quanta is fairly constant (8000 - 10000 ACH molecules). Quantal secretion causes the change in membrane potential that excite the postsynaptic cell, and transmission at a synapse is critically dependent upon this mechanism of release [32,51]. ACh concentrations depend on the volumes of both compartments. For *in vitro* experiments it is easy to control each compartment volume and consequently controlling of ACh concentration for both the two compartments. With respect to physiological situations, the synaptic clefts have variable constructions and volumes according to its locations and functions. Volume of synaptic clefts could be less than 0.1 femtoliter (10^{-15} liter) and could be greater than 150 femtoliter as shown by [52-55]. Releases of 1 quanta (10,000 molecules) correspond to the release of 0.1661×10^{-13} micromole of ACH taking Avogadro's number in consideration (6.02×10^{23} molecule/mole). So ACh concentrations may take very low values in micromole/lit units up to values of hundred micromole/lit according to the nature and size of the cleft.

In general, excitability means that the response of a system on external perturbations is "all" or "none" depending on whether the strength of the stimulus is above or below a critical threshold. Prominent examples of excitable systems are the spiking of neurons [56], the cardiac muscle [57], the dynamics of life populations [58], or nonlinear chemical reactions [59,60].

Two different forcing mechanisms have been investigated in this work:

1) Continuous oscillating ACh feeding which obeys the function $s1_f = s1_f0 \times (1 + \mathrm{am} \times \sin(\omega \times \tau))$, where (am) is the amplitude of forcing oscillations and ω is the

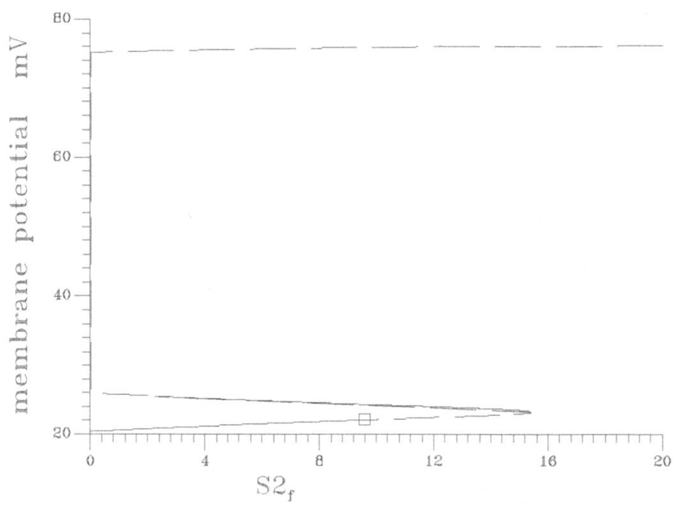

Figure 11. Bifurcation diagram [$S2_f$ vs membrane potential] extracted from Figure 9(c).

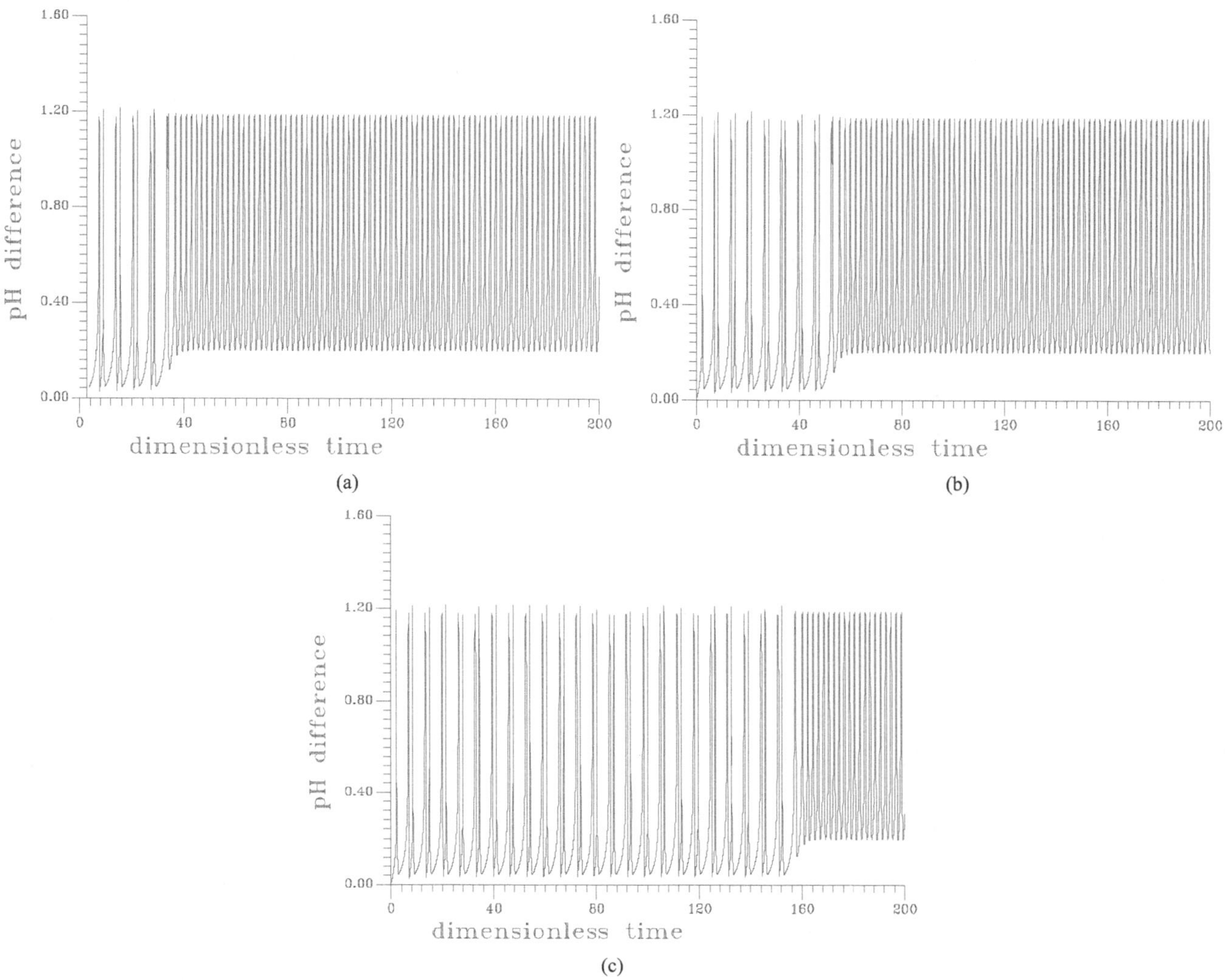

Figure 12. (a) Periodic transient at $s1_f$ = 14.40375 [this attractor is produced as the same set of parameters as in Table 1 but gama = 0.01, H_f = 0.0024]; (b) Periodic transient at $s1_f$ = 14.40373 [this attractor is produced as the same set of parameters as in Table 1 but gama = 0.01, H_f = 0.0024]; (c) Periodic transient at $s1_f$ = 14.403722 [this attractor is produced as the same set of parameters as in Table 1 but gama = 0.01, H_f = 0.0024].

frequency; $s1_f0$ is the starting value of ACh feed concentration ($\tau = 0$) and τ is the dimensionless time. **Figure 13** shows the one dimension excitation map for this case at $s1_f0 = 20$, am = 0.0015 and the natural period of the system before forcing is 11.2636099 and the rest of system parameters as in **Table 1**. It is clear that the bifurcation scenario obeys the so-called crises bifurcation where: period one attractor from left goes suddenly to chaotic attractor. Infinite periodic and chaotic patterns may obtain using **Figure 13**. **Figure 14** shows the phase plan of a chaotic attractor at $\omega = 3.5$.

2) Discontinuous constant feeding of ACh (square forcing): in this case the system response to long period forcing (120, 80) is shown in **Figures 15(a)** and **(b)**. The resulting patterns have the feature of actual physiological response (where the neural bursts involve one or more spikes (60)). **Figure 15** shows the oscillatory pattern of membrane potential at period equal to 80. **Figure 15(b)** shows the oscillatory pattern at period equal to 120.

5. Concluding Remarks

An eight-dimensional non-linear mathematical model (two-enzymes/two-compartments) is developed for a coupled AChE/ChAT enzymes system considering the physiological reality and importance of the transmembrane pH and electrical gradient (ΔpH and membrane potential differences respectively) of this neural system. The investigation has been based on a well established kinetic scheme and kinetic data. The model accounts for the effects of hydrogen ion concentrations on the kinetics and its role in creating membrane potential, assuming no other charged ions exist. Both autonomous and non-autonomous cases were investigated considering the two common mechanisms of applying acetylcholine in prac-

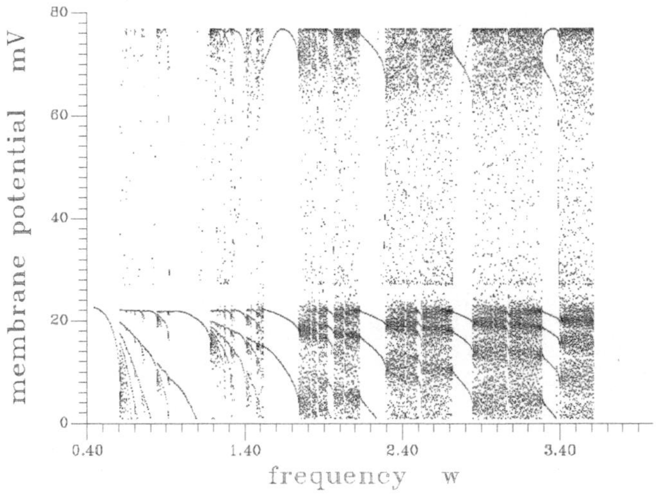

Figure 13. One dimension excitation map using the function $S1_f = S1_f0*(1 + Am*sin(\omega\tau))$-[Am = 0.015, natural period = 11.2636099].

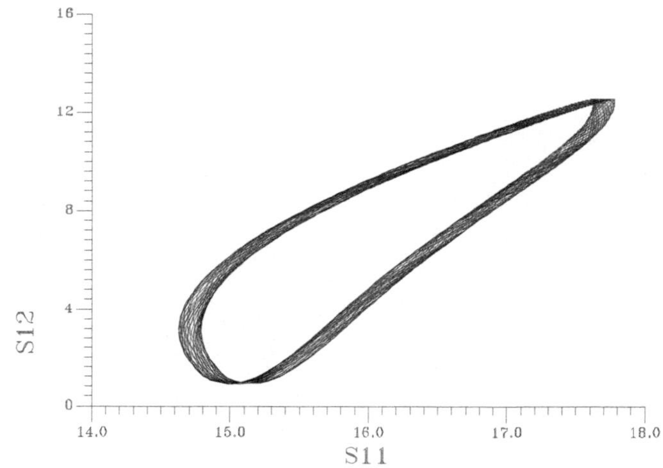

Figure 14. Phase plane of chaotic attractor at ω = 3.5 [Figure 13].

tical physiological situations (constant and quantal).

The investigation uncovered a wealth of static and dynamic bifurcations of the system including multiplicity of steady states, isola, periodic and aperiodic behavior. The use of two parameters continuation technique uncover different qualitative changes in bifurcations with respect to three of the system parameters (pH_f, γ, $S2_f$). Different bifurcations occurred at alkaline feed pH and a wide range of the other two parameters.

It was observed that all the state variables were very sensitive to the variations in the acetylcholine inlet concentration $S1_f$ as a bifurcation parameter. Their behavior was strongly dominated by the hysteresis and multiplicity phenomena throughout a large range of the bifurcation parameter. Different kinds of solutions existed: point (stationary state), periodic, chaotic bi-stability, tri-stability was found. Boundary crises bifurcations are the dominating scenario of bifurcation: periodic transient and boundary crises bifurcation to chaos were shown. Intrinsic bursts with one spike are the most common periodic and aperiodic behavior of the system.

It is noticed that the variation in pH and consequentially the membrane potential are a little bit bigger than those recorded in actual physiological situations; this is due to the fact that the model ignores any effects of other ions (generation and transport) but hydrogen ions, and the assumption of fully ionization of the acetic acid.

Different kinds of bursts may be obtained in response to the quantal feeding mechanism; bursts with more than one spike were obtained (spike train).

An experimental research work is recommended in order to use this model for simulating real physiological behavior. Availability of good experimental data for human brain in the future can help in improving this model. It can also help in deeper understanding of the physiological behavior and in planning better brain experiments and in linking the complex behavior investigated to the cholinergic disorders.

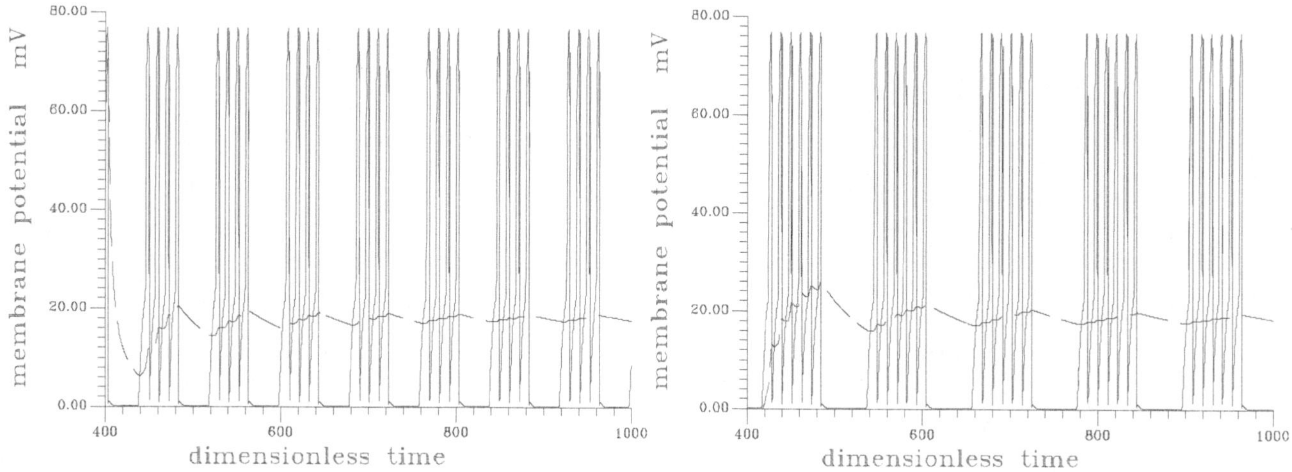

(a) (b)

Figure 15. (a) Spike train obtained at square input [$s1_f = 20$], period = 80 and frequency of the input = 0.078539816. [$\gamma = 0.032$, $H_f = 0.003$]; (b) Spike train at frequency of 0.0523598775, period of 120, and the same parameters as (a).

6. Acknowledgements

The authors would like to thank the Deanship of Scientific Research, Salaman Bin AbdulAziz University, Kingdom of Saudi Arabia for their support to this research.

REFERENCES

[1] D. M. Quinn, A. S. Balasubremanian, B. P. Doctor and P. Taylor, "Enzymes of the Cholinestrase Family," Plenum Press, London, 1995.

[2] S. Tucek, "Acetylcholine Synthesis in Neurons," Chapman and Hall, London, 1978.

[3] C. A. Guyton, "Medical Physiology," 7th Edition, W. B. Saunders Company, Philadelphia, 1986.

[4] T. E. Barman, "Enzyme Handbook," Vol. 1, Springer-Verlage, New York, 1969.

[5] R. R. Llinás, "The Squid Giant Synapse: A Model for Chemical Transmission," Oxford University Press, London, 1999.

[6] http://www.biology.ucsd.edu/classes/bibc102.WI99/neuro/Neurotransmitter. html

[7] H. Soreq and H. Zakut, "Human Cholinsterases and Anticholinesterases," Academic Press, San Diego, 1993.

[8] A. L. Koch, "The pH in the Neighborhood of Membranes Generating a Protomotive Force," *Journal of Theoretical Biology*, Vol. 120, No. 1, 1986, pp. 73-84.

[9] R. Friboulet, A. David and D. Thomas, "Excitability Memory and Oscillation in Artificial Acetylcholinesterase Membranes," *Journal of Membrane Science*, Vol. 8, No. 33, 1981, pp. 33-39.

[10] J. L. Hindmarsh and R. M. Rose, "A Model of Neuronal Bursting Using Three Coupled First Order Differential Equations," *Proceedings of the Royal Society*, Vol. 221, No. 1222, 1984, pp. 87-102.

[11] J. L. Hindmarsh and R. M. Rose, "A Model of the Nerve Impulse Using Two First-Order Differential Equations," *Nature*, Vol. 296, No. 5853, 1982, pp. 162-164.

[12] R. Fitzhugh, "Impulses and Physiological States in Theoretical Models of Nerve Membranes," *Biophysical Journal*, Vol. 1, No. 6, 1985, pp. 445-466.

[13] A. V. Holden and Y. S. Fan, "Crisis-Induced Chaos in the Rose-Hindmarsh Model for Neuronal Activity," *Chaos, Solitons and Fractals*, Vol. 2, No. 6, 1992, pp. 583-595.

[14] A. V. Holden and Y. S. Fan, "From Simple to Complex Oscillatory Behaviour via Intermittent Chaos in the Rose-Hindmarsh Model for Neuronal Activity," *Chaos, Solitons and Fractals*, Vol. 2, No. 4, 1992, pp. 349-369.

[15] A. V. Holden and Y. S. Fan, "From Simple to Simple Bursting Oscillatory Behaviour via Chaos in the Rose-Hindmarsh Model for Neuronal Activity," *Chaos, Solitons and Fractals*, Vol. 2, No. 3, 1992, pp. 221-236.

[16] Y. S. Fan and A. V. Holden, "Bifurcation, Bursting, Chaos and Crisis in the Rose-Hindmarsh Model for Neuronal Activity," *Chaos, Solitons and Fractals*, Vol. 3, No. 4, 1993, pp. 439-449.

[17] G. Ibrahim, O. Saleh, I. H. Mustafa, A. H. El Ahwany and S. S. E. H. El Nashaie, "Modeling Peiodic and Aperiodic Behavior of Acetylcholine Hydrolysis," *International Review of Chemical Engineering*, Vol. 2, No. 6, 2010, pp. 19-33.

[18] G. Ibrahim and S. S. E. H. Elnashaie, "Hyperchaos in Acetylcholinesterase Enzyme Systems," *Chaos, Solitons and Fractals*, Vol. 8, No. 12, 1997, pp. 1997-2007.

[19] I. H. Mustafa, G. Ibrahim, A. Elkamel, S. S. E. H. Elnashaie and P. Chen, "Non-Linear Feedback Modeling and Bifurcation of the Acetylcholine Neurocycle and Its Relation to Alzheimer's and Parkinson's Diseases," *Che-*

mical Engineering Science, Vol. 64, No. 1, 2009, pp. 69-90.

[20] I. H. Mustafa, A. Elkamel, G. Ibrahim, P. Chen and S. S. E. H. Elnashaie, "Effect of Choline and Acetate Substrates on Bifurcation and Chaotic Behavior of Acetylcholine Neurocycle and Alzheimer's and Parkinson's Diseases," *Chemical Engineering Science*, Vol. 64, No. 9, 2009, pp. 2096-2112.

[21] I. H. Mustafa, A. Elkamel, P. Chen, G. Ibrahim and S. S. E. H. Elnashaie, "Effect of Cholineacetyltransferase Activity and Choline Recycle Ratio on Diffusion-Reaction Modelling, Bifurcation and Chaotic Behavior of Acetylcholine Neurocycle and Their Relation to Alzheimer's and Parkinson's Diseases," *Chemical Engineering Science*, Vol. 68, No. 1, 2012, pp. 19-35.

[22] P. Garhyan, A. M. Botero and S. S. E. H. Elnashaie, "Complex Bifurcation/Chaotic Behavior of Acetylcholinesterase and Cholineacetyltransferase Enzymes System," *Applied Mathematical Modelling*, Vol. 30, 2006, pp. 824-853.

[23] L. B. Hersh and M. Peet, "Re-Evaluation of the Kinetic Mechanism of the Choline Acetyltransferase Reaction," *The Journal of Biological Chemistry*, Vol. 252, No. 14, 1977, pp. 4796-4802.

[24] E. J. Doedel, A. R. Champneys, T. F. Fairgrieve, Y. A. Kuznetsov, B. Sandstede and X. J. Wang, "AUTO97: Continuation and Bifurcation Software for Ordinary Differential Equations," Department of Computer Science, Concordia University, Montreal, 1997.

[25] S. S. E. H. Elnashaie, G. Ibrahim and F. A. Teymour, "Chaotic Behavior of an Acetylcholinesterase Enzyme System," *Chaos, Solitons and Fractals*, Vol. 5, No. 6, 1995, pp. 933-954.

[26] S. Karel and H. Milan, "Biotechnological Aspects of Membrane Function," *Critical Reviews in Biotechnology*, Vol. 17, No. 2, 1997, pp. 69-86.

[27] M. Obara, M. Szeliga and J. Albrecht, "Regulation of pH in the Mammalian Central Nervous System under Normal and Pathological Conditions: Facts and Hypotheses," *Neurochemistry International*, Vol. 52, No. 6, 2008, pp. 905-919.

[28] J. Antosdiewicz, J. A. McCammon and M. K. Gilson, "Prediction of pH-Dependent Properties of Proteins," *Journal of Molecular Biology*, Vol. 238, No. 3, 1994, pp. 415-436.

[29] D. M. Michaelson and I. Angel, "Determination of Delta pH in Cholinergic Synaptic Vesicles: Its Effect on Storage and Release of Acetylcholine," *Life Sciences*, Vol. 27, No. 1, 1980, pp. 39-44.

[30] H. H. Fuldner and H. Stadler, "^{31}P-NMR Analysis of Synaptic Vesicles, Status of ATP and Internal pH," *European Journal of Biochemistry*, Vol. 121, No. 3, 1982, pp. 519-524.

[31] R. Ahdu-Hacohen, D. Duridanova, H. Meiri and R. Rahamimoff, "Hydrogen Ions Control Synaptic Vesicle Ion Channel Activity in Torpedo Electromotor Neurons," *Journal of Physiology*, Vol. 556, No. 2, 2004, pp. 347-352.

[32] S. M. Parsons, "Transport Mechanisms in Acetylcholine and Monoamine Storage," *The FASEB Journal*, Vol. 14, No. 15, 2000, pp. 2423-2434.

[33] S. Heven, X. H. Li and G. P. Michael, "Energization of Plant Cell Membranes by H^+-Pumping ATPases: Regulation and Biosynthesis," *The Plant Cell*, Vol. 11, No. 4, 1999, pp. 677-689.

[34] A. L. Koch, "The pH in the Neighborhood of Membranes Generating a Protomotive Force," *Journal of Theoretical Biology*, Vol. 120, No. 1, 1986, pp. 73-84.

[35] C. Rae, R. Scott, C. H. Thompson, I. Dumughn, G. Kemp, P. Styles, I. M. Tracey and G. K. Radda, "Is Brain pH a Biochemical Marker of IQ?" *Proceedings of the Royal Society*, Vol. 263, No. 1373, 1996, pp. 1061-1064.

[36] A. Zauner and J. P. Muizelaar, "Brain Metabolism and Cerebral Blood Flow," In: P. Reilly and R. Bullock, Eds., *Head Injury: Pathophysiology and Management of Severe Closed Injury*, Chapman and Hall, London, 1997.

[37] I. Wessler, E. Roth, S. Schwarze, W. Weikel, F. Bittinger, C. J. Kirkpatrick and H. Kilbindger, "Release of Non-Neural Acetylcholine from the Human Placenta: Difference to Neural Acetylcholine," *Naunyn-Schmiedeberg's Archives of Pharmacology*, Vol. 364, No. 3, 2001, pp. 205-212.

[38] S. Kysela and J. Torok, "Histamine H1-Receptor Antagonists do not Prevent the Appearance of Endothelium-Dependant Relaxation to Acetylcholine in Rat Pulmonary Artery," *Physiological Research*, Vol. 45, No. 4, 1996, pp. 345-350.

[39] R. R. Chay and J. Rinzel, "Bursting, Beating and Chaos in an Excitable Membrane Model," *Biophysical Journal*, Vol. 47, No. 3, 1985, pp. 357-366.

[40] W. F. Boron and E. L. Boulpaep, "Medical Physiology," W.B. Saunders, Philadelphia, 2002.

[41] P. Constable, "A Simplified Strong Ion Model for Acid-Base Equilibria: Application to Horse Plasma," *Journal of Applied Physiology*, Vol. 83, No. 1, 1997, pp. 297-311.

[42] K. A. Gossett, D. D. French, B. Cleghorn and G. E. Church, "Effect of Acute Acidemia on Blood Biochemical Variables in Healthy Ponies," *American Journal of Veterinary Research*, Vol. 51, No. 9, 1990, pp. 1375-1379.

[43] L. P. Sil'nikov, "A Contribution to the Problem of the Structure of an Extended Neighbourhood of a Structurally Stable Equilibrium of Saddle-Focus Type," *Mathematics of the USSR-Sbornik*, Vol. 10, No. 1, 1970, pp. 91-102.

[44] O. E. Rossler, J. L. Hudsun and R. Rossler, "Homoclinic Hyperchaos—An Explicit Example," *Physica D: Nonlinear Phenomena*, Vol. 62, No. 1-4, 1993, pp. 80-86.

[45] C. Javier, I. A. García and J. Sorolla, "Resolution of the Poincaré Problem and Nonexistence of Algebraic Limit Cycles in Family (I) of Chinese Classification," *Chaos, Solitons & Fractals*, Vol. 24, No. 2, 2005, pp. 491-499.

[46] S. Wiggins, "Introduction to Applied Nonlinear Dynami-

cal Systems and Chaos," Springer-Velag, New York, 1990.

[47] J. Guckenheimer and P. Hohns, "Nonlinear Oscillations, Dynamical Systems and Bifurcntion of Vector Fields," Springer, New York, 1983.

[48] M. Franaszek, "Long-Lived Periodic Transients," *Acta Physica Polonica B*, Vol. 22, No. 9, 1991.

[49] P. Fatt and B. Katz, "Spontaneous Sub Threshold Activity at Motor Nerve Endings," *The Journal of Physiology*, Vol. 117, No. 1, 1952, pp. 109-128.

[50] M. R. Bennett, "The Origin of Gaussian Distributions of Synaptic Potentials," *Progress in Neurobiology*, Vol. 46, No. 4, 1995, pp. 331-350.

[51] Y. Dunant and M. Israel, "Neurotransmitter Release at Rapid Synapses," Biochimie, Vol. 82, No. 4, 2000, pp. 289-302.

[52] M. C. Rosales-Hernandez, J. E. Mendieta-Wejebe, J. Correa-Basurto, J. I. Vazques-Alcantara, E. Terres-Rojas and J. Trujillo-Ferrara, "Catalytic Activity of Acetylcholinesterase Immobilized on Mesoporous Molecular Sieves," *International Journal of Biological Macromolecules*, Vol. 40, No. 5, 2007, pp. 444-448.

[53] F. H. White, D. A. Thompson and K. Cohari, "Ultra Structural Morphometry of Gap Junctions during Differentiation of Stratified Squamous Epithelium," *Journal of Cell Science*, Vol. 69, 1984, pp. 67-85.

[54] W. Van Der Kloot, J. Molgo, R. Cameron and C. Colasante, "Vesicle Size and Transmitter Release at the Frog Neuromuscular Junction When Quantal Acetylcholine Content Is Increased or Decreased," *Journal of Physiology*, Vol. 541, 2002, pp. 385-393.

[55] L. P. Sartchenko and D. A. Rusakov, "The Optimal Height of Synaptic Cleft," *Proceedings of the National Academy of Sciences of the USA*, Vol. 104, No. 6, 2007, pp. 1823-1828.

[56] A. L. Hodgkin, "The Local Electric Changes Associated with Repetitive Action in a Non-Medulated Axon," *Journal of Physiology*, Vol. 107, 1948, pp. 165-181.

[57] J. D. Murray, "Mathematical Biology," Springer, New York, 1990.

[58] D. R. Clother and J. Brindley, "Excitability of an Age-Structured Plankton Ecosystem," *Journal of Mathematical Biology*, Vol. 39, No. 5, 1999, pp. 377-420.

[59] A. Zhabotinsky and A. Zaikin, "Concentration Wave Propagation in Two-Dimensional Liquid-Phase Self-Oscillating System," *Nature*, Vol. 225, 1970, pp. 535-537.

[60] E. M. Izhikevich, "Dynamical Systems in Neuroscience: The Geometry of Excitability and Bursting," MIT Press, Cambridge, 2007.

Notation

Symbol	Description
A_m	Membrane area, m^2
E_1, E_2	Enzyme concentration in kg/m^3-1 for ChAT enzyme-2 for AChE enzyme
H_j	Hydrogen ions concentration in $kmol/m^3$
$K_s, K_{s1}, K_{s2}, K_{s3}, K_i$	Equilibrium and inhibition constants
$K_{h1}, K_{h2}, K'_{h1}, K'_{h2}$	Equilibrium constants
Q	Volumetric flow rate in m^3/s
R_1, R_2	Rate of production and consumption of ACh respectively in s^{-1}
S_{1j}	ACh concentration in $kmol/m^3$
S_{2j}	Choline concentration in $kmol/m^3$
S_{3j}	Acetyl Coenzyme concentration in $kmol/m^3$
t	Time in s
V_j	Compartment volume in m^3
V_{m1}	Specific enzyme velocity (ChAT) in s^{-1}
V_{m2}	Specific enzyme velocity (AChE) in s^{-1}
α'_h	Membrane permeability for hydrogen ions, m/s
α'_{oh}	Membrane permeability for hydroxyl ions, m/s
α'_{s1}	Membrane permeability for ACh, m/s
α'_{s2}	Membrane permeability for Choline, m/s
α'_{s3}	Membrane permeability for acetyl Co enzyme, m/s

Abbreviations

AChE	Acetylcholinesterase
ChAT	Choline Acetyltransferase
CoA	Coenzyme A
ACh	Acetylcholine
Ch	Choline
CSTR	Continuously Stirred Tank Reactor
HB	Hopf Bifurcation
SB	Static Bifurcation
U	Periodic Limit Point
PD	Period Doubling
Pi	Periodicity *i* of the Periodic Orbit

Subscripts

1	Compartment 1
2	Compartment 2
f	Feed condition

Legend for Figures

———	Stable steady state branch
---------	Unstable steady state branch
•••••••	Stable periodic branch

Dimensionless Parameters and State Variables

Symbol	Description
$\alpha = \frac{K_{s1}}{K_i}$	Dimensionless substrate inhibition constant
$\alpha_h = \frac{\alpha'_h A_m}{q}$	Dimensionless permeability of hydrogen ions
$\alpha_{oh} = \frac{\alpha'_{oh} A_m}{q}$	Dimensionless permeability constant of hydroxyl ions
$\alpha_{s1} = \frac{\alpha'_{s1} A_m}{q}$	Dimensionless permeability c onstant of ACh
$\alpha_{s2} = \frac{\alpha'_{s2} A_m}{q}$	Dimensionless permeability constant of Choline
$\alpha_{s3} = \frac{\alpha'_{s3} A_m}{q}$	Dimensionless permeability constant of Acetyl Co-enzyme
$\beta = \frac{K_{s2}}{K_{s3}}$	Dimensionless equilibrium ratio
$\beta_1 = \frac{V_1 V_{m2} E_2}{K_{h2} q}$	Dimensionless reaction activity constant
$\beta_2 = \frac{V_1 V_{m1} E_1}{K_{s1} q}$	Dimensionless reaction activity constant
$\beta_3 = \frac{V_1 V_{m2} E_2}{K_{s1} q}$	Dimensionless reaction activity constant
$\beta_4 = \frac{V_1 V_{m1} E_1}{K_{s2} q}$	Dimensionless reaction activity constant
$\beta_5 = \frac{V_1 V_{m2} E_2}{K_{s2} q}$	Dimensionless reaction activity constant
$\beta_6 = \frac{V_1 V_{m1} E_1}{K_{s3} q}$	Dimensionless reaction activity constant
$\beta_7 = \frac{V_1 V_{m2} E_2}{K_{s3} q}$	Dimensionless reaction activity constant
$\gamma = \frac{K_w}{K_{h2}^2}$	Dimensionless equilibrium ratio
$\delta_1 = \frac{K_{h2}}{K'_{h2}}$	Dimensionless equilibrium ratio

$\delta_2 = \dfrac{K_{h2}}{K'_{h1}}$	Dimensionless equilibrium ratio
$\delta_3 = \dfrac{K_{h1}}{K_{h2}}$	Dimensionless equilibrium ratio
$\tau = \dfrac{qt}{V_1}$	Dimensionless time
$h_j = \dfrac{H_j}{K_{h2}}$	Dimensionless hydrogen ions concentration
h_f	
$k_d = \dfrac{k_s}{k_{S2}k_{S3}}$	Dimensionless equilibrium ratio
$S_{1,j} = \dfrac{S_{1,j}}{K_{S1}}$	Dimensionless ACh concentration
$S_{2,j} = \dfrac{S_{2,j}}{K_{S2}}$	Dimensionless Choline concentration
$S_{3,j} = \dfrac{S_{3,j}}{K_{S3}}$	Dimensionless Acetyl Co-enzyme concentration
S_{1f}	ACh feed concentration
S_{2f}	Choline feed concentration
S_{3f}	Acetate feed concentration
$V_r = \dfrac{V_1}{V_2}$	Compartment volume ratio

Dehydrocyclization of n-Hexane over Heteropolyoxometalates Catalysts

Abdellah Eid[1], Ouarda Benlounes[1], Hikmat S. Hilal[2], Chérifa Rabia[3], Smain Hocine[1]
[1]Laboratoire de Chimie Appliquée et de Genie Chimique, Universite Mouloud Mammeri, Tizi-Ouzou, Algeria
[2]Laboratory of Semiconductor and Solar Energy Research, Chemistry Department, An-Najah N. University, Nablus, Palestine
[3]Laboratoire de Chimie du Gaz Naturel, Institut de Chimie USTHB, Alger, Algeria

ABSTRACT

The catalytic dehydrocyclization of n-hexane was studied here for the first time using a number of compounds based on $H_3PMo_{12}O_{40}$. The described catalysts were prepared by either replacing the acidic proton with counter-ions such as ammonium or transition metal cations (NH_4^+, Fe^{3+}, K^+), or by replacing Mo^{6+} with (Ni^{3+}, Co^{3+}, Mn^{3+}) in the polyoxometalate framework, as reported earlier. For comparison purposes, the known $(TBA)_7PW_{11}O_{39}$ catalyst system was used. All reactions were conducted at different temperatures in the range 200˚C - 450˚C. The Keggin structure of these heteropolycompounds was ascertained by XRD, UV and IR measurements. ^{31}P NMR measurements and thermal behaviour of the prepared catalysts were also studied. These modified polyoxometalates exhibited heterogeneous super-acidic catalytic activities in dehydrocyclization of n-hexane into benzene, cyclohexane, cyclohexene and cyclohexadiene. The catalysts obtained by substituting the acidic proton or coordination atom exhibited higher selectivity and stability than the parent compound $H_3PMo_{12}O_{40}$. Catalytic activity and selectivity were heavily dependent on the composition of the catalyst and on the reaction conditions. At higher temperatures, the catalyst exhibited higher conversion efficiency at the expense of selectivity. Using higher temperatures (>400˚C) in the presence of hydrogen carrier gas, selectivity towards dehydrocyclization ceased and methane dominated. To explain the results, a plausible mechanism is presented, based on super-acidic nature of the catalyst systems.

Keywords: Dehydrocyclization; Heteropolyacid; Catalyst; n-Hexane; Keggin Ion

1. Introduction

Catalytic reforming of n-hexane is one of the most widely used processes in oil refining industry. Fuel octane number can be enhanced by increasing concentrations of aromatic compounds [1-3].

Catalytic reforming proceeds over bifunctional catalysts having both metallic and acidic functions are currently being considered [4,5]. Heteropoly acids (HPAs) are complex proton acids that incorporate polyoxometalate anions (heteropoly anions) having metal-oxygen octahedra as the basic structural units. HPAs have several advantages such as very strong Brönsted acidities approaching the super acid region [6] and high oxidant activities in multielectron redox reactions under mild conditions. Their acid-base and redox properties can be varied widely by changing their chemical compositions.

HPAs have a discrete ionic structure, containing the fairly mobile heteropoly anions and counter-cations (H^+, H_3O^+, $H_5O_2^+$, etc.). This unique structure results in significantly high proton mobility and pseudo liquid phase behaviour. In addition, HPAs are highly soluble in polar solvents and their thermal stabilities are quite high. These properties make HPAs potentially promising acid, redox, and bifunctional catalysts in both homogeneous and heterogeneous systems [7,8].

Heteropolyacids (HPA) usually act as strong acidic catalysts, with the ability to activate alkanes and produce carbonium ion intermediates [9,10]. Such intermediates are necessary for alkane cracking, alkylation and isomerization processes [11]. The large-scale industrial applications of these reactions drew attention to such catalyst systems for long time [12].

The transformation of n-hexane depends on the nature of the catalyst and on the operation conditions. The reactant may undergo three different competing reactions, namely: Hydrogenolysis to short linear chain alkanes such as methane, ethane, propane, butane and pentane in presence of H-[Al]-ZMS-5 leading [13]. Isomerization over Pt/zeolithe giving iso-alkanes, such as isobutane, isopentane, 2,2-dimethylbutane, 2,3-dimethylbutane, 2-methylpentane and 3-methylpentane [14]. Dehydrocyclization and aromatization giving cyclohexane, cyclohexene, and benzene on catalysts such as WO_3/ZrO_2 [15].

In our search for highly efficient, stable and selective catalyst systems for dehydrocyclization of hexane, synthesised and characterised a number of compounds based on modified polyoxomolybdates and polyoxytengestates. The investigated compounds are the heteropolyacid $H_3PMo_{12}O_{40}{\cdot}xH_2O$ (noted PMo_{12}), the ammonium salt of the (Co, Ni, Mn) substituted species $(NH_4)_4PMo_{11}O_{39}M$ $(H_2O){\cdot}xH_2O$ (M: Co, Ni, Mn) (noted $PMo_{11}Co$, $PMo_{11}Ni$, $PMo_{11}Mn$). $(TBA)_7PW_{11}O_{39}{\cdot}xH_2O$ (noted PW_{11}) and $KFePMo_{12}O_{40}{\cdot}xH_2O$ (noted $FePMo_{12}$). Such complexes have been intensively studied for different organic reactions [16-20], and can activate alkanes due to their superacidic nature [21-24].

In this work, the compounds have been used in hexane dehydrocyclization reaction for the first time. The reaction may involve other products such as cyclohexane, cyclohexene and cyclohexadiene, in addition to the unwanted methane product. The main goal is to maximize benzene production and minimize methane in the product mixture, by investigating different catalysts, varying temperature, and carrier gas and retention time.

2. Experimental Part

2.1. Preparation of Catalysts

Pure heteropolyacid $H_3PMo_{12}O_{40}{\cdot}xH_2O$ was prepared as described earlier [25]. The mixed potassium-iron salt $K_{0.5}Fe_{0.1}H_{0.2}PMo_{12}O_{40}{\cdot}xH_2O$ was prepared as follows: to an aqueous solution of $H_3PMo_{12}O_{40}$ (50 ml, 0.08 M) were added successively 9.0 ml of 0.10 M $Fe(NO_3)_3$ and 7.5 ml of 0.80 M KNO_3. The obtained precipitate was filtered over glass frit and dried at 50°C under vacuum for 5 h.

The ammonium salt $(NH_4)_4PMo_{11}MO_{39}$ $(H_2O){\cdot}xH_2O$ (M = Ni, Co, Mn) was prepared as described earlier [26]. A mixture of H_3PO_4 (5.0 ml; 1.00 M), H_2SO_4 (10.0 ml; 0.50 M) and MSO_4 (5.0 ml; 1.00 M) was slowly added to 250 ml of an aqueous solution of ammonium paramolybdate $[(NH_4)_6Mo_7O_{24}{\cdot}4H_2O]$ (48.8 g, 40 mmol) at 0°C. The ammonium salt completely precipitated after addition of solid ammonium nitrate (NH_4NO_3).

$(TBA)_7PW_{11}O_{39}$ was prepared by the method described in the literature [27]. A 15.00 g quantity of $H_3PW_{12}O_{40}{\cdot}14H_2O$ was dissolved in 20.0 mL of water; a 5.0 g quantity of $[(n\text{-}C_4H_9)_4N]Br$ (TBABr) was dissolved in 50.0 mL of CH_2Cl_2. The mixture was poured into the molybdophosphate solution with vigorous stirring. Three phases were formed: a yellow solid one, a lower yellow liquid one (CH_2Cl_2), and a poorly yellow coloured aqueous upper one. The solid was washed with Et_2O, providing 8.50 g of product. A further yield of 2.50 g was obtained by addition of Et_2O to the CH_2Cl_2 phase. Recrystalization was performed in CH_2Cl_2.

2.2. Equipment

The UV analysis of the catalyst samples (dissolved in acetonitrile/water 1:1 ratio, as dilute samples solutions 10-5-10-3 M) were performed on a UV-1601PC SHIMADZU spectrophotometer. IR spectra of the samples were obtained at room temperature with a BIO-RAD FTS 165 FTIR spectrometer using standard KBr pellet technique. Measurements were taken in the wave-number range of 4000 - 400 cm^{-1}, 256 scans, and resolution 2 cm^{-1}.

The X-ray powder diffraction (XRD) diagrams were recorded on a SIEMENS D5000 equipment in the 5 - 45° 2θ range, at a scanning speed of 1°C per minute, using CuK_α radiation. The ^{31}P NMR measurements were performed on a Bruker CXP10 spectrophotometer at 10 MHz. Chemical shifts were referenced to $Al(PO_3)_3$.

Thermal behaviour of the prepared catalysts was studied using a TGA-DTA SETARAM apparatus. For each experiment 20 mg of powdered sample were heated at a temperature ramp rate of 5°C/min under air flow. Differential scanning calorimetry experiments (DSC) were run on a DSC 111 SETARAM. Apparatus under similar conditions, with a nitrogen flow.

2.3. Catalytic Experiments

The conversion of n-hexane was carried out in a quartz fixed-bed flow reactor at different temperatures from 200°C to 400°C, using 0.5 g of catalyst. Catalysts were heated to the desired reaction temperature at 5°C/min under nitrogen flow and then maintained at this temperature for one hour under nitrogen. A mixture of n-hexane (50 torr) and hydrogen was flowed over the catalyst (0.1 l/h). The reaction mixture was analysed by gas chromatography on a Hewlett Packard 5830A apparatus equipped with two successive separation columns, filled respectively with ®Porapak QS for a TCD detector and 5Å-molecular sieve for FIDCG detector.

Reaction rates (r) were calculated from the following equation:

$$r\left(mol \cdot h^{-1} \cdot g^{-1}\right) = F/22.4 \times 273/T \times p/760 \times 1/m \quad (1)$$

where F is the flow rate of the vector gas (l/h), T the reaction temperature (K), p the partial pressure of the products (torr) and m the mass of the catalyst (g).

The activation energy (E_a) was obtained from the Arrhenius plot according to:

$$r = k \times \exp\left(-E_a/RT\right) \quad (2)$$

where R is the perfect gas constant and k the pre-exponential factor.

3. Results and Discussion

3.1. Characterization of Catalysts

In the UV, the Keggin heteropolyanions show two absorption bands, one around 200 nm and another around 280 nm. According to literature [28], when catalyst concentration inside the suspension decreases, the two bands disappear gradually indicating decomposition of the polyanion with dilution [29]. Therefore, intermediate concentrations of different catalyst systems were used here for characterization, as shown in **Figure 1**.

Values of absorption bands are summarized in **Table 1**. The Table shows main band observed for each prepared catalyst. The bands are assigned respectively to the vibrations of terminal Mo = Ot and bridging bonds. The bridging bonds are two types: the inter-bridges (M-Ob-M) that occur between two adjacent octahedra, and the intra-bridges (M-Oc-M) that occur within same octahedron.

Solid-state FT-IR spectra measured for different prepared catalysts are given in **Figure 2**. The spectra display similar patterns characteristic for the Keggin structure [30,31]. Values of IF bands are summarized in **Table 2**.

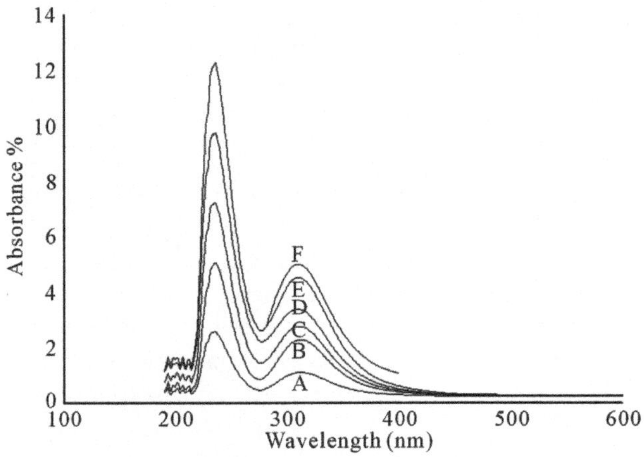

Figure 1. UV spectra for different catalyst systems (A: PMo_{12}; B: $FePMo_{12}$; C: $PMo_{11}Ni$; D: $PMo_{11}Co$; E: $PMo_{11}Mn$; F: PW_{11}). A suspension of each catalyst system in acetonitrile/water mixture was used for UV absorption spectra.

Table 1. Main UV bands observed for different HPA catalyst systems.

Catalyst	λ_{max} (nm)	
	M = Ot	M-Ob
PMo_{12}	237	310
PW_{11}	248	311
$PMo_{11}Co$	232	306
$PMo_{11}Ni$	236	306
$PMo_{11}Mn$	237	311
$KFeMo_{12}O_{40}$	235	306

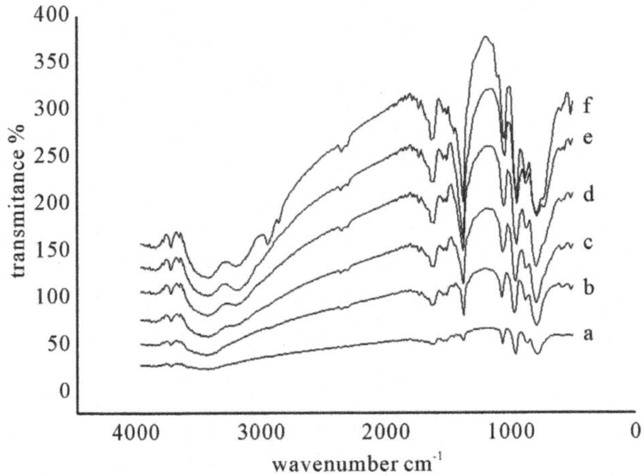

Figure 2. Solid state FT-IR spectra measured as KBr discs for a: PW_{11}; b: $PMo_{11}Mn$; c: $PMo_{11}Ni$; d: $PMo_{11}Co$; e: $KFePMo_{12}$; f: PMo_{12}.

Table 2. FT-IR absorption band values for different catalyst systems, together with their assignments.

Catalys	P-O_a	Mo-O_t	Mo-O_b-Mo	Mo-O_c-Mo
$HPMo_{12}$	1064	962	868	789
$PMo_{11}Ni$	1047	945	870	725
$PMo_{11}Mn$	1108	933	900	725
$PMo_{11}Co$	1047	943	806	737
PW_{11}	1059	962	884	797
$KFePMo_{12}$	1064	962	868	789

FT-IR spectra show that main peaks around 1064, 961, 861 and 786 cm^{-1} are attributed to P-O_a, Mo = O_t, Mo-O_b-Mo, and Mo-O_c-Mo bonds, respectively.

^{31}P NMR spectra were measured for only two catalyst types, as shown in **Figures 3(a)** and **(b)**. **Figure 3(a)** shows a single peak at −3.47 ppm for salt $Fe_{0.1}PMo_{12}O_{40}$, which indicates purity of the sample. The KPW_{11} heteropolyanion showed a signal at 10.35 ppm, **Figure 3(b)**.

The signal is characteristic for the lacunary species $K_7[\alpha\text{-}PW_{11}O_{39}]$ [32,33]. ^{31}P NMR measured and literature data are summarized in **Table 3**.

XRD patterns were measured for all solid catalysts systems in the powder form. **Figure 4** shows the diffractograms of $HPMo_{12}$, $FePMo_{12}$ and $PMo_{11}Ni$. The diffractogram of $H_3PMo_{12}O_{40}\cdot 13H_2O$ was consistent with literature [36] and corresponded to a triclinic structure.

Other diffractograms show that the salt $FePMo_{12}$ crystallises in a cubic structure with intense (222) peak [16, 37]. This corresponds to same diffraction plane for ammonium salts of ($PMo_{11}Ni$, $PMo_{11}Co$, $PMo_{11}Mn$) with a monoclinic structure.

Thermogravimetric (TGA) analysis of the $H_3PMo_{12}O_{40}$ acid shows two mass loss signals between 40°C and

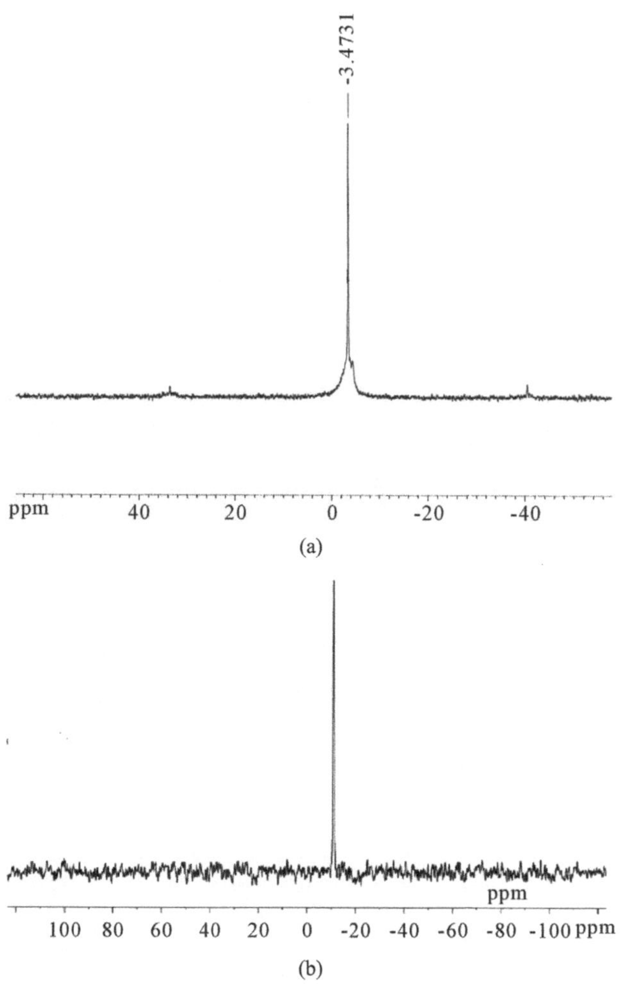

Figure 3. ^{31}P NMR spectra for: (a) $FePMo_{12}$; (b) PW_{11}.

Table 3. ^{31}P NMR measured and literature data for different catalyst systems.

No.	Catalysts	Chemical shift ppm	Ref.
1	PMo_{12}	−3.7	[34]
2	$PMo_{12}Ni$	−1.1551	[35]
3	$PMo_{11}Co$	−4826	[35]
4	PW_{11}	−10.35	This work
5	$FePMo_{12}$	−3.47	This work

140˚C corresponding to the departure of water of crystallization or hydration. A second mass loss at temperatures between 250˚C and 350˚C, was attributed to water content resulting from combination of H^+ ions and network oxygen. This mass loss leads to a reversible modification of the polyanion.

In differential thermal (DT) analysis, the loss of water corresponds to endothermic peaks. An exothermic peak was observed above 350˚C and was attributed to acid decomposition into P_2O_5 and MoO_3 oxides.

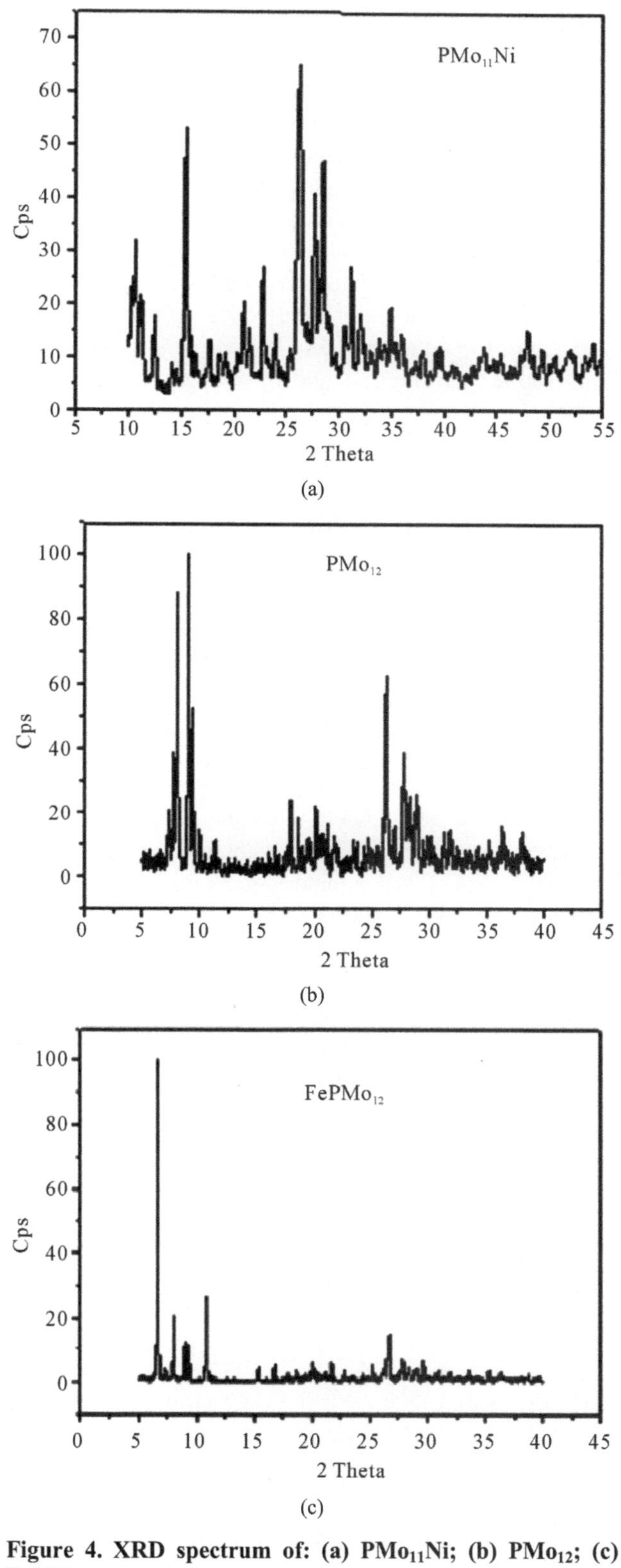

Figure 4. XRD spectrum of: (a) $PMo_{11}Ni$; (b) PMo_{12}; (c) $FePMo_{12}$.

TGA analysis for the salts $PMo_{11}Ni$, $PMo_{11}Co$, $PMo_{11}Mn$, $FePMo_{12}$, PW_{11} shows three mass losses. A first mass loss occurred around 100˚C, which is interpreted as the start of physisorbed water loss. A second departure is at around 240°C to 300°C, which is due to loss of constitution water molecule. Water molecules coordinated to nickel and cobalt ions are involved here. Finally, a loss around 300°C to 400°C, corresponding to departure of molecular ammonia or molecular nitrogen, was observed. Salt catalysts exhibit higher thermal stability than $H_3PMo_{12}O_{40}$ (**Figure 5**). The temperature at which their degradation starts depends on the position of the ion M^{3+}. When the counter ion involved metal ionic species, the decomposition temperature exceeded 500°C. On the other hand, the heteropolyacid $H_3PMo_{12}O_{40}$ is less thermally stable than the salt, and its decomposition started at 350°C.

3.2. Catalytic Study

3.2.1. Effect of Catalyst Type

Steady-state activity for catalytic n-hexane dehydrocyclization was reached within about 4 h. The observed products were benzene, (one most preferred product), cyclohexane, cyclohexene, cycloheadiene and methane (least preferred product). As expected, the catalytic activity and selectivity were dependent on the nature of the catalyst.

Figure 6 shows the conversion of n-hexane using different catalysts. The conversion of the solid catalyst was sensitive to the nature of the coordination ion (Mo^{6+}, W^{6+}, Ni^{3+}, Co^{3+}, Mn^{3+}) and counter ion H_3O^+, NH_4^+, Fe^{3+}). The replacement of mobile M (Ni^{3+}, Co^{3+}, Mn^{3+}) ions in the HPA framework increased the activity of the catalyst. Replacing H^+ ion with Fe^{3+} lowered the activity. PW_{11} exhibited relatively high activity as well.

Detailed studies on n-hexane isomerization catalysis with HPC are summarized in (**Tables 5** and **6**). The results show that benzene was the major product at low conversion level while using different catalysts. Cyclohexane, cyclohexene, cyclohexadiene, benzene and methane were also observed. As the conversion level increased benzene product ratio was lowered in ratio of methane product.

The 30% activity for $H_3PMo_{12}O_{40}$ (or even 100% for other catalysts excluding FePMo) shows that the protons in anhydrous HPA are accessible to the reactant molecules. The high catalytic activity of HPA may thus be related to high acid strength and high mobility of protons, as reported earlier [38]. The resulting protons are responsible of the aromatization of n-hexane. Cyclohexane, cyclohexene cyclohexadiene were also obtained, with variable ratios depending on type of catalyst. An important selectivity obtained by PMoNi catalyst involved cyclohexane 14%, cyclohexene 2%, cyclohexadiene 40%. A lower selectivity with PMoCo and PMoMn catalysts was observed. The results indicate that both relatively strong acidity and metallic propriety are responsible for

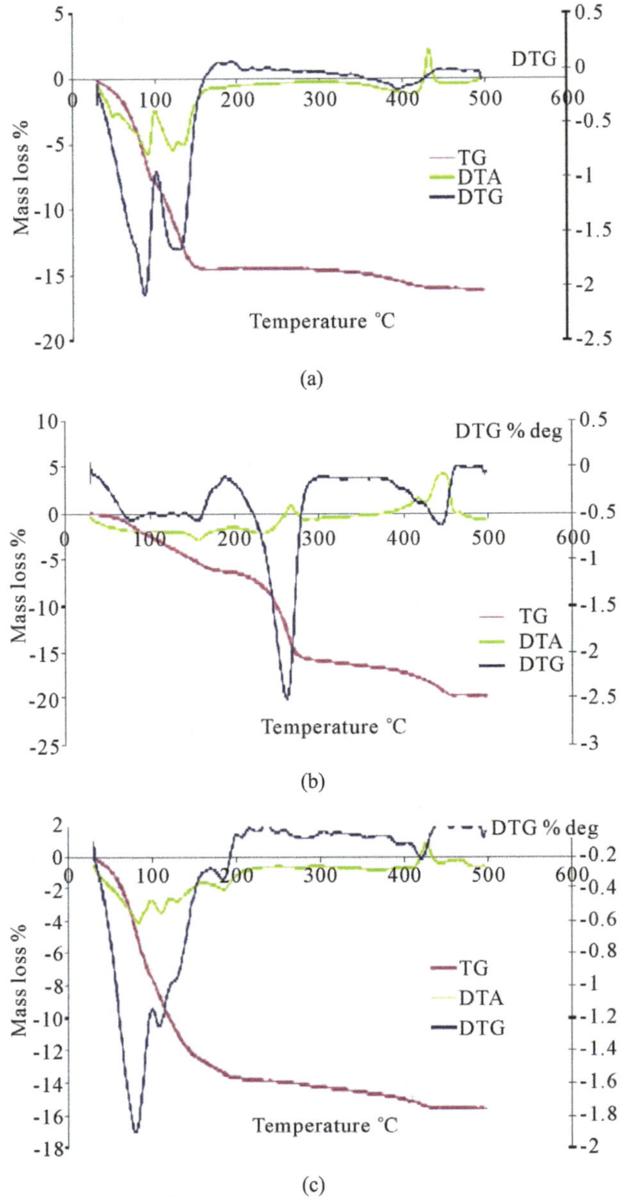

Figure 5. Thermal analysis of: (a) PMo_{12}; (b) $PMo_{11}Ni$; (c) $FePMo_{12}$.

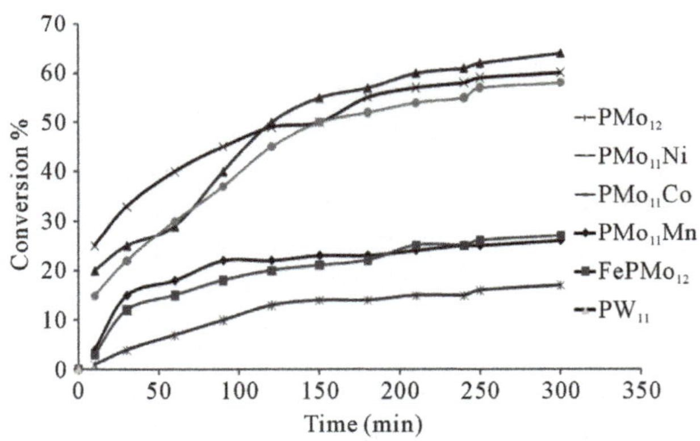

Figure 6. Conversion of n-hexane as function of time at 350°C, (using N_2 as carrier gas).

Table 4. Lattice parameters (A°) measured for different catalyst systems.

Heteropolycompounds	Parameters (A°)		Ref.
$H_3PMo_{12}O_{40}\cdot 13H_2O$	a = 14.10	α = 112.1°	[35,36]
	b = 14.13	β = 109.8°	
	c = 13.39	γ = 60.73°	
	triclinic		
$(NH_4)_4HPMo_{11}NiO_{39}$	a = 20.072	b = 11.55938	[35]
	c = 18.77828	β = 109.58°	
	monoclinic		
$(NH_4)_4HPMo_{11}CoO_{39}$	a = 21.01022	b = 7.8468	[35]
	c = 17.00056	β = 97,539°	
	monoclinic		
$(NH_4)_4PMo_{11}MnO_{39}$	a = 21.12322	b = 7.19688	[35]
	c = 17.00056	β = 97.653°	
	monoclinic		

Table 5. Catalytic performances of catalysts (Carrier gas H_2, flow rate: 0.1 l\h).

Catalyst	T °C	Con. %	Selectivity				
			CH_4	C_6H_6	C_6H_{12}	C_6H_{10}	C_6H_8
PMo_{12}	300	5	79	21			
	350	30	83	17			
	380	100	100				
$PMo_{11}Ni$	250	48	53	21	7	12	7
	300	77	56	21	8	9	6
	350	100	85	12	3		
	380	100	100				
$PMo_{11}Co$	250	35	57	23	20		
	300	50	86	8	6		
	350	100	98		2		
	380	100	100				
$PMo_{11}Mn$	250	60	90	7	3		
	300	80	98		2		
	350	100	100				
	380	100	100				
$Fe_{0.1}PMo_{12}$	250	56	52	20		12	12
	300	90	89	11			
	350	100	100				
	380	100	100				
PW_{11}	300	3	0	100			
	350	69	0	100			

Table 6. Catalytic performances of catalysts (carrier gas N_2, flow rate: 0.1 L\h).

Catalyst	T °C	Con. %	Selectivity				
			CH_4	C_6H_6	C_6H_{12}	C_6H_{10}	C_6H_8
PMo_{12}	300	2	0	100			
	350	23	20	80			
	380	100	60	40			
$PMo_{11}Ni$	250	1	0	78	12	2	8
	300	8	0	52	24	4	20
	350	24	7	37	14	2	40
	400	100	100	0			
$PMo_{11}Co$	250	4	0	100			
	300	11	0	100			
	350	60	58	42			
	400	100	100	0			
$PMo_{11}Mn$	250	3	100	0			
	300	11	0	100			
	350	58	53	47			
	400	100	100	0			
$Fe_{0.1}PMo_{12}$	300	6	0	100			
	350	15	20	80			
	380	50	80	20			
	400	86	100	0			
PW_{11}	300	3	0	100			
	350	57	0	100			
	400	67	0	100			

catalyst efficiency at lower temperature (lower than 350°C).

The PW_{11} catalyst showed specially high efficiency, selectivity and stability towards benzene formation. For example, PW_{11} showed high activity (69%) conversion and high selectivity (100%) at 350°C compared to PMo_{12} (30% conversion and 17% selectivity), PMoNi (100% and 11.5%), and PMoCo and PMoMn (100% and 0%). Such behaviour should be ascribed to the Lewis acid properties of the TBA cations, Also, we may conclude that Lewis acid species should be important components of the active site in the isomerization of n-hexane reaction over HPAs, in agreement with other studies [39,40].

3.2.2. Effect of Temperature

The effect of temperature on efficiency and benzene selectivity has been studied for different catalysts. **Figure 7** shows how conversion varies with reaction temperature. The temperature curves exhibit linear character. For dif-

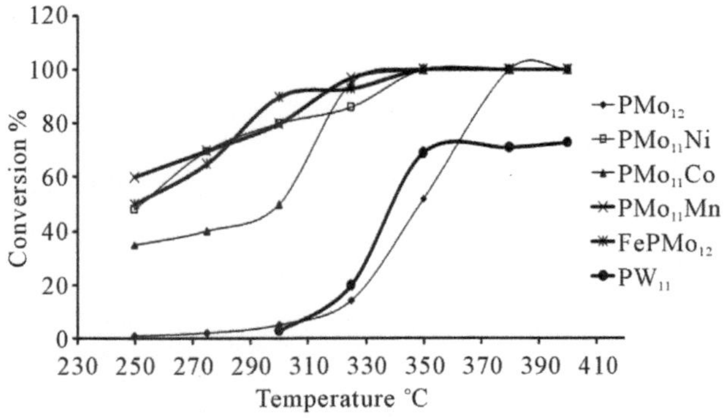

Figure 7. Conversion of n-hexane as function of temperature, (using N_2 as carrier gas).

Table 7. Reaction activation energy for different catalyst systems measured based on amount of n-hexane consumption.

Catalyst	Ea (KJ/mol)
$H_3PMo_{12}O_{40}$	95.2
$H(NH_4)_6PMo_{11}NiO_{40}$	83.2
$H(NH_4)_6PMo_{11}CoO_{40}$	63.8
$H(NH_4)_6PMo_{11}MnO_{40}$	43
$KFe_{0.1}PMo_{12}O_{40}$	83.2
$(TBA)_7PW_{11}O_{39}$	52.7

ferent catalysts, overall hexane conversion increased with temperature. The transformation of n-hexane on the tested catalysts started only above 250°C and the conversion reached 100% at 350°C - 400°C.

Until 300°C, benzene was the unique or major product. The selectivity to benzene decreased with increasing temperature (**Tables 4** and **5**). The conversion increased from 2 up to 30%, whereas benzene selectivity decreased from 100% to 80% when the reaction temperature was increased from 300°C to 350°C when using $HPMo_{12}O_{40}$ heteropolyacid catalyst.

Methane was obtained with all catalysts at temperature above 300°C. Methane production selectivity increased with temperature at the expense of benzene. At 400°C n-hexane was converted completely to methane with all catalysts. The high methane selectivity was with PMoMn catalysts.

At 400°C the H_3PMo_{12} heteropolyacid was decomposed into Mo_3 and P_2O_5. This result agrees with literature [41] which showed that the isomerization of n-hexane was due to surface Bronsted acid sites.

3.2.3. Activation Energy

n-hexane conversion rate were determined over the temperature range 200°C - 450°C. From (**Figures 8(a)-(f)** n-hexane dehydrocyclization rate constants were determined over the temperature range 200°C - 450°C. From Arrhenius plots (**Figure 9**), Activation energies for the transformation of n-hexane over acidic catalysts were calculated as shown in **Table 7**.

Values of activation energies for the overall process of dehydrocyclization and aromatization of n-hexane using the acidic catalyst systems (43 to 95.2 $kJ \cdot mol^{-1}$) are much lower than those known for metal type catalysts (typically 230 to 293 $kJ \cdot mol^{-1}$) [42]. Acidic catalytic activity is expected to occur by the highly acidic heteropoly compounds. Actually the activation energies are comparable to those found for the transformation of the n-hexane in benzene using acidic catalysts. [43].

Furthermore, the selectivity of the benzene decreases, while the conversion increases, with higher temperatures or longer retention times. In case of all catalysts, the benzene is the secondary product of the reaction.

3.2.4. Effect of Time of the Reaction

The conversion of n-hexane was increase with time for all of the catalysts at 350°C as it is clear from **Figure 10**. The selectivity of benzene only slightly decreased with time on all catalysts **Figure 8**, when keeping temperature constant at 350°C. This due to high thermodynamic stability of benzene.

3.2.5. Effect of Type of Carrier Gas

The type of carrier gas affected both conversion and selectivity of the reaction. The overall conversion increased by using H_2 as a carrier gas, while benzene selectivity increased when using N_2 as a carrier gas. This applied to all catalytic systems, as shown in **Tables 5** and **6**. Comparison between the two Tables shows that for each catalyst system, using H_2 gas gives higher conversion than using N_2. On the other hand, the Tables show using N_2 gave higher benzene selectivity than using H_2, for each catalyst.

3.2.6. Mechanism

A plausible mechanism has been suggested to explain the observations discussed above. As shown in Scheme I, two possible products can be expected by the HPA catalyst systems. The Scheme shows that with excess H_2 (carrier gas, as discussed above) the carbonium ion formed by super-acid activation of hexane will be cracked into CH_4 [44]. Such a process involves complete saturation of all carbon atoms with hydrogen, via complete cracking of the hydrocarbon chain. On the other hand, in case of nitrogen carrier gas (with no hydrogen) dehydrocylization process is favored to dominate, with no significant carbon chain cracking, as shown in the Scheme. The process is multi-stage involving β-H elimination in each step. This explains the production of cyclohexane, cyclohexene and cyclohexadiene as ac-

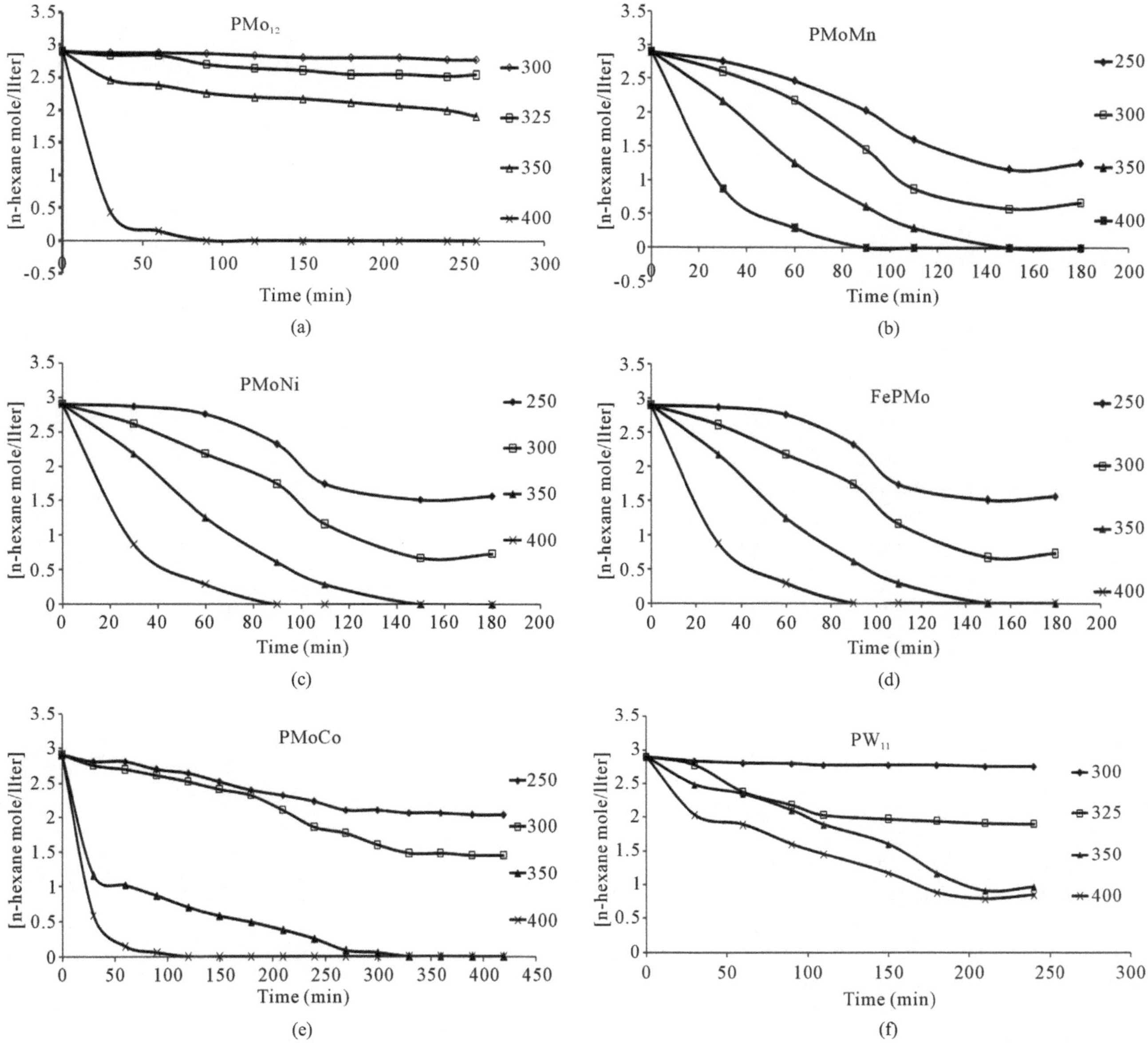

Figure 8. n-Hexane conversion as a function of time over: (a) PMo_{12} catalyst; (b) $PMo_{11}Mn$ catalyst; (c) $PMo_{11}Ni$ catalyst; (d) $FePMo_{12}$ catalyst; (e) $PMo_{11}Co$ catalyst; (f) PW_{11} catalyst.

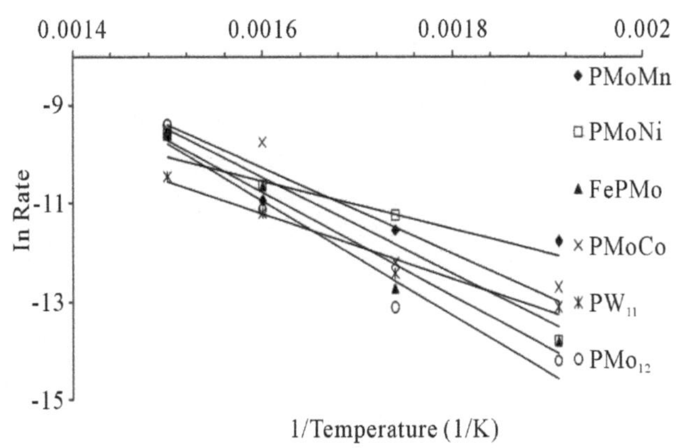

Figure 9. Arrhenuis plots for different catalyst systems (using N_2 as carrier gas).

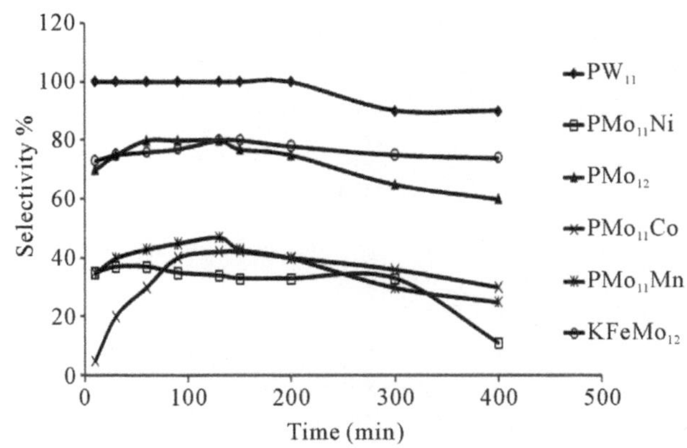

Figure 10. Selectivity of benzene as function of time at 350°C, (using N_2 as carrier gas).

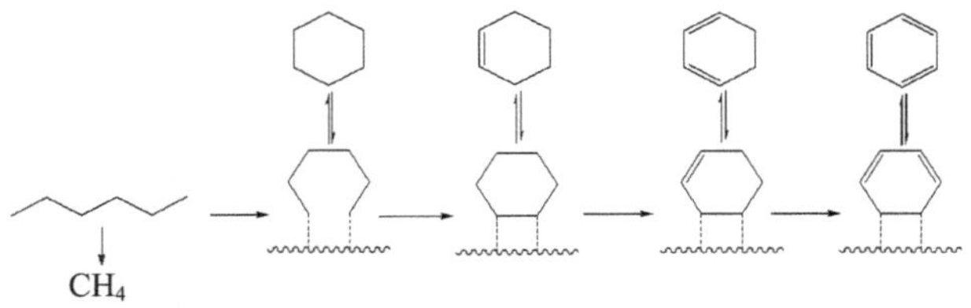

$$\begin{array}{ccccc} C_6H_{12} & C_6H_{10} & C_6H_8 & & \\ \uparrow -H_2 & \uparrow -H_2 & \uparrow -H_2 & \uparrow -H_2 & -H_2 \end{array}$$

$$C_6H_{14} \rightarrow [C_6H_{14}]_{ads} \rightarrow [C_6H_{12}]_{ads} \rightarrow [C_6H_{10}]_{ads} \rightarrow [C_6H_{18}]_{ads} \rightarrow C_6H_6$$

$$\downarrow$$

$$CH_4$$

Figure 11. Reaction scheme of n-hexane dehydrocylization over heteropolycompounds catalysts.

companying products to benzene.

It should also be noted that this logic does not fully work under higher temperatures, where CH_4 production dominates even in the absence of hydrogen gas. In such a case, high temperature experiments without hydrogen yielded high molecular weight aromatics and tars, in parallel to CH_4 formation, which caused some technical difficulties such as reactor blockage during experiments. Therefore, If benzene is the desired product from n-hexane, hydrogen gas should not be used, and the reaction must be conducted under mild temperatures (350˚C or lower).

4. Conclusion

A number of molybdenum-based compounds $H_3PMo_{12}O_{40}$, $KFePMo_{12}O_{40}$, $(TBA)_7PW_{12}O_{40}\cdot xH_2O$, $(NH_4)_4PMo_{11}CoO_{40}$, $(NH_4)_4PMo_{11}NiO_{40}$, $(NH_4)_4PMo_{11}MnO_{40}$ have been prepared and characterized by UV, IR, NMR Spectra, TGA and XRD. The compounds were of Keggin type structure. Due to their super-acidic nature, the compounds showed catalytic efficiency in the dehydrocyclization of n-hexane. The product distribution (benzene, cyclohexane, cyclohexene, cyclohexadiene and methane) strongly depends on the nature of the catalyst, the type of carrier gas and the reaction temperature. The observed differences in the behaviour of the studied catalysts should be firstly ascribed to the superacidic nature of the catalysts, which can activate saturated alkanes.

REFERENCES

[1] B. C. Gates, "Catalytic Chemistry," John Wiley & Sons, Inc., New York, 1992.

[2] A. Corma, J. M. Serra and A. Chica, "Discovery of New Paraffin Isomerization Catalysts Based on SO_4^{2-}/ZrO_2 and WO_x/ZrO_2 Applying Combinatorial Techniques," *Catalysis Today*, Vol. 81, No. 3, 2003, pp. 495-506.

[3] Y. Ono, "A Survey of the Mechanism in Catalytic Isomerization of Alkanes," *Catalysis Today*, Vol. 81, No. 1, 2003, pp. 3-16.

[4] B. Bachiller-Baeza, J. Alvarez-Rodríguez, A. Guerrero-Ruiz and I. Rodríguez-Ramos, "Support Effects on Ru-HPA Bifunctional Catalysts: Surface Characterization and Catalytic Performance," *Applied Catalysis A: General*, Vol. 333, No. 2, 2007, pp. 281-289.

[5] M. H. Jordao, V. Simoes and D. Cardoso, "Zeolite Supported Pt-Ni Catalysts in n-Hexane Isomerization," *Applied Catalysis A: General*, Vol. 319, 2007, pp. 1-6.

[6] Y. Gucbilmez, A. S. Yargic and I. Calis, "A Comparative Characterization of the HPA-MCM-48 Type Catalysts Produced by the Direct Hydrothermal and Room Temperature Synthesis Methods," *Journal of Nanomaterials*, Vol. 2012, 2012, Article ID: 210437.

[7] I. V. Kozhevnikov, "Heteropoly Acids and Related Compounds as Catalysts for Fine Chemical Synthesis," *Catalysis Reviews*, Vol. 37, No. 2, 1995, pp. 311-352.

[8] I. V. Kozhevnikov, K. R. Kloetstra, A. Sinnema, H. W. Zandbergen and H. Van Bekkum, "Study of Catalysis Comprising Keteropoly Acid $H_3PW_{12}O_{40}$ Supported on MCM-41 Molecular Sieve and Amorphous Silica," *Journal of Molecular Catalysis A: Chemical*, Vol. 114, No. 1-3, 1996, pp. 287-298.

[9] H. Hayashi and J. B. Moffat, "Methanol Conversion over Metal Salts of 12-Tungstophosphoric Acid," *Journal of Catalysis*, Vol. 81, No. 1, 1983, pp. 61-66.

[10] T. Baba, H. Watanabe and Y. Ono, "Generation of Acidic Sites in Metal Salts of Heteropoly Acids," *Journal of Physical Chemistry*, Vol. 87, No. 13, 1983, pp. 2406-2411.

[11] K. Nowińska, "Catalytic Activity of Supported Heteropoly Acids for Reactions Requiring Strong Acid\Centres," *Journal of Chemical Society, Faraday Transactions*, Vol. 87, No. 5, 1991, pp. 749-753.

[12] V. Haensel, "Process of Reforming a Gasoline with an Alumina-Platinum-Halogen Catalyst," US Patent No. 2479109 and 2479110, 1949.

[13] H. E. Kluksdahl, "Reforming a Sulfur-Free Naphtha with a Platinum-Rhenium Catalyst," US Patent No. 3415737, 1968.

[14] M. H. Jordão, V. Simões and D. Cardoso, "Zeolite Supported Pt-Ni Catalysts in n-Hexane Isomerization," *Applied Catalysis A: General*, Vol. 319, 2007, pp. 1-6.

[15] C. Rocchioccioli-Deltcheff, M. Fournier, R. Franck and R. Thouvenot, "Vibration Investigatios of Polyoxometalates. 2. Evidence for Anion-Anion Interaction in Molybdenum (VI) and Tungsten(VI) Compounds Related to the Keggin Structure," *Inorganic Chemistry*, Vol. 22, No. 2, 1983, pp. 207-216.

[16] T. Mazari, S. Hocine, N. Salhi and C. Rabia, "Oxidation of Propane over Ammonium-Transition Metal Mixed Ke-

ggin Phosphomolybdate Salts," *Journal of Natural Gas Chemistry*, Vol. 19, No. 1, 2010, pp. 54-60.

[17] O. Benlounes, S. Cheknoun, S. Mansouri, C. Rabia and S. Hocine, "Catalytic Activation of C-H Bonds of Hydrocarbons by Heteropolycompounds," *Journal of the Taiwan Institute of Chemical Engineers*, Vol. 42, No. 1, 2011, pp. 132-137.

[18] F. M. Zhang, M. P. Guo, H. Q. Ge and J. Wang, "Hydroxylation of Benzene with Hydrogen Peroxide over Highly Efficient Molybdovanadophos Phoric Heteropoly Acid Catalysts," *Chinese Journal of Chemical Engineering*, Vol. 15, No. 6, 2007, pp. 895-898.

[19] O. Benlounesa, S. Mansouria, C. Rabiab and S. Hocine, "Direct Oxidation of Methane to Oxygenates over Heteropolyanions," *Journal of Natural Gas Chemistry*, Vol. 17, No. 3, 2008, pp. 309-312.

[20] H. Kima, P. Kima, K. Leeb, S. H. Yeomc, J. Yia and I. K. Song, "Preparation and Characterization of Heteropoly Acid/Mesoporous Carbon Catalyst for the Vapor-Phase 2-Propanol Conversion Reaction," *Catalysis Today*, Vol. 111, No. 3-4, 2006, pp. 361-365.

[21] J. N. Beltramini, "Studies in Surface Science and Catalysis," *Proceedings of the 3rd International Mesostructured Materials Symposium, Nanotechnology in Mesostructured Materials*, Jeju, 8-11 July 2002, pp. 653-656.

[22] R. S. Drago, J. A. Dias and T. O. Maier, "An Acidity Scale for Brönsted Acids Including $H_3PW_{12}O_{40}$," *Journal of American Chemical Society*, Vol. 119, No. 33, 1997, pp. 7702-7710.

[23] K. Nowinska, R. Fiedorow and J. Adamiec, "Catalytic Activity of Supported Heteropoly Acids for Reactions Requiring Strong Acid Centres," *Journal of the Chemical Society, Faraday Transactions*, Vol. 87, No. 5, 1991, pp. 749-753.

[24] K. J. Nowinska, "Evidence for Superacid Sites on the Ammonium Salt of 12-Tungstophosphoric Acid from a Catalytic Test Reaction," *Journal of the Chemical Society, Chemical Communications*, 1990, pp. 44-45.

[25] C. Marchal-Roch, J. M. Millet and C. R. Acad, "Phosphomolybdic Heteropolycompounds as Oxidation Catalysts. Effect of Transition Metals as Counter-Ions," *Science Chemistry*, Vol. 4, No. 5, 2001, pp. 321-329.

[26] S. Hocine, C. Rabia, M. M. Bettahar and M. Fournier, "Oxidative Dehydrogenation of Cyclohexane over Heteropolymolybdates," *Studies in Surface Science and Catalysis*, Vol. 130, 2000, pp. 1895-1900.

[27] R. Tayebee, "Simple Heteropoly Acids as Water-Tolerant Catalysts in the Oxidation of Alcohols with 34% Hydrogen Peroxide," *Journal of the Korean Chemical Society*, Vol. 52, No. 1, 2008, pp. 23-29.

[28] T. Okuhara, "Catalytic Chemistry of Heteropoly Compounds," *Advances in Catalysis*, Vol. 41, 1996, pp. 113-252.

[29] Y. He, W. Sang, J. Wang, R. Wu and J. Min, "Vertically Well-Aligned ZnO Nanowires Generated with Self-Assembling Polymers," *Materials Chemistry and Physics*, Vol. 94, No. 1, 2005, pp. 29-33.

[30] C. L. Hill and C. M. Prosser-Mccartha, "Homogeneous Catalysis by Transition Metal Oxygen Anion Clusters," *Coordination Chemistry Reviews*, Vol. 143 ,1995, pp. 407-455.

[31] A. V. Churakov, E. A. Legurova, A. A. Dutov, P. V. Prikhodchenko and T. A. Tripol'skaya, "Peroxide Derivatives of Heteropoly Compounds with Keggin Anions $[PW_{12}O_{40}]^{3-}$ and $[SiW_{12}O_{40}]^{4-}$: Synthesis and Structure," *Russian Journal of Inorganic Chemistry*, Vol. 53, No. 8, 2008, pp. 1187-1192.

[32] T. J. R. Weakley and S. A. Malik, "Triheteropolyanions Containing Copper(II), Manganese(II), or Manganese (III)," *Journal of Inorganic and Nuclear Chemistry*, Vol. 32, No. 12, 1970, pp. 3875-3890.

[33] O. Benlounesa, "Oxidation of Methane on Heteropoly Compounds," Ph.D. Dissertation, University of Mouloud Mammeri-Tizi-Ouzou, Tizi-Ouzou, 2010.

[34] S. S. Lima, G. I. Park, I. K. Song and W. Y. Lee, "Heteropolyacid (HPA)-Polymer Composite Films as Catalytic Materials for Heterogeneous Reactions," *Journal of Molecular Catalysis A: Chemical*, Vol. 182-183, 2002, pp. 175-183.

[35] S. Hocine, "Heteropolyphosphomolybdates. Preparation, Charactarization, Catalytic Activity of Cyclohexane Oxydehydrogenation," Ph.D. Dissertation, Houari Boume Diene University, Algiers, 2003.

[36] A. Popa, V. Sasca, M. Stefanescu, E. Kis and R. Marinkovic-Neducin, "The Influence of the Nature and Textural Properties of Different Supports on the Thermal Behavior of Keggin Type Heteropolyacids," *Journal of the Serbian Chemical Society*, Vol. 71, 2006, pp. 235-249.

[37] G. A. Tsigdinos and C. J. Hallada, "Molybdovanadophosphoric Acids and Their Salts Investigation of Methods of Preparation and Characterization," *Inorganic Chemistry*, Vol. 7, No. 3, 1968, pp. 437-441.

[38] T. Okuhara, N. Mizuno and M. Misono, "Catalysis by HeteroPolycompounds—Recent Developments," *Applied Catalysis A: General*, Vol. 222, No. 1-2, 2001, pp. 63-77.

[39] R. Hubaut, B. Ouled Ben Tayeb, W. Kuang, A. Rives and M. Fournier, "Mechanical Mixtures of Me(Ni, Pd)Ce Oxides and Silica-Supported Heteropolyacids: Role and Optimal Content of Each Active Species for n-Hexane Isomerization," *Kinetics and Catalysis*, Vol. 47, No. 1, 2006, pp. 20-24.

[40] V. V. Brei, O. V. Melezhyk, S. V. Prudius, M. M. Levechuk and K. I. Patrylak, "Superacid WO_x/ZrO_2 Catalysts for Isomerization of n-Hexane and for Nitration of Benzene," *Studies in Surface Science and Catalysis*, Vol. 143, 2002, pp. 387-395.

[41] B. Demirel and E. N. Givens, "Transformation of Phosphor Molybdic Acid into an Active Catalyst with Potential Application in Coal Liquefaction," *Catalysis Today*,

Vol. 50, No. 1, 1999, pp. 149-158.

[42] F. Garin and F. G. Gault, "Mechanisms of Hydrogenolysis and Isomerization of Hydrocarbons on Metals. VIII. Isomerization of Carbon-13 Labeled Pentanes on a 10% Platinum-Aluminum Oxide Catalyst," *Journal of the American Chemical Society*, Vol. 97, No. 16, 1975, pp. 4466-4476.

[43] F. R. Ribeiro, C. Marcilly and M. Guisnet, "Hydroisomerization of n-Hexane on Platinum Zeolites," *Journal of Catalysis*, Vol. 78, No. 2, 1982, pp. 267-280.

[44] S. Kotrel, H. Knozinger and B. C. Gates, "The Haag-Dessau Mechanism of Protolytic Cracking of AlKanes," *Microporous and Mesoporous Materials*, Vol. 35-36, 2000, pp. 11-20.

4

Elucidation of Reaction Behaviors in Sonocatalytic Decolorization of Amaranth Dye in Water Using Zeolite Y Co-Incorporated with Fe and TiO_2

Atheel Hassan Alwash[1,2*], Ahmad Zuhairi Abdullah[1], Norli Ismail[3]
[1]School of Chemical Engineering, Universiti Sains Malaysia, Penang, Malaysia
[2]Department of Chemistry, College of Science, Al-Nahrain University, Baghdad, Iraq
[3]School of Industrial Technology, Universiti Sains Malaysia, Penang, Malaysia

ABSTRACT

The Zeolite Y as a support was modified by the incorporating of Fe and TiO_2 in one single step using impregnation method. The synergistic effects between those components enhance the catalytic activity for the degradation of amaranth dye under ultrasonic irradiation with output power of 50 W and 40 kHz frequency. Different characterization techniques were used to elucidate the physical and chemical properties of the produced catalysts. The XRD results indicated that the type of titanium precursor significantly effects on the crystallinity of 0.4% Fe/15%TiO_2-NaY catalyst. The AFM results detected that the TiO_2 formed a layer covered the surface of zeolite while Fe clusters were located close to TiO_2. The influence of reaction parameters such as TiO_2 and Fe content, pH values, amount of hydrogen peroxide used, catalyst loading and the initial dye concentration were investigated for the decolorization efficiency of amaranth. The maximum decolorization efficiency for 0.4%Fe/15%TiO_2-NaY was 100% after 120 min of reaction time with an initial dye concentration of 30 mg/L, 2 g/L of catalyst loading, natural pH about 5.5 and 0.65 mM H_2O_2.

Keywords: Dye Degradation; Impregnation Method; Incorporation of Metals

1. Introduction

Dyes with their structural variety have been used as important raw materials for colorants in different industries, such as textiles, rubber, leather, paper, food and plastic. Large quantities of aqueous waste and dye effluents are produced during processing operations [1]. The high stability of the dyestuffs makes them resistance to the conventional treatment processes and limited their biodegradation. Thus advanced oxidation processes (AOPs) such as ultrasonic process have been widely investigated to overcome the drawbacks of the traditional methods.

The synergetic effect between ultrasonic irradiation and heterogeneous catalyst such as titanium oxide enhance the activity of the process due to the generation of more hydroxyl radicals species. However, the separation difficulties for the superfine TiO_2 particles after the reaction makes the efforts focused on the immobilization of TiO_2 on different types of support [2]. Zeolite support has been investigated extensively for the immobilization of TiO_2 especially for photocatalytic application due to its special properties such as the presence of Lewis acid and Lewis base sites in the framework that play an important role in charge transfer and electron transfer processes [3,4].

Our previous work [5] reported the successful results for the catalytic activity of Fe loaded on TiO_2 encapsulated zeolite for the degradation of dye under ultrasonic reaction. The position of TiO_2 in the previous study was encapsulated into the pores of zeolite using ion-exchange method while the Fe was located at the external surface of zeolite using impregnation method. Therefore, it was of great interest to examine the variations occurs in characteristic and catalytic activity of the produced catalyst by replacing the position of TiO_2 to be located at the external surface of zeolite together with Fe under ultrasonic reaction. The selection of ultrasonic-assisted process should provide an efficient activity to the waste water treatment process due to the strong penetration ability of ultrasound in water which can usually reach 20 - 30 cm from the energy source [6].

*Corresponding author.

The co-incorporation of TiO_2 and Fe with zeolite structure should enhance the sonocatalytic activity since more active sites will be provided in the reaction. Furthermore, using zeolite as a support for these metals was ascribed to its high surface area and its ability to act as a host for more than one active component. Different characterization techniques have been used to elucidate the physical and chemical properties of the produced catalysts. The sonocatalytic activity has been investigated as a function of different reaction parameters such as solution acidity, catalyst loading, the addition of H_2O_2 and the effect of initial dye concentration for the decolorization of amaranth dye which has been used as the probe molecule in this study.

2. Experimental

2.1. Materials

H-Y zeolite with Si/Al 15 was obtained from Zeolyst International while amaranth dye was purchased from Sigma. Meanwhile, potassium titanooxalate dehydrate and sodium chloride crystals were supplied by J. T. Baker while iron (III) nitrate nonhydrate was obtained from Merck Sdn. Bhd. Hydrogen peroxide (35%) used to accelerate the reaction was supplied by R&M Chemicals.

2.2. Catalyst Preparation Method

2.2.1 Preparation of NaY

Dried H-Y zeolite was first converted to its sodium form by means of an ion-exchange method [7]. The zeolite was washed with 1 M NaCl at 80˚C for 4 h in such a way that the ratio of zeolite to sodium chloride solution was 1:80. The slurry was then filtered and washed with deionized water to remove the excess amount of NaCl. In order to achieve complete exchange of sodium into the zeolite, the above cycle was repeated twice. The sample was then dried for 12 h at a temperature of 80˚C and then calcined in a muffle furnace at 550˚C for 5 h.

2.2.2. Incorporation of Fe and TiO_2 with NaY

The heterogeneous catalyst was prepared using the procedure reported by Li *et al.* [8]. One gram of NaY was suspended in 50 mL water and known amounts of $Fe(NO_3)_3$ (Fe^{3+} loading from 0.2 - 0.6 wt%) and fixed amount of titanium precursor (the theoretical proportion of TiO_2 in the catalyst is 15 wt% TiO_2) was added into the suspended solution and mixed by mechanical stirrer for 24 h at ambient temperature. The resulting samples were then dried using a rotary evaporator and placed in an oven at 100˚C for 4 h. Then, the produced solids were calcined at 550˚C for 5 h. The produced catalysts were denoted as *x* Fe%/15%TiO_2-NaY where *x* is the wt% of Fe. In order to investigate the effect of TiO_2 with fixed amount of Fe, the same preparation method was repeated with TiO_2 loading from 5 - 20 wt% and fixed amount of Fe.

2.3. Catalytic Characterization

Characterization of the catalysts with different analytical techniques was performed to elucidate the incorporation of Ti and Fe species with the framework of Na-Y. Among various catalysts loadings, the best catalyst was demonstrated by activity of 0.4% Fe/15%TiO_2-NaY. Thus it was chosen to completely characterize. The crystalline structure of the obtained catalysts was investigated by means of X-ray diffraction using an XRD system (Philips Goniometer PW 1820).The Fourier transforms infrared spectrometry (Perkin Elmer) was carried out with wave number range between (400 - 4000) cm^{-1}. Transmission electron microscope (TEM) images were obtained using a Phillips CM 12 transmission electron microscope equipped with an image analyzer and operated at 120 kV. Finally, the roughness of the catalyst surface was investigated using SPA 400 atomic force microscopy (AFM) technique.

2.4. Sonocatalytic Reaction

All sonocatalytic experiments were carried out in a 250 mL cylindrical vessel and placed in an ultrasonic bath (Cole-Parmer) for up to 120 min. Before starting the ultrasonic reaction, the solution was stirred for 30 min at room temperature to maintain the good dispersion of catalyst with the dye solution. The experimental conditions were set at an initial amaranth dye concentration of 10 mg/L, a catalyst amount of 1.5 g/L and the solution pH was kept at its original level without any adjustment. The ultrasonic output power was fixed at 50 W with a frequency of 40 kHz. The decolorization efficiency was calculated based on UV-vis spectrophotometer results. The concentration of the dye solution was determined at its maximum absorbance wavelength of 521 nm. The decolorization efficiency of the catalyst was calculated based to the following equation;

$$\text{Decolorization}(\%) = \left[1 - \frac{C_t}{C_0}\right] \times 100\% \quad (1)$$

where, C_0 (mg/L) is the initial concentration of dye and C_t (mg/L) is the concentration of dye at certain reaction time, *t*(min).

3. Results and Discussion

3.1. Characterization of the Catalyst

3.1.1. XRD

The XRD pattern of NaY, potassium titanium oxalate (PTO) calcined at 550˚C, 0.4% Fe-NaY, 15% TiO_2-NaY and 0.4% Fe/15%TiO_2-NaY are shown in **Figure 1**. The

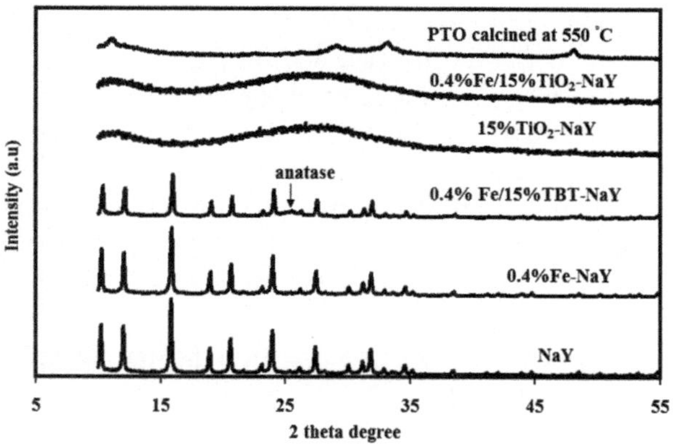

Figure 1. X-ray diffraction patterns of different catalysts used in this study.

parent NaY shows high crystalline structure while the calcined PTO showed an amorphous structure with small anatase peak at 2θ 48.1°. However, after the incorporation of Fe and TiO_2 with NaY for the case of 0.4% Fe/15% TiO_2-NaY, the crystallinity of zeolite disappeared and a single wide peak with low intensity appeared at 2θ of 27.7° ascribed to rutile phase. The formation of new material after the incorporation of Fe/TiO_2 with zeolite was not suggested since the diffraction pattern of 0.4% Fe/15% TiO_2-NaY was similar to that of 15% TiO_2-NaY catalyst.

No characteristic peak for iron oxide phase that could be formed after the calcination step was detected either for 0.4% Fe-NaY or 0.4% Fe/15% TiO_2-NaY. This could be ascribed to the low loading of Fe used in this study. Castano *et al.* [9] reported that the unclear diffraction for some metals could be attributed to the small crystal size (<4 nm) or to the well-dispersion of amorphous phases that caused unclear detection by XRD. Wang *et al.* [10] did not indicate any clear peak for Fe phases after the doping of TiO_2 and Fe into a natural zeolite even at higher concentrations up to 7% mol of Fe.

It has been reported in literature that the transformation phase temperature from anatase to rutile usually occurs at temperature between 700°C - 800°C [11,12]. Therefore, the appearance of the rutile phase for both catalysts *i.e.* 15% TiO_2-NaY and 0.4% Fe/15% TiO_2-NaY should not be attributed to the calcination temperature used in this study. However, the titanium precursor could be the reason behind the appearance of rutile phase since no anatase peak at 25.1° (highest peak for anatase phase) was detected in the bare potassium titanium oxalate (PTO) diffraction calcined at 550°C.

In order to investigate the effect of titanium precursor on the crystallinity of zeolite, the titanium precursor was changed with another organic precursor *i.e.* tetra butyl orthotitanate (TBT) using the same preparation method and the catalyst was denoted as 0.4% Fe/15% TBT-NaY as shown in **Figure 1**. The result indicated that zeolite support was able to retain its crystallinity even after the loading of both Fe and TiO_2 and a small anatase peak was detected at 25.1°. As a result, it seemed that the dispersion of TiO_2 clusters with zeolite depend on the type of titanium precursor and consequently affecting on the crystallinity of the support. The partial incorporation between Ti species and the framework of zeolite might be occurred during the preparation method since the zeolite structure have super cages and pore size of 1.3 nm and 0.74 nm, respectively. The excess amount of Ti species could be anchored on the surface of the support after the calcination step thus lead to a significant reduction in the crystallinity of catalyst due to the accumulation of TiO_2 phase on it.

3.1.2. Fourier Transforms Infrared Spectrometry

The FTIR spectra for the parent NaY, 0.4% Fe-NaY, 15% TiO_2-NaY and 0.4% Fe/15% TiO_2-NaY are presents in **Figure 2**. The absorption features of NaY resulted in different absorption peaks in the range between (4000 - 400) cm^{-1}. The peak at 3439 cm^{-1} and 1626 cm^{-1} represents the presence of water molecules or hydroxyl groups attached to zeolite structure. Hassan and Hameed [13] reported that the region from 1600 to 3700 cm^{-1} is ascribed to the existence of zeolite water molecules. These water molecules attached to the silicate minerals can be found in two types *i.e.* intact molecules or as OH groups formed from the dissociation of water interact with metals such as (Si^{4+}, Al^{3+}). The peak at 1138 cm^{-1} was ascribed to the asymmetric stretching modes of internal tetrahedral. While the absorption peak at 514 and 450 cm^{-1} represent the T-O bending vibrations of internal tetrahedral in zeolite [14]. The zeolite framework also present peak around 1050 cm^{-1} ascribed to Si-O-Si asymmetric stretching mode of external linkages bond in zeolite framework. It should be mentioned that the terms internal and external vibrations have been used in FTIR spectroscopy for zeolites materials to describe the vibrations in tetrahedral building units and between them [15].

A significant change in stretching vibration for 15%

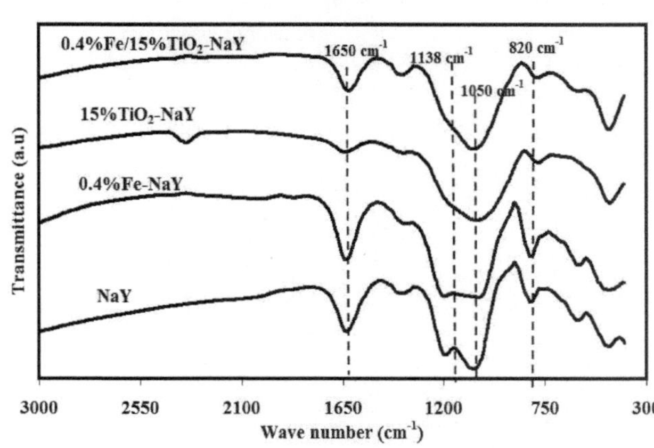

Figure 2. FTIR spectrum of different types of catalysts.

TiO_2-NaY and 0.4% Fe/15% TiO_2-NaY in comparative to NaY was observed at peak value around 1650 cm^{-1} ascribed to the existence of water molecules in zeolite [13]. Nikazar *et al.* [16] reported that the changes occurs in this regions after the doping of metal indicated that a certain amount of OH was demolished by this metal *i.e.* Ti and Fe. The peaks at 1138 cm^{-1} and 1050 cm^{-1} for NaY were attributed to the asymmetric stretching modes of internal tetrahedral and asymmetric stretching of external linkage for (T-O-T) respectively [17]. However, the peak at 1138 cm^{-1} (NaY) was disappeared in the case of 15% TiO_2-NaY and 0.4% Fe/15% TiO_2-NaY due to the effect of TiO_2 and Fe loading. Accordingly, the change occurred in peak at 1050 cm^{-1} was ascribed to the sensitivity of this region to the structure changes.

No peak could be detected at 960 cm^{-1} for 15% TiO_2-NaY and 0.4% Fe/15% TiO_2-NaY after the loading of TiO_2 which was assigned to the anti-symmetric stretching vibration of the Ti-O-Si [18]. This peak was ascribed to the interaction of Ti species with the internal framework of zeolite. However, the absence of this peak in this catalyst suggests the dispersion of TiO_2 to be more on the external surface of zeolite. This result is in agreement with the results obtained by XRD.

Quintero Ortíz *et al.* [19] reported that the peak in NaY at 820 cm^{-1} was assigned to the stretching vibrations of external bond between tetrahedral of zeolite. However, this peak showed a reduction in stretching vibration and shifting to lower wavelength after the incurporation of TiO_2 for the case of 15% TiO_2-NaY and 0.4% Fe/15% TiO_2-NaY. While no changes were detected for the case of 0.4% Fe-NaY so, this change was ascribed to the incorporation of TiO_2. On the other hand, the stretching vibration from 500 to 700 cm^{-1} was ascribed to the sensitive change in the pseudo- lattice vibration of zeolite after the loading of Metal/Zeolite [20]. From the above analysis, the differences in FTIR spectra were evident especially for TiO_2 loading due to its higher amount in comparative to Fe.

3.1.3. AFM Technique

The morphological characteristics was investigated based on the values of mean roughness that were presented in two and three dimensional AFM images of NaY, 15% TiO_2-NaY, 0.4% Fe-NaY and 0.4% Fe/15% TiO_2-NaY as in **Figure 3**. The variation occurred in surface morphology of the prepared catalysts suggested the significant effect of Ti and Fe incorporated with zeolite. The results for the degree of roughness was in the sequence of NaY > 0.4% Fe-NaY > 0.4% Fe/15% TiO_2-NaY > 15% TiO_2-NaY with roughness values of 74 nm, 72 nm, 48 nm and 44 nm, respectively. The similarity in the roughness of the catalyst surface between NaY and 0.4% Fe-NaY seemed to be due to the low loading of the Fe_2O_3 which might be formed after the calcination step. Thus, showed this slight effect on the uniformity on the catalyst surface.

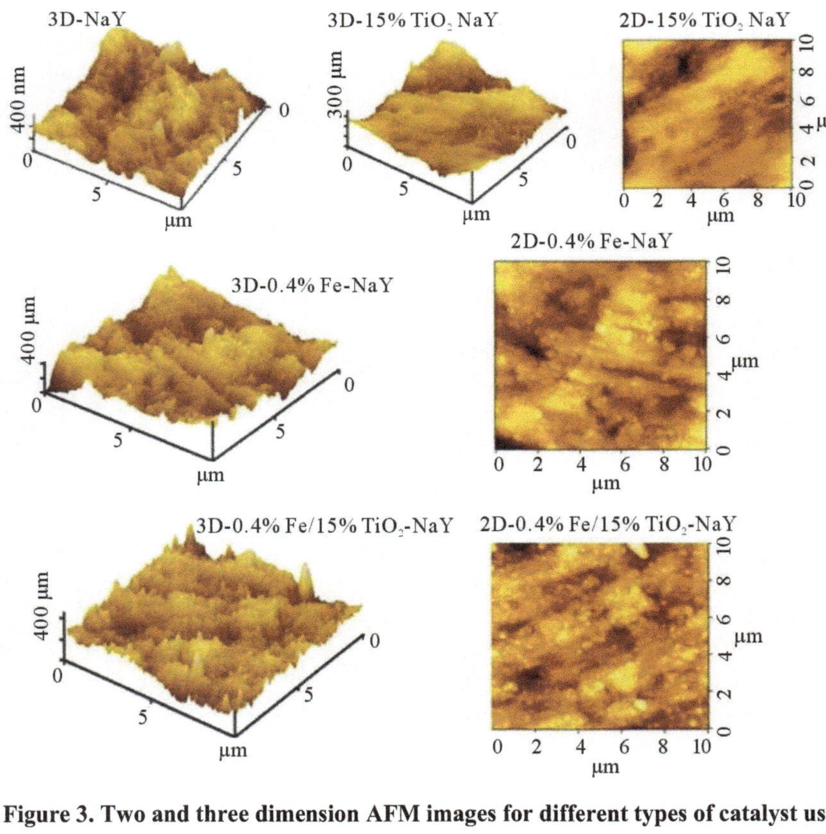

Figure 3. Two and three dimension AFM images for different types of catalyst used.

However, it was clear that the roughness of the catalyst was significantly decreased after the loading of TiO_2 for both cases of 15% TiO_2-NaY and 0.4% Fe/15% TiO_2-NaY. This extreme reduction in the roughness of the catalysts surfaces were ascribed to the small crystal size of the supported TiO_2 which was in agreement with XRD results (low intensity and wide peak). Although, the AFM technique can not differentiate between the existing metal on the surface of the catalyst. However, small clusters were significantly detected for the case of 0.4% Fe/15% TiO_2-NaY which could be ascribed to Fe oxide formed after the calcination step. While for the case of 15% TiO_2-NaY, it was clear that TiO_2 phase formed a layer covered the surface of zeolite and no small clusters were observed on its surface comparing to that of 0.4% Fe/15% TiO_2-NaY.

3.1.4. Transmission Electron Microscope

The TEM imaging was carried out for parent NaY, and 0.4% Fe/15% TiO_2-NaYcatalysts. The unmodified zeolite structure can be observed in **Figures 4(a)** and **(b)** for two different magnifications *i.e.* 8 kX and 45 kX, respectively. The hexagonal crystalline geometry of the zeolite remained virtually unchanged for the case of 0.4% Fe/ 15% TiO_2-NaY using impregnation method as was observed in **Figure 4(c)**. These results proved that the zeolite was generally able to maintain its original structure after the loading of TiO_2 and Fe since no significant damages were detected on its hexagonal structure. The TEM image suggested that the position of TiO_2 clusters were mainly positioned on the external surface of zeolite due to appearance of many white clusters accumulated on the external surface of zeolite structure and it was not shielded by zeolite particles. However, it was not possible to detect the Fe metal due to the higher loading of TiO_2 that was significantly occupied the surface of zeolite.

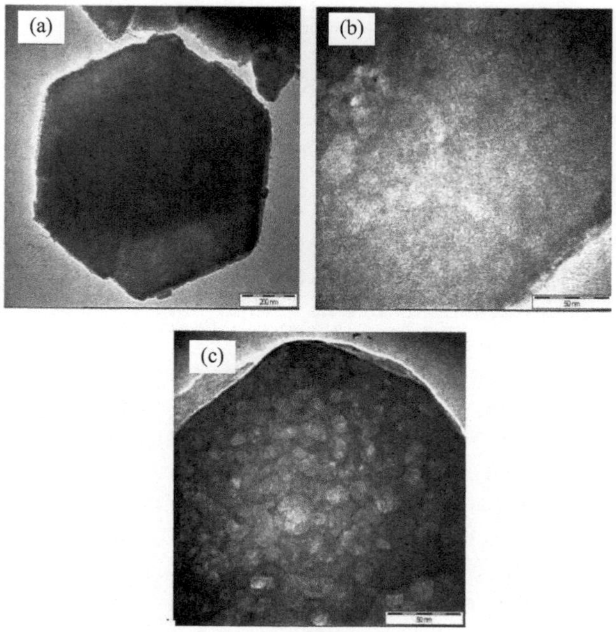

Figure 4. TEM images for (a) and (b) NaY and (c) 0 .4% Fe/ 5% TiO_2-NaY.

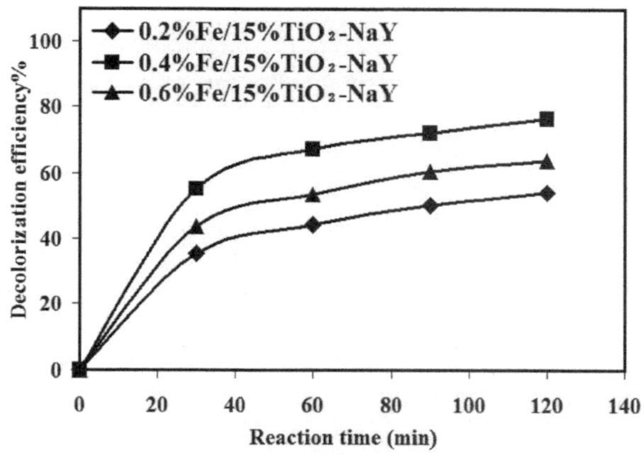

Figure 5. Effect of interaction between different loadings of Fe and fixed amount of TiO_2 with zeolite.

3.2. Sonocatalytic Degradation Process

3.2.1. Effect of Incorporation of Fe and Fixed Amount of TiO_2 with Zeolite

In order to obtained higher decolorization activity of the catalyst, preliminary experiments were carried out to investigate the synergistic effect between different wt% of Fe with fixed amount of TiO_2 15 wt% as shown in **Figure 5**. The experimental runs were carried out with 10 mg/L initial dye concentration, catalyst loading of 1.5 g/L and natural pH about 5.5 without the addition of H_2O_2 in the reaction. The results showed that the catalytic activity significantly increased with increasing the wt% of Fe loading from 0.2% to 0.4% Fe with catalytic activities of 54% and 76%, respectively. In this reaction, the Fe acted as electron acceptor from the e^--h^+ pairs generated through the activation of TiO_2 in the zeolite. Increasing the loading amount of Fe would provide more electron traps to hinder e-h^+ ions recombination leading to an increase in the sonocatalytic activity. In a complementary role, the zeolite support worked as a hole trap thus enhance efficiently the inhibition of the electron hole recombination. Therefore, the lower catalytic activity for the catalyst at 0.2% Fe loading was ascribed to the insufficient amount of Fe ions to inhibit the $e^- - h^+$ recombination. However, with increasing loading of Fe up to 0.6 wt%, the catalytic activity was found to decrease to 64%. In this case, as the wt% Fe was further increased, the $e^- - h^+$ recombination rate would increase also due to the existence of excess amount of Fe which would act as a charge carrier to cause this reduction in the catalytic activity. Bajamundi *et al.* [21] investigated the effect of Fe loading from 0 - 3.0 wt% Fe on the activity of TiO_2/Al_2O_3 and the higher degradation efficiency of methylene blue was at 0.1% Fe. As a conclusion it seems that increasing the concentration of Fe up to 0.4 wt% showed significant enhance in catalytic activity and the addition of higher concentration affected negatively on catalytic activity.

3.2.2. Effect of Incorporation of Tio_2 and Fixed Amount of Fe with Zeolite on Sonocatalytic Activity

As the optimum loading of Fe was 0.4 wt%, further experiment runs were carried out to investigate the interacttion between fixed loading of 0.4% Fe with different loadings of TiO_2 from 5 - 20 wt% and the results are shown in **Figure 6**. The sonocatalytic activity was found to increase with increasing the loading of TiO_2 from 5 - 15 wt% reaching its maximum activity of 76% at the highest TiO_2 loading. Increasing the loading of TiO_2 would provide more active sites to enhance the generation of $^{\bullet}OH$ radicals during the sonocatalytic reaction [11]. Further increase in TiO_2 loading above the optimum value *i.e.* 20 wt%, the decolorization efficiency of amaranth was significantly decreased to 34%. This drastic reduction in catalytic activity might be ascribed to the aggregation of TiO_2 clusters on the surface of zeolite [20]. Furthermore, with increasing amount of TiO_2, the amount of available Fe sites would not be sufficient to inhibit the e^--h^+ recombination thus, reduced the catalytic activity of the catalyst.

Another factor that can negatively effect on the catalytic activity might be referring to the formation of a separate TiO_2 phase with increasing the TiO_2 loading. This phase will no longer be incorporated with zeolite framework consequently, cause this reduction in catalytic activity. This result was suggested based on the results obtained from the AFM analysis since, there was a thin layer covered the surface of zeolite after the loading of 15 wt% of TiO_2. However, the incorporation of 0.4% of Fe/15% TiO_2 with zeolite showed the higher catalytic activity and further increase in TiO_2 was negatively affected.

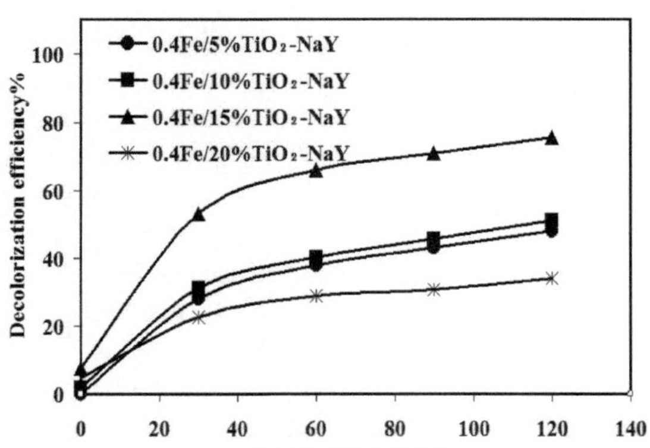

Figure 6. Effect of interaction of different loadings of TiO_2 and fixed amount of Fe with zeolite.

3.2.3. Control Experiment

A preliminary study was carried out to investigate the role of catalyst under both silent (mechanical stirring only) and ultrasonic reaction. The adsorption study without the presence of ultrasonic irradiation was carried out to check the adsorption capacity of 0.4% Fe/15% TiO_2-NaY catalyst for amaranth dye under continuous mechanical stirring for 120 min. The experimental runs were at an initial dye concentration of 10 mg/L, a catalyst loading of 1.5 g/L and at original pH of the dye solution. The results presented in **Figure 7** showed that only 4% of amaranth dye was adsorbed after 120 min of continuous stirring.

The decolorization efficiency under ultrasonic irradiation without the presence of catalyst was also investigated under the same reaction condition and the efficiency of the process was only about 12%. This low process efficiency was mainly ascribed to the limited generation of the hydroxyl radicals by the dissociation of water molecules during the ultrasonic reaction. In addition, Mahamuni *et al.* [18] suggested that the recombinetion of the $^{\bullet}OH$ radicals within the cavity during the sonolysis process could eliminate the generation of the active species thus reduce the efficiency of the sonolysis process.

On the other hand, the decolorization efficiency visibly increased after the addition of catalyst under ultrasonic irradiation *i.e.* 17.4%, 30%, 68% and 76% for PTO (potassium titanium oxalate) calcined at 550°C, 0.4% Fe-NaY, 15% TiO_2-NaY and 0.4% Fe/15% TiO_2-NaY, respectively. The synergistic between the catalysts and the ultrasonic process led to increase the production of more hydroxyl radicals and enhance the decolorization efficiency. From **Figure 7**, it is clear that the incorporation

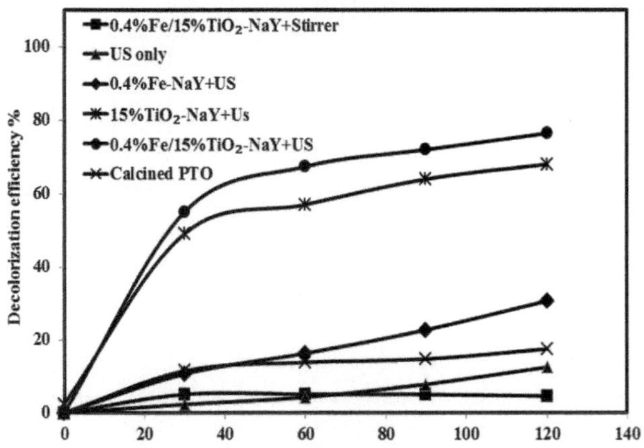

Figure 7. Preliminary study for the decolorization of amaranth dye under various conditions.

of Fe (III) and Ti species with zeolite could enhance the decolorization efficiency of amaranth as compared to that of 0.4 % Fe-NaY and 15% TiO_2-NaY catalysts.

3.2.4. Effect of pH

The effect of reaction acidity on the sonocatalytic degradation of amaranth was investigated within the pH range of 2.5 to 9.0 under ultrasonic reaction and the results are present in **Figure 8** . Before starting the sonocatalytic reaction, all the reactions were adjusted to the desire pH using 0.1 M NaOH or HCl with an initial dye concentration of 10 mg/L, 0.65 mM of H_2O_2 and 1.5 g/L catalyst loading for up to 120 min. It was noticed that the decolorization efficiency of the catalyst in the first 60 min was increased from 9% to 76% for pH 2.5 and 5.5, respectively. After longer reaction time, the color removal at different pH diminished and the reaction was almost complete with decolorization efficiency of about 98% for pH ranging from 3.5 to 9.0. Meanwhile, very low sonocatalytic activity of about 22.4% was obtained at the acidic condition of pH 2.5 after 120 min of reaction. The potassium (K) which was originally exists in the chemical structure of the titanium precursor was the reason behind this low catalytic activity at pH 2.5. The trace of such highly basic element exists at the catalyst in addition to the Fe and Ti will significantly affect the environment of the catalyst surface. Huang *et al.* [2] reported that the decolorization efficiency for methyl orange dye using Pt/TiO_2-zeolite catalyst was higher in high acidic condition.

An experiment run was carried out to investigate the effect of titanium precursor on the decolorization efficiency of amaranth at pH value of 2.5. The TiO_2 precursor (potassium titanium oxalate) used in this study was replaced with tetrabutyl orthotitanate using the same preparation method used in this study and the produced catalyst was denoted as 0.4% Fe/15% TBT-NaY. The dark experiment run showed an adsorption rate of 68% after 30 min of mechanical stirring without ultrasonic irradiation. However, the absorbed dye was desorbed

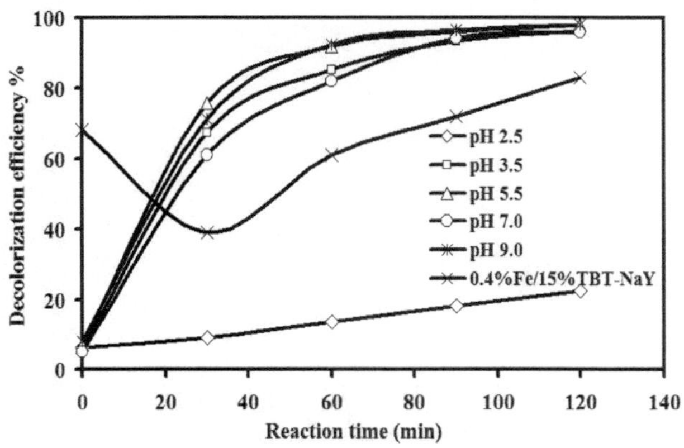

Figure 8. Effect of pH on decolorization efficiency of amaranth.

from the catalyst surface at the beginning of ultrasonic reaction due to the mechanical vibration provided by ultrasonic. The catalytic activity was 83% after 120 min of reaction without any addition of H_2O_2. It was clear that this large difference in catalytic activity between the original catalyst 0.4% Fe/15% TiO_2-NaY and 0.4% Fe/15% TBT-NaY at pH 2.5 was ascribed to the difference in chemical nature of TiO_2 precursor.

It is important to highlight that the zero point charge of the catalyst is depend on dispersion nature and the difference in catalyst diameter [6]. Furthermore, the surface properties are affected by the synergistic between Fe, TiO_2 and zeolite in addition to the trace of potassium oxide. This properties is complicated to be understand due the different component interact together. As a result, one can observe that the synthesized catalyst had high catalytic activity in a wide range of pH from 3.5 to 9.0 which could affect by the chemical nature of the catalyst.

3.2.5. Effect of Catalyst Loading

The decolorization efficiency of amaranth was demonstrated to obtain the optimum loading with respect to the amount of catalyst used from 1.0 to 2.5 g/L at an initial dye concentration of 10 mg/L, 0.65 mM of H_2O_2 and pH of 5.5 for up to 120 min. In **Figure 9**, it was notice that the difference in catalytic activity with different loading of catalyst was insignificant after 120 min of reaction and the activity was about 98% for the catalyst loadings variation from 1.0 to 2.5 g/L. This behavior could be attributed to the synergistic effect between the heterogeneous catalyst and the ultrasonic irradiation that provide mechanical mixing between the generated radicals and the dye molecules which enhance the catalytic activity [6]. However, there was a significant difference in decolorization efficiency at the first 30 min. The catalytic efficiency was found to increase with increasing catalyst loading to about 83% at catalyst loading of 2.0 g/L. Further increase in the amount of catalyst up to 2.5 g/L caused an activity drop to 74%.

The difference in sonocatalytic activity with different loading of catalyst at the first 30 min of reaction was

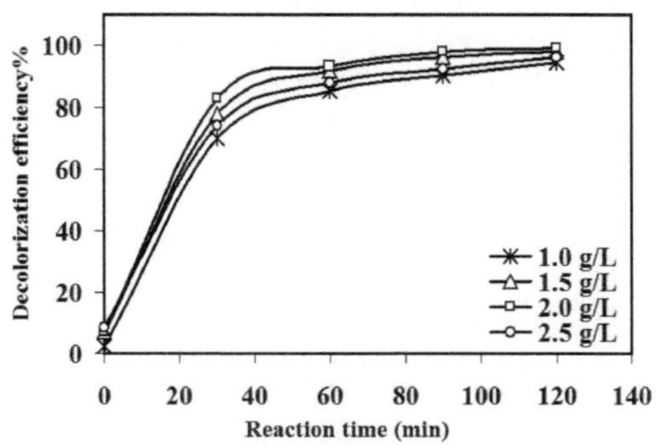

Figure 9 . Effect of catalyst loading on decolorization efficiency of amaranth.

ascribed to multiple reason such as increase the number of the active sites of such catalyst that generate more ˙OH radicals during the sonocatalytic reaction. Furthermore, the presence of Fe ions provided higher decomposition rate for H_2O_2 to generate more hydroxyl radicals during the reaction [22]. Therefore, the reaction becomes faster with increasing the time of reaction. As such, the difference in catalytic activity with different loading of catalyst was diminished after 120 min of reaction. Wang *et al.* [23] reported that the optimum amount of catalyst loading for the Fe doped TiO_2 was 1 g/L for the degradation of azo fuchsine under ultrasonic reaction. Meanwhile, Zhong *et al.* [24] reported that the maximum catalyst loading of Fe_2O_3/Al_2O_3 for the sonocatalytic degradation of acid orange 7 was 0.3 g/L. As a conclusion, the optimal amount of catalyst loading varied depending on the process conditions, the nature of dye molecules and the properties of catalyst.

3.2.6. Effect of H_2O_2 Concentration

The decolorization efficiency of amaranth dye as a function of different H_2O_2 concentration was investigated at initial dye concentration of 10 mg/L, 2.0 g/L of catalyst loading, original pH of the dye about 5.5 and with different concentrations of H_2O_2 from (0.32 - 2.62) mM. The decolorization efficiency was found to increase from 72 % to 83 % at the first 30 min with increasing the concentration of H_2O_2 from 0.32 to 0.65 mM as presented in **Figure 10**. However, by increasing the concentration of H_2O_2 above 0.65 mM, the decolorization efficiency was decreased to 69% at 2.61 mM of H_2O_2 concentration. This behavior could be ascribed to the excess amount of H_2O_2 added to the reaction solution that led to the generation of hydroperoxyl radicals (HO_2•) which was considered as the less reactive radicals compared to ˙OH radicals [25]. Aravindhan *et al.* [26] reported that excess amount of H_2O_2 added into the reaction system can negatively affect the efficiency of the process due to the scavenging of ˙OH.

Pang *et al.* [27] reported that the optimum amount of H_2O_2 needed to add into the ultrasonic reaction is depends on the type of the organic compound to be degraded and the ultrasonic operating condition. However, the presence of heterogeneous catalyst in synergistic with ultrasonic irradiation can be consider another factor that effect on the amount of H_2O_2 added into the reaction. In section 3.2.2 the sonocatalytic activity of 0.4% Fe/15% TiO_2-NaY without the addition of H_2O_2 was 74% thus, only 0.65 mM was needed to enhance the catalytic activity to 97% after 120 min of reaction and that no desirable effect in the color removal was observed with the addition of higher concentration of H_2O_2.

3.2.7. Effect of Initial Dye Concentration

The decolorization efficiency of 0.4% Fe/15% TiO_2-NaY catalyst as a function of initial dye concentration was also investigated. The experimental runs were carried at 2 g/L of catalyst loading, original pH of the dye solution *i.e.* about 5.5 and 0.65 mM of H_2O_2. It was noted in **Figure 11** that nearly complete decolorization was obtained at the end of the reaction at different initial dye concentrations from 10 to 100 mg/L. However, the rate of decolorization in the first 30 min of reaction was directly proportional to the decrease in initial dye concentration.

As the time of reaction was further increased, the color removal was relatively fast for the case of 30 mg/L of initial dye concentration. This behavior can be explained based on the collusion theory [28]. With increasing the concentration of dye molecules per unit volume of reaction, the chance of collisions between these reactants molecules increased thus, the decolorization efficiency steadily increased. Above the optimum value of 30 mg/L of the initial dye concentration, the decolorization efficiency after 60 min decreased from 100% to 85% for 30 and 100 mg/L of initial dye concentration, respectively. Jamalluddin Abdullah [29] reported the low decolorizetion efficiency with increasing the concentration of dye could be affected by the mutual screening between the

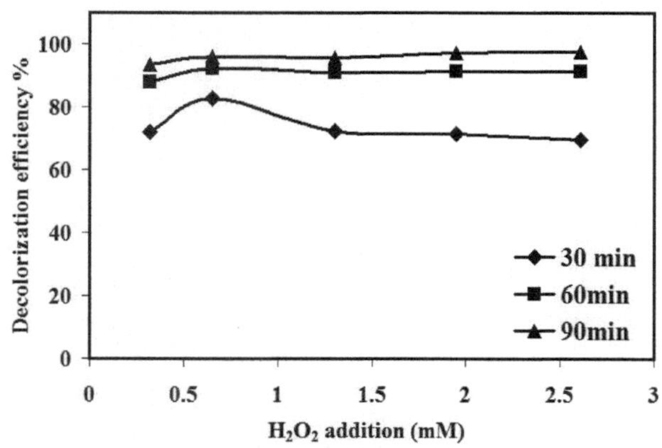

Figure 10. Effect of H_2O_2 on decolorization efficiency amaranth.

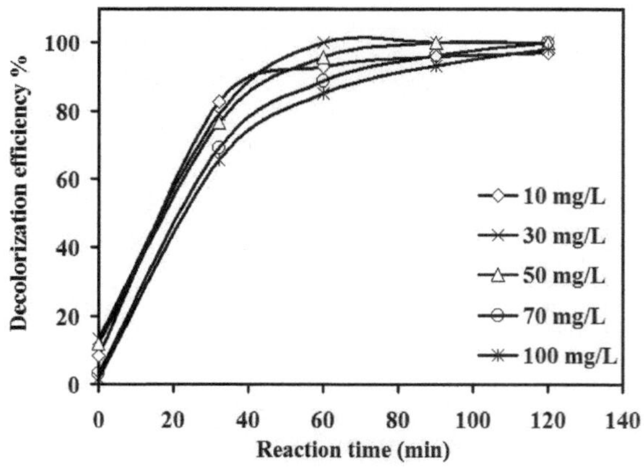

Figure 11. Effect of initial dye concentration on decolorizetion efficiency of amaranth.

organic dyes molecules at higher concentration of dye. However, with increasing the time of reaction more decolorization rate was observed and this mutual screening with higher dye concentration was significant only in earlier time of reaction.

It was found that the leaching of Fe^{3+} ions increased with increasing the concentration of dye molecules due to the generation of more acidic-by products that were formed during the reaction thus enhance the dissolving of Fe^{3+} ions during the reaction. The concentrations of Fe^{3+} found in the solution after 120 min of reaction were 0.011, 0.017, 0.033, 0.051 and 0.06 mg/L for 10, 30, 50, 70 and 100 mg/L, respectively. The more Fe^{3+} leached out from the catalyst, the more loss in active sites of catalyst was expected [30]. However, this small amount of Fe^{3+} ions separated from the catalyst did not affect the decolorization efficiency even for the high concentration of dye as it was clear from **Figure 11**.

Our previous report [5] investigates the degradation of amaranth using the heterogeneous catalyst of Fe loaded on encapsulated TiO_2. The higher decolorization efficiency of amaranth was 98% at 10 mg/L initial concentration, 2.5 pH, 0.2 M H_2O_2 and 2 g/L catalyst loading. However, in this catalyst the maximum decolorization efficiency was obtained at 30 mg/L initial dye concentration, 5.5 pH, 0.65 mM H_2O_2 and 2 g/L catalyst loading. The difference in optimum conditions between the two catalysts was significant especially for amount of H_2O_2 and pH of reaction solution. It seems that position of TiO_2 with respect to zeolite structure, preparation method and the amount of TiO_2 affect extremely on chemical environment of the catalyst and consequently effect on the optimum condition of the reaction.

4. Conclusion

A heterogeneous catalyst was successfully synthesized in one simple step using a wet impregnation method. Incorporation between Ti and Fe into the zeolite successfully enhanced the sonocatalytic reaction. The XRD results showed that chemical nature of titanium precursor affected significantly on the crystallinity of the catalyst. The AFM results showed a significant reduction in the surface roughness of the catalyst after the loading of 15% TiO_2 and 0.4% Fe/15% TiO_2 into the zeolite. Effects of different reaction variables were elucidated to achieve the higher decolorization of amaranth. The maximum decolorization efficiency of the 0.4% Fe/15% TiO_2-NaY was 100% at an initial dye concentration of 30 mg/L, 2 g/L of catalyst loading, natural pH of 5.5 and H_2O_2 concentration of 0.65 mM.

5. Acknowledgements

The Research University grant from Universiti Sains Malaysia to support this work is gratefully acknowledged.

REFERENCES

[1] H. Wang, Y. Shen, C. Shen, Y. Wen and H. Li, "Enhanced Adsorption of Dye on Magnetic Fe_3O_4 via HCl-Assisted Sonication Pretreatment," *Desalination*, Vol. 284, 2012, pp. 122-127.

[2] M. Huang, C. Xu, Z. Wu, Y. Huang, J. Lin and J. Wu, "Photocatalytic Discolorization of Methyl Orange Solution by Pt Modified TiO_2 Loaded on Natural Zeolite," *Dyes and Pigments*, Vol. 77, No. 2, 2008, pp. 327-334.

[3] S. S. Rayalu, N. Dubey, R. V. Chatti, M. V. Joshi, N. K. Labhsetwar and S. Devotta, "Effect of Heteropolyacid and Its Route of Incorporation on the Photocatalytic Properties of Zeolite-Based Materials," *Current Science*, Vol. 93, No. 10, 2007, pp. 1376-1382.

[4] A. N. Soon and B. H. Hameed, "Heterogeneous Catalytic Treatment of Synthetic Dyes in Aqueous Media Using Fenton and Photo-Assisted Fenton Process," *Desalination*, Vol. 269, No. 1-3, 2011, pp. 1-16.

[5] A. H. Alwash, A. Z. Abdullah and N. Ismail, "Zeolite Y Encapsulated with Fe-TiO_2 for Ultrasound-Assisted Degradation of Amaranth Dye in Water," *Journal of Hazardous Materials*, Vol. 233-234, 2012, pp. 184-193.

[6] Y. L. Pang, S. Bhatia and A. Z. Abdullah, "Process Behavior of TiO_2 Nanotube-Enhanced Sonocatalytic Degradation of Rhodamine B in Aqueous Solution," *Separation and Purification Technology*, Vol. 77, No. 3, 2011, pp. 331-338.

[7] R. J. Tayade, R. G. Kulkarni and R. V. Jasra, "Enhanced Photocatalytic Activity of TiO_2-Coated NaY and HY Zeolites for the Degradation of Methylene Blue in Water," *Industrial and Engineering Chemistry Research*, Vol. 46, No. 2, 2007, pp. 369-376.

[8] X. Li, B. Shen and C. Xu, "Interaction of Titanium and Iron Oxide with ZSM-5 to Tune the Catalytic Cracking of Hydrocarbons," *Applied Catalysis A: General*, Vol. 375, No. 2, 2010, pp. 222-229.

[9] P. Castaño, B. Pawelec, J. L. G. Fierro, J. M. Arandes and J. Bilbao, "Aromatics Reduction of Pyrolysis Gasoline (PyGas) over HY-Supported Transition Metal Catalysts," *Applied Catalysis A: General*, Vol. 315, 2006, pp. 101-113.

[10] C. Wang, H. Shi and Y. Li, "Synthesis and Characteristics of Natural Zeolite Supported Fe^{3+}-TiO_2 Photocatalysts," *Applied Surface Science*, Vol. 257, No. 15, 2011, pp. 6873-6877.

[11] C. C. Wang, C. K. Lee, M. D. Lyu and L. C. Juang, "Photocatalytic Degradation of C.I. Basic Violet 10 Using TiO_2 Catalysts Supported by Y Zeolite: An Investigation of the Effects of Operational Parameters," *Dyes and Pigments*, Vol. 76, No. 3, 2008, pp. 817-824.

[12] A. Z. Abdullah and P. Y. Ling, "Heat Treatment Effects

on the Characteristics and Sonocatalytic Performance of TiO_2 in the Degradation of organic Dyes in Aqueous Solution," *Journal of Hazardous Materials*, Vol. 173, No. 1-3, 2010, pp. 159-167.

[13] H. Hassan and B. H. Hameed, "Oxidative Decolorization of Acid Red 1 Solutions by Fe-Zeolite Y Type Catalyst," *Desalination*, Vol. 276, No. 1-3, 2011, pp. 45-52.

[14] G. Li, "FT-IR Studies of Zeolite Materials: Characterization and Environmental Applications," The University of Iowa's Institutional Repository, 2005.

[15] C. Li and Z. Wu, "Microporous Materials Characterized by Vibrational Spectroscopies," In: S. M. Auerbach, K. A. Carrado and P. K. Dutta, Eds., *Handbook of Zeolite Science and Technology*, CRC Press, Boca Raton 2003, pp. 423-514.

[16] M. Nikazar, K. Gholivand and K. Mahanpoor, "Photocatalytic Degradation of Azo Dye Acid Red 114 in Water with TiO_2 Supported on Clinoptilolite as a Catalyst," *Desalination*, Vol. 219, No. 1, 2008, pp. 293-300.

[17] H. G. Karge, "Characterization by Infrared Spectroscopy," *Microporous Mesoporous Materals*, Vol. 22, No. 4-6, 1998, pp. 547-549.

[18] N. N. Mahamuni and A. B. Pandit, "Effect of Additives on Ultrasonic Degradation of Phenol," *Ultrasonic Sonochemistry*, Vol. 13, No. 2, 2006, pp. 165-174.

[19] F. A. Q. Ortíz, J. T.Valenzuela and C. A. R. Reyes, "Zeolitisation of Neogene Sedimentary and Pyroclastic Rocks Exposed in Paipa (Boyacá), in the Colombian Andes: Simulating Their Natural Formation Conditions," *Earth Sciences Research Journal*, Vol. 15, No. 2, 2011, pp. 89-100.

[20] H. Faghihian and A. Bahranifard, "Application of TiO_2-Zeolite as Photocatalyst for Photodegradation of Some Organic Pollutants," *Iranian Journal of Catalysis*, Vol. 1, No. 1, 2011, pp. 45-50.

[21] C. Bajamundi, M. Dalida, K. Wantala, P. Khemthong and N. Grisdanurak, "Effect of Fe^{3+} Doping on the Performance of TiO_2 Mechanocoated Alumina Bead Photocatalysts," *Korean Journal of Chemical Engineering*, Vol. 28, No. 8, 2011, pp. 1688-1692.

[22] J. Herney-Ramirez, M. A. Vicente and L. M. Madeira, "Heterogeneous Photo-Fenton Oxidation with Pillared Clay-Based Catalysts for Wastewater Treatment: A Review," *Applied Catalysis B: Environmental*, Vol. 98, No. 1-2, 2010, pp. 10-26.

[23] J. Wang, W. Sun, Z. Zhang, Z. Jiang, X. Wang, R. Xu, R. Li and X. Zhang, "Preparation of Fe-Doped Mixed Crystal TiO_2 Catalyst and Investigation of Its Sonocatalytic Activity during Degradation of Azo Fuchsine under Ultrasonic Irradiation," *Journal of Colloid and Interface Science*, Vol. 320, No. 1, 2008, pp. 202-209.

[24] X. Zhong, L. Xiang, S. Royer, S. Valange, J. Barrault and H. Zhang, "Degradation of C.I. Acid Orange 7 by Heterogeneous Fenton Oxidation in Combination with Ultrasonic Irradiation," *Journal of Chemical Technology & Biotechnology*, Vol. 86, No. 7, 2011, pp. 970-977.

[25] M. Neamţu, C. Zaharia, C. Catrinescu, A. Yediler, M. Macoveanu and A. Kettrup, "Fe-Exchanged Y Zeolite as Catalyst for Wet Peroxide Oxidation of Reactive Azo Dye Procion Marine H-EXL," *Applied Catalysis B: Environmental*, Vol. 48, No. 4, 2004, pp. 287-294.

[26] R. Aravindhan, N. N. Fathima, J. R. Rao and B. U. Nair, "Wet Oxidation of Acid Brown Dye by Hydrogen Peroxide Using Heterogeneous Catalyst Mn-Salen-Y Zeolite: A Potential Catalyst," *Journal of Hazardous Materials*, Vol. 138, No. 1, 2006, pp. 152-159.

[27] Y. L. Pang, A. Z. Abdullah and S. Bhatia, "Review on Sonochemical Methods in the Presence of Catalysts and Chemical Additives for Treatment of Organic Pollutants in Wastewater," *Desalination*, Vol. 277, No. 1-3, 2011, pp. 1-14

[28] M. B. Kasiri, H. Aleboyeh and A. Aleboyeh, "Degradation of Acid Blue 74 Using Fe-ZSM5 Zeolite as a Heterogeneous Photo-Fenton Catalyst," *Applied Catalysis B: Environmental*, Vol. 84, No. 1-2, 2008, pp. 9-15.

[29] N. A. Jamalluddin and A. Z. Abdullah, "Reactive Dye Degradation by Combined Fe(III)/TiO_2 Catalyst and Ultrasonic Irradiation: Effect of Fe(III) Loading and Calcination Temperature," *Ultrasonic Sonochemistry*, Vol. 18, No. 2, 2011, pp. 669-678.

[30] N. A. Jamalluddin and Z. A. Ahmad, "Effect of pH and Catalyst Dosage on the Leaching of Fe from Fe(III) Doped Zeolite Y Used for Sonocatalytic Degradation of Acid Red B," *Research Journal of Chemistry and Environment*, Vol. 15, No. 2, 2011, pp. 860-865.

Application of Solution Calorimetry to the Prediction of 20.15[(2.038 + x)SiO_2-(1.457 − x)Na_2O]-2.6-P_2O_5-26.95CaO Glass Bioactivity

Amadou Kouyate[1,2], Ange Privat Ahoussou[1*], Jacques Rogez[2], Pierre Benigni[2]
[1]Laboratoire de Chimie-Physique, UFR-SSMT, Université de Cocody, Abidjan, Côte d'Ivoire
[2]Faculté des Sciences et Techniques, Campus de Saint-Jérôme, Aix-Marseille Université, Marseille, France

ABSTRACT

Bioactive glasses in the system SiO_2-Na_2O-CaO-P_2O_5 were prepared by the conventional melting process at a high temperature in an electric furnace. The densities of the glasses as well as their glass transition temperatures were measured experimentally. Also, the glasses were characterized by solution calorimetry at 298 K in an acidic solvent consisting of HF (6 M) and HNO_3 (4 M). A simulation of the bioactivity properties of these obtained glasses was carried out by thermodynamic calculations, with the purpose of understanding the *in vitro* results obtained.

Keywords: Bioactive Glass; Solution Calorimetry; Free Enthalpy of Formation; Reaction Number

1. Introduction

In recent decades, the applications of bioactive glasses as bioimplants to repair damaged or diseased bones have escalated, particularly in the context of their ability to form a strong interfacial bond with living tissues. These glasses have been used in many clinical operations during the last twenty years including surgery of human body parts such as ears, teeth, knees and hips. The choice of bioactive glasses for implant applications requires design optimized compositions [1]. Thus, preparing bioactive glasses with different compositions, and evaluating their physicochemical properties, structural properties and chemical stabilities is crucial for the subsequent optimization of their biocompatibility [2,3]. Some compositions of glasses in the system SiO_2-Na_2O-CaO-P_2O_5, used as biomaterials, are able to form a bond with bone once they are implanted, such as the Bioglass [4,5]. The main advantages of these synthetic biomaterials are availability, reproducibility and reliability [6].

On the other hand, the synthetic biomaterials must be fully biocompatible, they should not generate any complication or inflammatory response, and also, they must be suitable in their mineral compositions. A common characteristic of all known bioactive materials is the formation of a biologically active HCA [7-9] layer in the presence of *in vivo* body fluids or *in vitro* simulated body fluids (SBF). SBF solution mimics the inorganic composition of human body fluids. The kinetics of the dissolution of the glass in the biological environment and consequently the bioactivity of the upper thin layer depend on the chemical composition of the glass [10,11].

In this work, synthesis of bioactive glasses was carried out by using the melting and rapid quenching technique. Samples in the system 20.15[(2.038 + x)SiO_2-(1.457 − x)-Na_2O]-2.6-P_2O_5-26.95CaO were studied over the whole composition range, using X-ray diffraction, pycnometry, differential scanning calorimetry and solution calorimetry according to the composition. The properties such as density, glass transition temperature and chemical stability were closely studied and correlated to changes in the structure and the evolution of the index of bioactivity by Andersson.

2. Experimental

Nine different glassy compositions belonging to the system 20.15[(2.038 + x)SiO_2-(1.457 − x)Na_2O]-2.6-P_2O_5-26.95CaO were prepared by the melt quenching technique, as shown in **Table 1**.

High purity reagent grade silica (SiO_2, Merck product),

*Corresponding author.

Table 1. Compositions in mol% (wt%) of glasses investigated.

Sample	x	SiO_2	P_2O_5	Na_2O	CaO
40S5	0	41.07 (40)	2.6 (6)	29.36 (29.5)	26.95 (24.5)
42.5S5	0.125	43.58 (42.5)	2.6 (6)	26.84 (27)	26.95 (24.5)
45S5	0.25	46.10 (45)	2.6 (6)	24.32 (24.5)	26.95 (24.5)
47.5S5	0.375	48.62 (47.5)	2.6 (6)	21.80 (22)	26.95 (24.5)
50S5	0.5	51.14 (50)	2.6 (6)	19.28 (19.5)	26.95 (24.5)
52.5S5	0.625	53.66 (52.5)	2.6 (6)	16.76 (17)	26.95 (24.5)
55S5	0.75	56.18 (55)	2.6 (6)	14.25 (14.5)	26.95 (24.5)
57.5S5	0.875	58.70 (57.5)	2.6 (6)	11.73 (12)	26.95 (24.5)
60S5	1	61.22 (60)	2.6 (6)	9.21 (9.5)	26.95 (24.5)

calcium carbonate ($CaCO_3$, Prolabo product), sodium carbonate (Na_2CO_3, Aldrich product) and di-ammonium hydrogen phosphate ($(NH_4)_2HPO_4$, Prolabo product) were used to obtain the glass compositions. These reagents were mixed up in the desired proportions and thoroughly ground in an agate mortar. The mixtures were fired in air in a platinum crucible using a programmed kiln, in the following manner:

a) Reaching 500°C at the end of 80 min and maintaining this temperature for 30 min to allow the evaporation of water from the powders to be as complete as possible;

b) Reaching 1000°C in 85 min and maintaining this temperature for 30 min to allow all the gases removal (CO_2, NH_3, H_2O);

c) Reaching 1500°C in 85 min and maintaining this temperature for 90 min to allow a complete homogenization of the melt.

The melt was quenched by dipping the bottom of the crucible into cold water. The resulting glass fragments were transparent and free from bubbles. The samples were then kept in vacuum desiccators in the presence of P_2O_5 to avoid possible moisture sorption before testing.

The amorphous state of our samples was analyzed by X-ray diffraction. The diffractograms were recorded between 10°C and 60° (2θ) at a scanning speed of 2°/min and a fixed step-size of 0.02° using a D5000 SIEMENS diffractometer with λ = 1.5406 Å and a tube voltage of 40 kV. The samples were all analyzed in powder form.

The glass transition temperatures (Tg) were measured by a differential scanning calorimeter (DSC 111 SETERAM). The mass of each analyzed glass sample was ~50 to 60 mg so that the curve obtained by DSC is operative. The handling was carried out under an atmosphere of argon in a crucible made of stainless steel seam. The heating rate was 5°C/min. The rise in temperature is limited to 700°C in order to keep the apparatus in good working condition. The values of the temperature (Tg) were given with an accuracy of ±3°C.

Densities of glass samples were estimated at 296.15 K by using a pycnometer with diethylphtalate (d = 1.118) as the immersion fluid. The pycnometer consists in glass flask with a close-fitting ground glass stopper surmounted by a capillary indexed tube. This tube releases a part of the liquid after closing the top-filled pycnometer and allows to measure with a very high accuracy, masses of constant volumes of working liquid or liquid and solid. Care was taken to make sure that all of the cast samples were bubble free and that the glass samples were fully wetted when immersed in the diethylphtalate. All the weight measurements have been made using a digital balance Sartorius, having an accuracy of ±0.1 mg. The reported values of density are averages of the four to five independent measurements. Experimental accuracy in the density measurement is ±0.03 $g \cdot cm^{-3}$.

As the reaction between the glassy oxides cannot be carried out directly into the calorimeter at ambient temperature due to kinetic limitations, the well-known dissolution method is applied. This indirect method based on a Hess thermodynamic cycle consists in the successive dissolutions in the same solvent and at the same temperature of all the compounds involved in the reaction of formation [12-14]. HF calorimetric experiments were performed with a swinging calorimeter. The calorimeter and standard procedures have been described previously [14, 15] and so only a brief description will be given below. Thermopiles consisting of hundreds of chromel-alumel thermocouples were inserted into an aluminum block isotherm. The high stability of the device allows very long period of experimentation around room temperature (293 - 350 K). The whole measurement cell, made of Poly Tetra Fluoro Ethylene (PTFE) resin allows dissolutions in all mineral acidic or basic aqueous solutions. The stirring rate was 3 rpm. The calorimeter is in a thermally regulated room at ± 1 K and the dissolutions are carried out at 298.15 ± 0.01 K. Before dissolution, the solute and the solvent are whole isolated each other in two distinct compartments separated by a platinum cap and a n-decane liquid joint. The acidic vapor cannot react with the solute during the thermal stabilization time of the calorimeter. After thermal equilibrium during about 4 h, the solute is put in touch with the solvent which consists in a 4 mL HF (6 M) and 4 mL HNO_3 (4 M) mixture by rotating the calorimeter. The dissolution starts with the first rotation. The rotation of the calorimeter is stopped as soon as the flow of energy produced by the reaction is cancelled. The data acquisition is stopped about an hour after the end of the dissolution so the final thermal and mechanical state of the calorimeter is identical to the initial state. The calorimeter was calibrated by using the heat effect of the well-known enthalpy of dissolution of tris-hydroxy-methylaminomethane in hydrochloric solution [15].

3. Results and Discussion

3.1. Results

X-ray diffraction measurements at room temperature confirmed the amorphous state of the samples. DSC thermograms showed endothermic peaks in the temperature range 489˚C - 613˚C (**Figure 1**). These endothermic peaks are attributed to the glass transition (Tg), at which atoms start to be reorganized towards crystallization. The Tg data are gathered in **Table 2** and shown as a function of the SiO_2 content in **Figure 2**. The line is only drawn as a guide to the eye. It has been found that Tg increased from 489˚C to 613˚C with the increase in the SiO_2 mole fraction from nought to one.

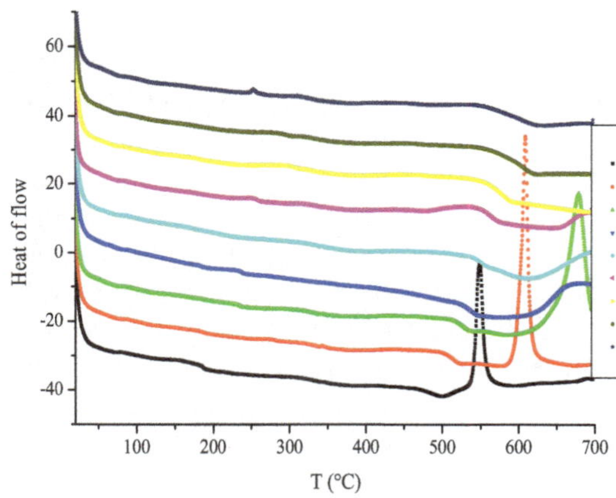

Figure 1. DSC curves for studies glasses.

Table 2. Measurements of the glass transition temperature (Tg), density (*d*) and molar solution enthalpy at infinite dilution $\left(\Delta_{diss}H^{\infty}\right)$ at 298.15 K of the 20.15[(2.038 + *x*)SiO_2-(1.457 − *x*)Na_2O]-2.6-P_2O_5-26.95CaO bioactive glasses.

x	Tg (˚C)	*d* (g·cm^{-3})	$\Delta_{diss}H^{\infty}$ (kJ·mol^{-1})
0	489	2.79	−149.36
0.125	515	2.78	−146.81
0.25	529	2.76	−144.27
0.375	535	2.74	−143.34
0.5	550	2.73	−139.61
0.625	564	2.72	−137.97
0.75	576	2.70	−137.56
0.875	607	2.69	−134.71
1	613	2.67	−135.98
	± 3	± 0.03	± 0.64

Measurements of the densities of the studied bioactive glasses are also listed in **Table 2**. Density is a quantity sensitive to the spatial arrangement and the nature of atoms [16]. Therefore, it depends on the composition. We observe the same decrease monotonically, from 2.79 to 2.66 g·cm^{-3} with the addition of silica content in the mixture (**Figure 3**).

For each glass composition, the enthalpy of dissolution was measured five times at different solute concentrations. The results are summarized in **Table 3**. The enthalpies of dissolution are strongly negative and strongly exothermic. **Table 2** shows the values of the enthalpy of dissolution at infinite dilution at 298 K. They were deduced by linear extrapolation to zero concentration enthalpies of dissolution for each glass composition. The accuracy of these estimations was about ±0.64 kJ·mol^{-1}.

Classically, the enthalpy of formation determined by the dissolution method is obtained with the algebraic sum of the enthalpies of solution: Let $A_xB_{1-x}C$ be a mixture of the A, B and C compounds in the same physical state at the same temperature:

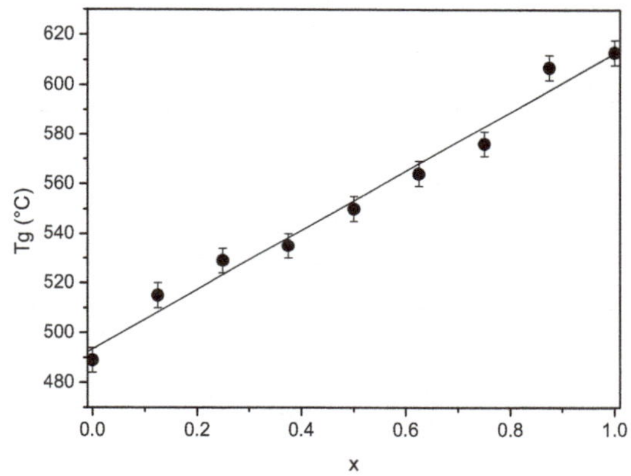

Figure 2. The variation of the glass transition temperature (Tg) of 20.15[(2.038 + *x*)SiO_2-(1.457 − *x*)Na_2O]-2.6-P_2O_5-26.95CaO bioactive glasses with the silica content.

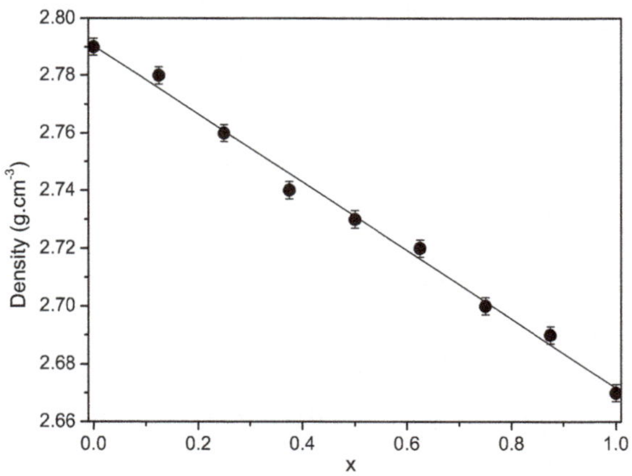

Figure 3. Variations of experimental densities with the composition (*x*) of the 20.15[(2.038 + *x*)SiO_2-(1.457 − *x*)Na_2O]-2.6-P_2O_5-26.95CaO bioactive glasses.

Table 3. Measurements of enthalpies of dissolution of 20.15[(2.038 + x)SiO_2-(1.457 − x)Na_2O]-2.6-P_2O_5-26.95CaO bioactive glasses at 298K in 4 mL HF (6 M) and 4 mL HNO_3 (4 M).

x	Glasses	m (mg)	ΔH_{diss} (kJ.mol^{-1})
		5.88	−149.62
		4.38	−149.20
0	40S5	4.13	−149.86
		5.40	−149.60
		6.04	−148.52
		5.95	−147.77
		4.30	−146.11
0.125	42.5S5	4.54	−147.60
		5.53	−146.08
		5.74	−146.51
		4.58	−143.32
		6.04	−143.81
0.25	45S5	5.27	−144.45
		5.90	−144.82
		5.82	−144.96
		5.71	−143.62
		5.81	−143.71
0.375	47.5S5	5.95	−143.59
		5.97	−143.01
		5.27	−142.75
		5.71	−139.93
		4.99	−139.06
0.5	50S5	5.14	−139.69
		5.48	−140.30
		5.98	−139.07
		4.99	−138.02
		5.79	−137.26
0.625	52.5S5	4.80	−138.24
		4.58	−139.20
		5.67	−137.13
		4.73	−137.84
		4.38	−138.13
0.75	55S5	5.14	−137.01
		5.05	−137.21
		5.00	−137.62
		4.54	−134.29
		5.57	−135.54
0.875	57.5S5	4.72	−134.74
		4.49	−134.77
		4.60	−134.23
		5.15	−135.32
		4.59	−135.88
1	60S5	4.63	−135.46
		5.19	−137.44
		5.21	−135.81

$$xA + (1-x)B + C \rightarrow A_xB_{1-x}C \qquad \Delta H_f^\circ$$

Let be the dissolution reactions extrapolated at infinite dilution at the same temperature and in the same solvent S, the $\Delta_{diss}H_i^\infty$ being the enthalpies at infinite dilution previously calculated.

$$A + S \rightarrow ((A, S, \infty)) \qquad \Delta_{diss}H_A^\infty$$
$$B + S \rightarrow ((B, S, \infty)) \qquad \Delta_{diss}H_B^\infty$$
$$C + S \rightarrow ((C, S, \infty)) \qquad \Delta_{diss}H_C^\infty$$
$$A_xB_{1-x}C + S \rightarrow ((A_xB_{1-x}C, S, \infty)) \qquad \Delta_{diss}H_{A_xB_{1-x}C}^\infty$$

The enthalpy of formation is obviously calculated by:

$$\Delta H_f^\circ = x\Delta_{diss}H_A^\infty + (1-x)\Delta_{diss}H_B^\infty + \Delta_{diss}H_C^\infty - \Delta_{diss}H_{A_xB_{1-x}C}^\infty \qquad (1)$$

Because the interaction between the solutes at infinite dilution in the solvent S is null or:

$$((A, S, \infty)) + ((B, S, \infty)) + ((C, S, \infty)) \rightarrow ((A_xB_{1-x}C, S, \infty)) \qquad \Delta H = 0$$

Experimental measurements or estimated enthalpies of dissolution values of SiO_2, P_2O_5, CaO and Na_2O oxides necessary to the calculation of the enthalpies of formation are indicated to the **Table 4**.

These data are used to calculate the corresponding enthalpies of formation using Equation (1). The uncertainties in the enthalpies of formation are extracted from the errors in the enthalpies of solution. For condensed phases and ambient temperature, the entropic contribution $(T\Delta S_f^\circ)$ is small and negligible in comparison with that of the enthalpy of formation (ΔH_f°). Therefore, the free enthalpy of formation (ΔG_f°) can be equated to the enthalpy of formation $(\Delta G_f^\circ \approx \Delta H_f^\circ)$. The results are summarized in **Table 5**. The so calculated free enthalpies of formation are listed in **Table 5** and reported on **Figure 4**.

The materials studied (**Table 1**) showed interesting biological properties largely developed in the literature [17,18]. Our approach is part of a process of pre-evaluation of the bioactivity of a glass on the basis of a study by solution calorimetry with its thermodynamic stability.

Table 4. Experimental measurements or estimated enthalpies of dissolution ΔH_{diss}^∞ (kJ·mol^{-1}) values of SiO_2, P_2O_5, CaO and Na_2O oxides at 298 K in 4 mL HF (6 M) and 4 mL HNO_3 (4 M).

SiO_2	P_2O_5	Na_2O	CaO
−129.09 ± 2.02	−404.73 ± 4.65	−385.84 ± 1.60	−163.77 ± 2.20

Table 5. Index of bioactivity (RN) and free enthalpies of formation at 298 K bioactive glasses system 20.15[(2.038 + x)-SiO_2-(1.457 − x)Na_2O]-2.6-P_2O_5-26.95CaO.

x	Glasses	Reaction number (RN)	ΔG_f° (kJ·mol^{-1})
0	40S5	6.12	−71.39 ± 2.13
0.125	42.5S5	6.17	−67.77 ± 2.18
0.25	45S5	6.08	−63.75 ± 2.19
0.375	47.5S5	5.84	−59.40 ± 2.13
0.5	50S5	5.46	−54.96 ± 2.21
0.625	52.5S5	4.93	−50.26 ± 2.31
0.75	55S5	4.25	−45.23 ± 2.23
0.875	57.5S5	3.43	−39.88 ± 2.35
1	60S5	2.47	−34.00 ± 2.30

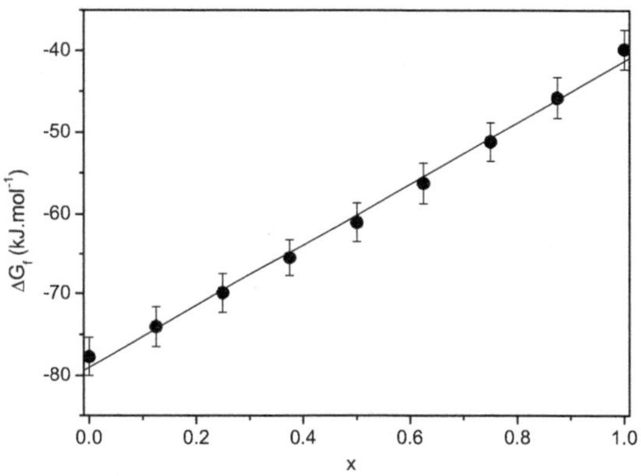

Figure 4. Variation of the free enthalpy of formation at 298 K as a function of the silica content of bioactive glasses 20.15[(2.038 + x)SiO_2-(1.457 − x)Na_2O]-2.6-P_2O_5-26.95CaO.

To achieve this, we calculated the free enthalpies of formation (ΔG_f°) of glasses in a solution of HF (6 M) + HNO_3 (4 M) at 298 K and obtained evidence of bioactivity following Andersson's equation [19,20]:

$$\begin{aligned} RN = 88.3875 - 0.0116272[SiO_2]^2 - 1.12306[CaO] \\ - 0.980188[Na_2O] - 1.20556[P_2O_5] \end{aligned} \quad (2)$$

with the proportions of the oxides in weight percent. RN is the "Reaction Number", a value permitting to define a domain of bioglasses. If RN is superior or equal to 5, then the glass develops apatite and bonds to bone. The results are also summarized in **Table 5**.

After eight weeks of *in vivo* experiments in rabbits' shins, Andersson states that when: RN = 1, the lenses are virtually inert, and it was found that a slight change in the surface composition occured. For RN = 2, there is contact with the bone but no real connection. Also, a layer of silica gel was formed on the surface but there was no precipitation of CaP. As for RN = 3, the glasses are very soluble, there is contact with the bone but again no real connection. The silica-rich layer is well formed but the precipitation of CaP is low. Then, for RN = 4, there was the formation of a CaP-rich layer but not enough to create a strong bond with the bone. Finally, for RN = 5, there was the formation of a layer of hydroxyapatite and consequently bone binding. The bioactive glasses have an RN value equal to or greater than 5.

According to **Table 5** glasses 40S5 (x = 0) and 42.5S5 (x = 0.125) are their RN higher than the Hench glass [4], reference glass called "Bioglass" noted 45S5 whose molar composition are 46.1% SiO_2, 26.9% CaO, 24.4% Na_2O and 2.6% P_2O_5, developing its hydroxyapatite layer after 12 hours immersion in contact with a physiological fluid. It is likely that these glasses are more bioactive than the Hench glass. Beyond the 52.5S5 composition, the link with the bone is weakened although in all cases, there is contact with the bone. In this study, the most bioactive glasses are poorer in silica and rich in sodium, in accordance with the work of Lebecq *et al.* [21].

With the results of the free enthalpies of formation given also in **Table 5**, we can assert that all compositions are very stable compared to their simple oxides at 298 K and their stabilities diminish gradually as the silica content increases in the system considered (**Figure 4**). **Figure 5** reports the change in reactivity index based on the free enthalpy of formation of each glass. The physicochemical reactions that occur on the surface when tested *in vitro* follow a mechanism common to almost all bioactive glasses [22]. However, the chemical reactivity of the glass in contact with the simulated biological fluid may be strongly influenced by the free enthalpy of formation of the latter. The shape of the curve of **Figure 5** for RN can be schematically divided into two zones, extensive variables defined by a change in chemical reactivity (slope change) on which the thermodynamic and physi-

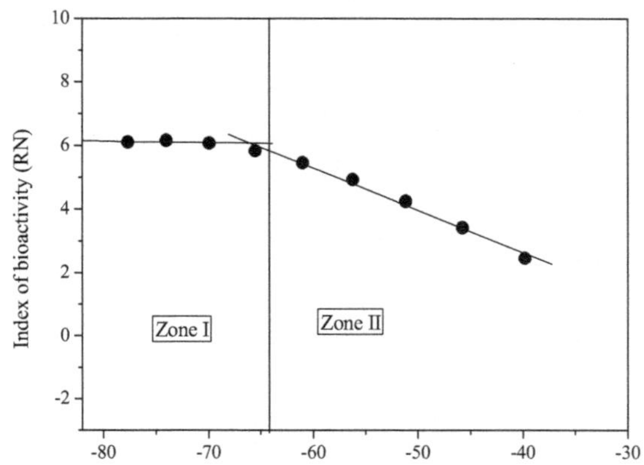

Figure 5. Variation of the index of bioactivity (RN) based on the free enthalpy of formation at 298 K bioactive glasses system 20.15[(2.038 + x)SiO_2-(1.457 − x)Na_2O]-2.6-P_2O_5-26.95CaO.

cochemical parameters can act and optimize the performance of bioactivity.

3.2. Discussion

The crystallinity of each obtained sample was systematically controlled by X-ray diffraction before any physicochemical analysis. The XRD results revealed that the samples are amorphous over the whole composition range in accordance with the work of L. L. Hench [23].

The glass transition temperature (Tg) plays an essential role in understanding the physical properties of bioactive glasses. In bioactive glasses, the glass transition temperature is particularly sensitive to the optimization of the composition of the bioactive glasses. All the samples studied herein were of the same thermal history and the same heating rate. Therefore, the cause of the changes in the Tg can be attributed to changes in the structure of the glass (**Figure 2**). Indeed, when the modifier oxide (Na_2O) is replaced by the forming oxide (SiO_2), molecular mobility is likely reduced because of the increase in the average degree of crosslinking. This increase is most likely a reflection of the building structure due to changes in composition. The creation of a network covalent character is favored by most orthophosphate groups which show no preferential association with Na^+ and Ca^{2+}, not used by non-bridging oxygens of the SiO_4 tetrahedra.

Density is an effective tool for exploring the evolution of the structure, coordination and cross-linking of glasses. Indeed, the density of a glass is directly related to the density of the elements involved in the system. According to **Figure 3**, the density of the bioactive glass system studied decreases almost linearly with the molar concentration of SiO_2 from 2.79 to 2.67 $g{\cdot}cm^{-3}$ without showing any abnormality. This result may suggest a change in the quasi-monotone structure due to the incorporation of SiO_2 at the expense of Na_2O in the glass. The quasi-linearity of the curve of variation of the enthalpy of formation may be indicative of a very weak interaction of the silica with the polymer chain.

Changes in the proportions of RN revealed the various stages leading to the formation of a hydroxyapatite layer in a simulated physiological environment. The slopes for these variations may be associated with the reactivity (clashes) of the different glasses via their reactive surfaces. The area in which curve I based RN is almost constant indicates a maximum threshold of bioactivity in the system studied. In this field of enthalpy of about −80 to −65 kJ/mol, the superposition of the silica-rich layer and the layer of hydroxyapatite is clearly observable above the glass matrix (in solution). They are predisposed to generate a series of physico-chemical reactions that can promote the growth of apatite in contact with a biological medium, allowing interfacial chemical bond with a bone marrow. The hydroxyapatite layer is intimately connected with the bone. As for zone II, it marks a nearly linear decrease (drastically) of RN index with increasing free enthalpy of formation. It therefore appears that in this case, the stable bond between the material and the host tissue is reduced or inhibited.

Several reasons explain the biological activity of a glass. We can distinguish kinetic energy and morphological root causes. Justifications based on kinetics and morphology have been widely put forward to explain and improve the bioactivity of glasses. Our study also shows that the nature and composition of the glass determine the biological characteristics (index of bioactivity) of the material. It has been shown, the importance of the thermodynamic parameter as well as the morphology and the kinetics. The nature and the composition of the glass affect the thermodynamic properties of the glass and thus the heat transfer. A glass, of low-silica content, induces a substantial increase in the enthalpy of formation in reference to constitutive single oxides, which promotes the simultaneous formation of a CaP layer and a layer rich in silica with a highspeed reaction. For glasses containing between 48 and 52 wt% of SiO_2, both layers are well separated: the layer of silica gel is formed first, and then the CaP surface develops [5,7]. This results in a decrease in the enthalpy of formation in the two systems. For glasses containing between 53 and 60 wt% silica, the enthalpy of formation becomes smaller in absolute value. In this case, the CaP layer does not form substantially.

4. Conclusion

It has been demonstrated in this work that in sodium and calcium phosphosilicate glasses given by $20.15[(2.038 + x)SiO_2\text{-}(1.457 - x)Na_2O]\text{-}2.6P_2O_5\text{-}26.95CaO$ from the Hench bioactivity diagram domain, the addition of the SiO_2 content leads to a gradual increase in Tg and a decrease in density. The enthalpy of mixing is negative and weak in the whole composition range. The reaction for the formation of the glass is favored thermodynamically. Silica-poor compositions are more stable. However, with the Silica-poor compositions, the products may form the HCA quickly, kinetic (the fastest product to form, training time short HAC) and thermodynamic (product thermodynamically most stable) controls being achieved simultaneously. Indeed, bioactivity is strongly dependent on both kinetic and thermodynamic control of the starting glass. Thermodynamic control is really crucial for bioactive glasses; the more stable the sample, the higher the ability to bind strongly to bone tissues. It reflects a weak chemical interaction between the components of the oxides. This study contributes to the enrichment of existing data on bioactive glasses.

REFERENCES

[1] T. Westerland, L. Hatakka and K. H. Karlsson, "A Model for Optimizing Glass Batch Compositions," *Journal of the American Ceramic Society*, Vol. 66, No. 8, 1983, pp. 574-579.

[2] K. J. L. Burg, S. Porter and J. F. Kellam, "Biomaterial Developments for Bone Tissue Engineering," *Biomaterials*, Vol. 21, No. 23, 2000, pp. 2347-2359.

[3] J. C. Knowles, K. Franks and I. Abrahams, "Investigation of the Solubility and Ion Release in the Glass System K_2O-Na_2O-CaO-P_2O_5," *Biomaterials*, Vol. 22, No. 23, 2001, pp. 3091-3096.

[4] L. L. Hench, "A Forecast for the Future. Biomaterials," *Biomaterials*, Vol. 19, No. 16, 1998, pp. 1419-1423.

[5] L. L. Hench, R. J. Splinter, W. C. Allen and T. K Greenlee Jr., "Bonding Mechanism at the Interface of Ceramics Prosthetic Materials," *Journal of Biomedical Materials Research*, Vol. 5, No. 6, 1971, pp. 117-141.

[6] L. L. Hench and J. Wilson, "Biocompatibility of Silicates for Medical Use," *Ciba Foundation Symposium*, Vol. 121, No. 2, 1986, pp. 231-246.

[7] T. Kokubo, H. Kushitani, C. Ohtsuki, S. Sakka and T. Yamamuro, "Effects of Ions Dissolved from Bioactive Glass-Ceramic on Surface Apatite Formation," *Journal of Materials Science*: *Materials in Medicine*, Vol. 4, No. 1, 1993, pp. 1-4.

[8] M. Vallet-Regi, I. Izquierdo-Barbo and A. J. Salinas, "Influence of P_2O_5 on Crystallinity of Apatite Formed *in Vitro* on Surface of Bioactive Glasses," *Journal of Biomedical Materials Research*, Vol. 46, No. 4, 1999, pp. 560-565.

[9] L. L. Hench and J. K. West, "Biological Applications of Bioactive Glasses," *Life Chemistry Reports*, Vol. 13, 1996, pp. 187-241.

[10] E. Dietrich, H. Oudadesse, A. Lucas-Girot, Y. Le Gal, S. Jeanne and G. Cathelineau, "Effects of Mg and Zn on the Surface of Doped Melt-Derived Glass for Biomaterials Applications," *Applied Surface Science*, Vol. 225, No. 2, 2008, pp. 391-395.

[11] E. Dietrich, H. Oudadesse, A. Lucas-Girot and M. Mami, "*In Vitro* Bioactivity of Melt-Derived Glass 46S6 Doped with Magnesium," *Journal of Biomedical Materials Research Part A*, Vol. 88A, No. 4, 2009, pp. 1087-1096.

[12] A. P. Ahoussou, J. Rogez and A. Kone, "Enthalpy of Mixing in 0.8[xB_2O_3-(1−x)P_2O_5]-0.2Na_2O Glasses at 298 K," *Thermochimica Acta*, Vol. 447, No. 1, 2006, pp. 96-100.

[13] A. P. Ahoussou, J. Rogez and A. Kone, "Thermodynamical Miscibility in 0.8[xB_2O_3-(1−x)P_2O_5]-0.2K_2O Glasses," *Journal of Non-Crystalline Solids*, Vol. 353, No. 3, 2007, pp. 271-275.

[14] A. P. Ahoussou, J. Rogez and A. Kone, "Solution Calorimetric Study of Mixing Enthalpy in 0.8[xB_2O_3-(1−x)-SiO_2]-0.2K_2O Glasses at 298 K," *Materials Research Bulletin*, Vol. 42, No. 8, 2007, pp. 1577-1581.

[15] M. Ganteaume, M. Coten and M. Decressac, "Unnouveau Calorimètre de Solution: Le Calsol," *Thermochimica Acta*, Vol. 178, 1991, pp. 81-98.

[16] M. B. Volf, "Mathematical Approach to Glass," Glass Science and Technology 9, Ed Elsevier, Amsterdam, 1988.

[17] Y. Yamasaki, Y. Yoshida, M. Okazaki, A. Shimazu, T. Uchida, T. Kubo, Y. Akagawa, Y. Hamada, J. Takahashi and N. Matsuura, "Synthesis of Functionally Graded $MgCO_3$ Apatite Accelerating Osteoblast Adhesion," *Journal of Biomedical Materials Research*, Vol. 62, No. 1, 2002, pp. 99-105.

[18] Y. Yamasaki, Y. Yoshida, M. Okazaki, A. Shimazu, T. Kubo, Y. Akagawa and T. Uchida, "Action of FG$MgCO_3$-Ap-Collagen Composite in Promoting Bone Formation," *Biomaterials*, Vol. 24, No. 27, 2003, pp. 4913-4920.

[19] O. H. Andersson, G. Liu, K. H. Karlsson, L. Niemi, J. Miettinen and J. Juanoja, " *In Vivo* Behaviour of Glasses in the SiO_2-Na_2O-CaO-P_2O_5-Al_2O_3-B_2O_3 System," *Journal of Materials Science*: *Materials in Medicine*, Vol. 1, No. 4, 1990, pp. 219-227.

[20] O. H. Andersson, K. H. Karlsson, K. Kangasniemi and A. Yli-Urpo, "Models for Physical Properties and Bioactivity of Phosphate Opal Glasses," *Glastechnische Berichte*, Vol. 61, 1998, pp. 300-305.

[21] I. Lebecq, F. Désanglois, A. Leriche and C. Follet-Houttemane, "Compositional Dependence on the *in Vitro* Bioactivity of Invert or Traditional Bioglasses in the Si-Ca-Na-P System," *Journal of Biomedical Materials Research Part A*, Vol. 83A, No. 1, 2007, pp. 156-168.

[22] P. Sepulveda, J. R. Jones and L. L. Hench, "*In Vitro* Dissolution of Melt-Derived 45S5 and Sol-Gel Derived 58S Bioactive Glasses," *Journal of Biomedical Materials Research*, Vol. 61, No. 2, 2002, pp. 301-311.

[23] L. L. Hench, "Bioceramics: From Concept to Clinic," *Journal of the American Ceramic Society*, Vol. 74, No. 7, 1991, pp. 1487-1510.

Copper Recovery from Barren Cyanide Solution by Using Electrocoagulation Iron Process

José R. Parga[1*], Guillermo Tiburcio Munive[2], Jesús L. Valenzuela[2], Víctor V. Vazquez[2], Gregorio González Zamarripa[3]
[1]Institute Technology of Saltillo, Saltillo, Mexico
[2]Departament of Chemistry and Metallurgy, University of Sonora, Hermosillo, Mexico
[3]Faculty of Metallurgy, University of Coahuila, Monclova Coah, México

ABSTRACT

This paper is a brief overview of the role of inducing the nucleated electro winning of copper by using iron electrodes in electrocoagulation (EC) process. Cyanide compounds are widely used in gold ore processing plants in order to facilitate the extraction and subsequent concentration of the precious metal. Owing to cyanide solution employed in gold processing, effluents generated have high contents of free cyanide as well as copper cyanide complexes, which lend them a high degree of toxicity. In this regard, two options for the treatment of cyanide barren solutions has been used; in two ways; first for cyanide destruction by oxidation with the use of the EC process, in theory, has the advantage of decomposing cyanide at the anode and collecting copper simultaneously by a sludge of copper magnetic iron. In both cases excellent performance can be achieved using the high capacity of the bipolar iron EC technology. We found that it is possible to reduce the copper cyanide complex from 720 $mg \cdot l^{-1}$ to below 10 $mg \cdot l^{-1}$ within 20 minutes.

Keywords: Copper Recovery; Electrocoagulation Process; Bipolar Iron Electrodes; Cyanide

1. Introduction

Due to the dwindling resources of simple cyanide extractable gold deposits, a large proportion of the gold processed in the 21st century will be recovered from complex gold ores, many of which will contain soluble copper minerals. It has been estimated that about 20% of all gold deposits have significant copper mineralization commonly associated with chalcopyrite, tetrahedrite, tennantite, enargite as well as bornite and chalcocite in certain ores [1]. It has also been found that the majority of copper minerals including copper oxides, carbonates, sulfides (with the exception of chalcopyrite) and native copper are highly soluble in cyanide solutions [2]. These copper containing minerals are problematic because, when ores containing such minerals are leached with cyanide to recover the gold, copper also dissolves to form stable copper cyanide complexes. The dissolution of copper consumes a substantial quantity of cyanide and thus, if not recovered imposes a significant financial cost on the gold mine. The presence of copper also causes other problems such as competition with gold to adsorb on carbon unless a sufficient free cyanide concentration is maintained, depletion of gold electrowinning cell efficiency, and gold losses by cementation onto certain copper minerals. Ores containing greater than 0.5% reactive copper may be generally considered uneconomical to process via conventional cyanidation due to the high reagent cost. Therefore, it is necessary to reduce the amount of copper, especially in leaching circuits, in order to increases the dissolution of gold and silver, to achieve this goal it is possible to remove the copper from barren solution after Merril-Crowe process using the electrocoagulation process (EC). This process can be interesting for the copper cyanide removal from processed solutions after the Merrill Crowe process, also has low cost of operation and investment.

2. Copper Cyanide Chemistry

The major challenges to the processing of gold-copper ores using cyanidation is that of the high cyanide consumptions that are typically experienced, along with effective control of the leach, particularly when there is variable cyanide-soluble copper in the ore. It is widely accepted that gold dissolution in cyanide solutions occurs as sequence of two reactions shown in Equations (1) and

*Corresponding author.

(2), Elsner's equation shows that oxygen is critical for the dissolution of gold.

$$2Au + 4NaCN + O_2 + 2H_2O \rightarrow 2Na\left[Au(CN)_2\right] + 2NaOH + H_2O_2 \quad (1)$$

$$2Au + 4NaCN + H_2O_2 \rightarrow 2Na\left[Au(CN)_2\right] + 2NaOH \quad (2)$$

The stoichiometry of the process shows that 4 moles of cyanide are needed for each mole of oxygen present in solution. At room temperature and standard atmospheric pressure, approximately 8.2 mg of oxygen are present in one liter of water. This corresponds to 0.27×10^{-3} mol/L accordingly, the sodium cyanide concentration (molecular weight of NaCN = 49) should be equal to $4 \times 0.27 \times 10^{-3} \times 49 = 0.05$ g/L or approximately 0.01%. This was confirmed in practice at room temperature by a very dilute solution of NaCN of 0.01% - 0.5% for ores, and for concentrates rich in gold and silver of 0.5% - 5% [3]. Also, lime is added to keep the system at an alkaline pH of 10.5 - 11.0. Other factors affecting gold leaching kinetics are grain size, agitation speed, temperature, pressure, foreign ions and cyanicides.

Free cyanide exists as the uncomplexed cyanide ion, CN^-, and molecular hydrogen cyanide, HCN. These species are related by the acid dissociation of HCN:

$$HCN_{(aq)} = CN^- + H^+ \quad (3)$$

The concentration of free cyanide is the sum of the CN^- and HCN concentrations, and the equilibrium diagram shown in **Figure 1** illustrates the distribution.

This figure shows the proportions of free cyanide as CN^-, and HCN as a function of pH at 25°C. At pH values below 7, cyanide is predominantly present as the unionized HCN molecule, which is easily volatilized because of its high vapor pressure. The equilibrium is displaced in favor of cyanide ion formation at pH values above 7.

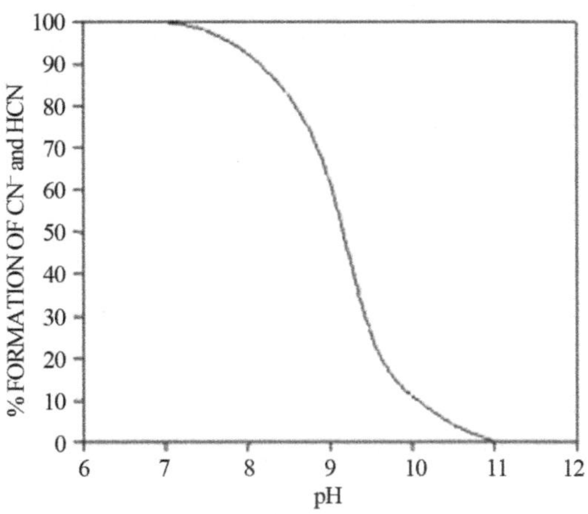

Figure 1. Equilibrium distribution diagram for cyanide as a function of pH.

Hydrogen cyanide (HCN), also known as hydrocyanic acid, is a colorless gas or liquid with a boiling point of 25.7°C, a vapor pressure of 100 kPa at 26°C and Henry's Law constant of 6.4 atm/mole [4], this makes HCN very volatile. Thus, low pH, high temperature, low pressure, and intimate contact with air, all tend to increase the rate of dissipation of cyanide from solution as hydrogen cyanide. In addition to free cyanide, other complexes such as the metal cyanide complexes formed with gold, silver, copper, nickel, iron and cobalt must be considered.

In cyanidation plants all around the world, the concentration of cyanide used to dissolve gold in ores is typically higher than the stoichiometric ratio, due to the solubility of other minerals. Free cyanide produces complexes with several metallic species, especially transition metals, which show a broad variation in both stability and solubility. Many common copper minerals are soluble in the dilute cyanide solution under typical of leach conditions found in the gold cyanidation process. Minerals such as azurite and malacite, are rapidly leached and are soluble in dilute cyanide solutions.

Enargite and chalcopyrite leach more slowly but are sufficiently soluble to cause excessive cyanide loss and contamination of the pregnant leach solutions. In reactions in aqueous solutions the cupric ion is rapidly converted to cuprous form and then copper forms a series of extremely stable soluble complexes in cyanide such as:

$$Cu^+ + CN^- = CuCN \quad (4)$$

$$CuCN + CN^- = Cu(CN)_2^- \quad (5)$$

$$Cu(CN)_2^- + CN^- = Cu(CN)_3^{2-} \quad (6)$$

$$Cu(CN)_3^{2-} + CN^- = Cu(CN)_4^{3-} \quad (7)$$

Under typical gold cyanidation conditions $Cu(CN)_3^{2-}$ has been shown to be the dominant species from the Eh-pH diagram for the copper-cyanide-water system [4,5]. The high consumption of cyanide during the cyanidation of copper-gold ores is due to the fact that copper forms complexes of high coordination numbers with cyanide (Reaction 3 to 6), $Cu(CN)_3^{2-}$ in particular. Therefore, hydrometallurgical treatment of these ores by cyaniding as a rule gives rise to a series of difficulties associated with increase in the cyanide consumption and decrease in the dissolution rate of gold and silver, and in the cementation process. This precipitate is of low quality, because the copper is precipitated along with gold and silver, resulting in a higher consumption of zinc dust, fluxes in the smelting of the precipitate and shorter life for crucibles.

In this regard a study is proposed to remove copper cyanide ions with, a very promising electrochemical treatment technique, which does not require chemical additions. This process is electrocoagulation (EC). The EC process operates on the principle that coagulation of copper cyanide ions from barren solutions from the Merrill-Crowe process is caused by the combined effects of electrolysis gases (H_2 and O_2) and the electrolytic production of cations from the iron anodes that corrode during electrolysis.

3. Electrocoagulation Fundamentals

The EC process operates on the principle that the cations produced electrolytically from iron and/or aluminum anodes enhance the coagulation of contaminants from an aqueous medium. Electrophoretic motion tends to concentrate negatively charged particles in the region of the anode and positively charged ions in the region of the cathode. The consumable, or sacrificial, metal anodes are used to continuously produce polyvalent metal cations in the vicinity of the anode. These cations neutralize the negative charge of the particles carried toward the anodes by electrophoretic motion, thereby facilitating coagulation. In the flowing EC techniques, the production of polyvalent cations from the oxidation of the sacrificial anodes (Fe and Al) and the electrolysis gases (H_2 and O_2) works in combination to flocculate the coagulant materials [6], the gas bubbles produced by the electrolysis carry the pollutant to the top of the solution where it is concentrated, collected and removed. **Figure 2** illustrates the schematic diagram of the process.

Generally, in the EC process bipolar electrodes are used. Pretorius [7] and Mameri [8] have reported on the use of cells with bipolar electrode arranged in series. These cells, operated at relatively low current densities, produce iron or aluminum coagulant in a more effective, fast and economical manner when compared to chemical coagulation. The savings result from a reduction in the time necessary for the treatment and an increase in the anodic area improved the efficiency of the electrolysis [7,8]. An electrocoagulation cell with bipolar electrodes

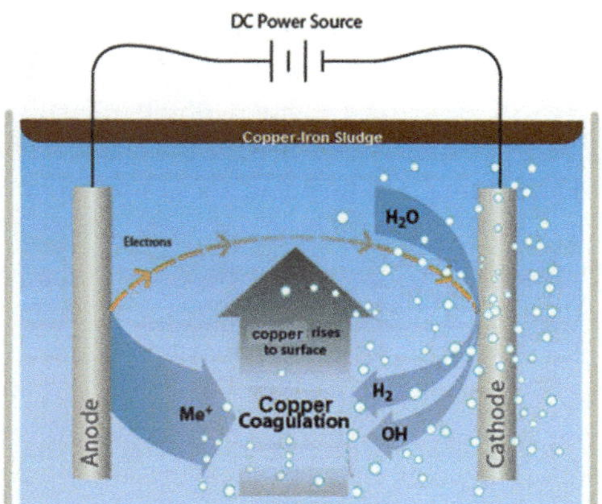

Figure 2. Schematic diagram of the EC process.

connected in series is shown in **Figure 3**. The inner electrodes are bipolar, that is they carry both positive and negative charges on opposing faces. These charges develop on the electrode surfaces and are opposite in sign of the charge carried by the parallel electrode surface (refer to **Figure 3**). During electrolysis the positive side of these bipolar electrodes undergoes anodic reactions, while on the negative side, cathodic reactions occur.

The released ions neutralize the charge of the particles and thereby initiate coagulation. The bipolar arrangement reduces the time needed for the treatment due to the increase in surface area mentioned above. This arrangement also has the practical advantage of simplified set-up in that only two monopolar electrodes are connected to the electric power source with no interconnections between the inner bipolar electrodes.

The Chemical Reactions of the Electrocoagulation Process

The chemicals reactions that have been proposed to describe the mechanism of EC for the production of $H_{2(g)}$ and $OH^-_{(ac)}$ (cathode) and $H^+_{(ac)}$ (anode) [9] are:

$$Fe \rightarrow Fe^{3+} + 3e^- \quad (8)$$

$$Fe(OH)_2^+ + H_2O \leftrightarrow Fe(OH)_3 + H^+ \quad (9)$$

$$H^+ + 2e^- \leftrightarrow H_{2(g)} \uparrow \quad (10)$$

$$Fe(OH)_2^+ + e^- \leftrightarrow Fe(OH)_{2(aq)} \quad (11)$$

$$Fe(OH)_{2(aq)} + H_2O \leftrightarrow Fe(OH)_3^- + H^+ \quad (12)$$

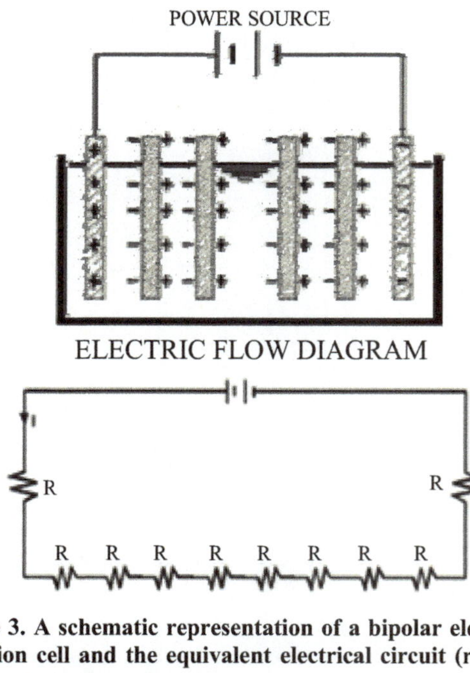

Figure 3. A schematic representation of a bipolar electrocoagulation cell and the equivalent electrical circuit (resistors reflects electrode surfaces).

$$Fe(OH)_3^- \leftrightarrow Fe(OH)_{3(aq)} + e^- \quad (13)$$

Overall reaction.

$$6Fe + (12+x)H_2O \leftrightarrow \frac{1}{2}(12-x)H_{2(g)} \uparrow + xFe(OH)_3 \cdot (6-X)Fe(OH)_{2(s)} \quad (14)$$

The pH of the medium usually rises as a result of this electrochemical process and the Green Rust formed $\left[xFe(OH)_3 * (6-X)Fe(OH)_{2(s)}\right]$ remains in the aqueous stream as a gelatinous suspension, which can remove the gold and silver from pregnant cyanide rich solutions, either by complexation or by electrostatic attraction followed by coagulation and flotation.

Formation of rust (dehydrated hydroxides) occurs a while after the process, as shown in the following:

$$2Fe(OH)_3 \leftrightarrow Fe_2O_3 + 3H_2O \text{ (hematite)} \quad (15)$$

$$Fe(OH)_2 \leftrightarrow FeO + H_2O \quad (16)$$

$$2Fe(OH)_3 + Fe(OH)_2 \leftrightarrow Fe_3O_4 + 4H_2O \text{ (magnetite)} \quad (17)$$

$$Fe(OH)_3 \leftrightarrow FeO(OH) + H_2O \text{ (goethite, lepidocrocite)} \quad (18)$$

A schematic representation of these reactions in an EC process, using iron electrodes, is shown in **Figure 4**. As mentioned above, the gas bubbles produced by electrolysis carry the copper along with the sludge to the top of the solution where it is collected and removed [10]. However, it is the reactions of the metal ions that enhance the formation of the coagulant. The metal cations are hydrolyzed, releasing hydrogen ions that result in hydrogen evolution at the cathode, to yield both soluble and insoluble hydroxides that will react with or adsorb copper from the cyanide solution and also contribute to coagulation by neutralizing the negatively charged colloidal particles that may be present at neutral or alkaline pH.

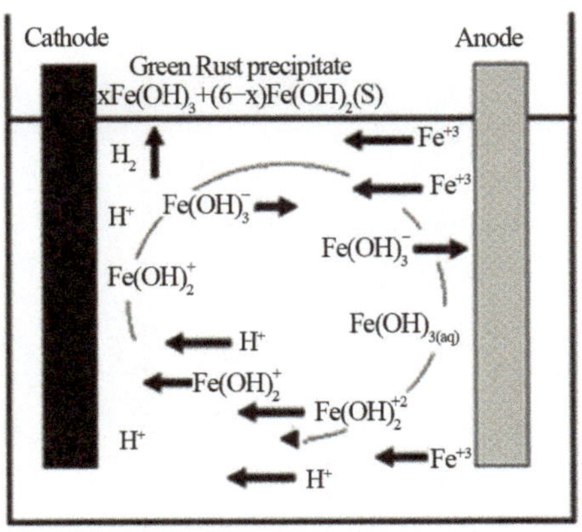

Figure 4. An illustration of the EC mechanism (arrow indicate the migration of ions, the H_2 evolution and the formation of green rust).

This enables the particles to approach closely and agglomerate under the influence of Van der Waals attracttive forces. The pH of the medium rises as a result of this electrochemical process and the $Fe(OH)_{n(s)}$ formed remains in the aqueous stream as gelatinous suspension, which can remove the $Cu(CN)_3^{2-}$ from the barren solution, either by complexation or by electrostatic attraction followed of coagulation and flotation [11]. Generally, in the EC process, bipolar electrodes are used. It has been reported that cells with bipolar electrodes, connected in series and operating at relatively low current densities [12], produce iron or aluminum coagulant more effecttively, more rapidly and more economically when compared to chemical coagulation.

4. Experimental Details

The experimental work was performed using a barren solution from the Merrill Crowe process containing an average in the range of 660 - 712 ppm of copper. EC experiments were performed using a 600 ml Pyrex beaker glass (**Figure 5**), equipped with two iron electrodes (10 cm × 2.5 cm) with an electrode separation of 5 mm. A regulated power supply (model Steren PRL-25) to provide the necessary energy to the electrocoagulation cell, magnetic stirrer (Model PC-310, Corning) with 100 rpm was used, and the initial and final pH was taken with a pH meter (VWR Scientific 8005), filtration was performed with Whatman filter paper No.42. To determine the adsorption copper in iron hydroxide species generated at the anode, a barren solution after the cementation process was used that was provided by the William Mine Co. This solution contained an average of 0.02 mg/L gold and 0.1 mg/L silver along with amounts of zinc, lead. Among others, the analysis was performed according to EPA 200.7 (EPA 200.7 is an analytical method for

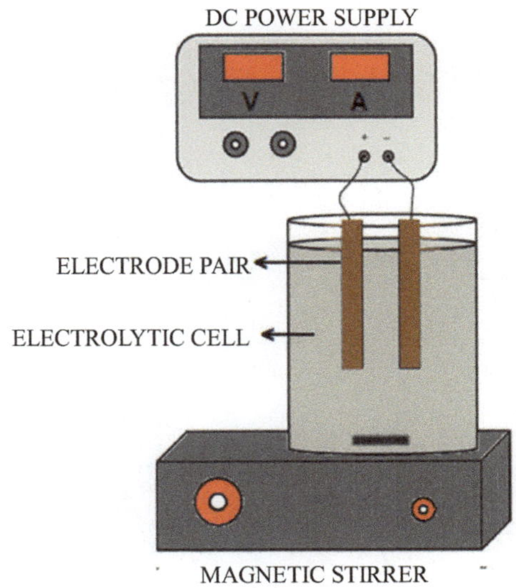

Figure 5. EC schematic diagram of the experimental setup.

detection of metals and trace elements by ICP/atomic emission spectrometry). The solution and the solids were separated by filtration through filter paper, the sludge from the EC was dried for 8 hrs in an oven at 80°C. The tables below show the initial conditions of the EC test to remove copper.

Optimization of Parameters

In order to find the optimum parameters of the EC process for the removal of copper, experiments were carried out by changing the pH of the solution, residence time in the EC cell and voltage and amperage.

5. Results

Tables 1-3 show the results for tests performed with 4 grams/liter of sodium chloride.

From these results it was determined that the optimum parameters were: pH 8, residence time of 20 minutes and 4 grams/liter of NaCl, this achieved 99% copper removal. Also, when the time increased from 15 to 20 minutes the removal of copper increased from 92% to 99%, this occurs in the pH range from 8 to 9 approximately, this coincides with the production of the magnetic iron, Fe_3O_4, which has magnetic properties that accelerates the process of adsorption of metals, the adsorption rate is then physically, because it is caused by the magnetic forces of the magnetite into copper, without altering their chemical composition. This removal of copper also can explain with a decrease in the zeta potential on iron hydroxides which causes a decrease in repulsive forces between the particles, generated collision between particles thus favors the formation of flocs which float to the water surface through micro bubbles generated from oxygen and hy-

Table 1. Results obtained for pH 8 and 4 g/liter of NaCl.

Initial Conditions of EC Test					Sludge Chemical Essay		
Time (min)	Voltage (volt)	Current (A)	%Cu	%Fe	$[Cu]_{initial}$ (ppm)	$[Cu]_{final}$ (ppm)	%Elimination
5	12.1	10.3	4.6	21.3	660	328	50
10	12.3	10.5	6.2	23	664	188	72
15	11.5	10.9	9.4	24	712	70	90
20	12.4	10.1	13	27	716	5	99

Table 2. Results obtained for pH 9 and 4 g/liter of NaCl.

Initial Conditions of EC Test					Sludge Chemical Essay		
Time (min)	Voltage (volt)	Current (A)	%Cu	%Fe	[Cu]initial (ppm)	[Cu]final (ppm)	%Elimination
5	12.1	10.3	4.4	22	673	428	36
10	12.3	10.5	7	23.5	677	190	76
15	11.5	10.9	10	24	679	80	88
20	12.4	10.1	12.8	27.5	701	10	98

Table 3. Results obtained for pH 10 and 4 g/liter of NaCl.

Initial Conditions of EC Test					Sludge Chemical Essay		
Time (min)	Voltage (volt)	Current (A)	%Cu	%Fe	$[Cu]_{initial}$ (ppm)	$[Cu]_{final}$ (ppm)	%Elimination
5	12.1	10.3	4.8	21.5	667	424	36
10	12.3	10.5	6.8	22	686	186	73
15	11.5	10.9	10.8	24	680	92	86
20	12.4	10.1	13	28	705	15	98

drogen from the iron electrodes. Also, the advantages of the EC process is the decomposing of cyanide at the anode, where the anodic oxidation of cyanide is proportional to the alkalinity of the electrolyte and consistent with the following mechanism:

$$CN^- + 2OH \rightarrow CNO^- + H_2O + 2e^- \quad (18)$$

$$2CNO^- + 4OH^- \rightarrow 2CO_2 + N_2 + 2H_2O + 6e^- \quad (19)$$

$$CNO^- + 2H_2O \rightarrow NH_4 + CO_3^- \quad (20)$$

Figure 6 shows the graph of time- elimination where it is observed the behavior of the removal of copper.

6. Product Characterization

In order to identify the iron species present, Scanning Electron Microscope (SEM/EDX) was used to characterize the solid products formed during the EC process for removal of copper with iron electrodes.

6.1. Scanning Electron Microscopy (SEM/EDAX)

Figure 7 shows SEM image and **Figure 8** shows EDAX of copper adsorbed on iron species. These SEM and EDAX results show that the surfaces of these iron oxide/

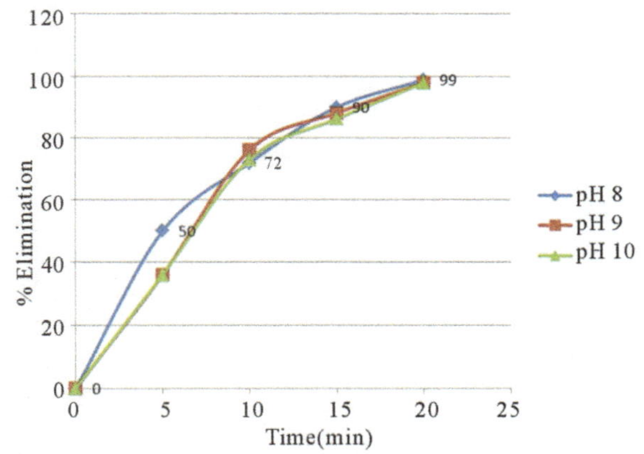

Figure 6. % Elimination of copper at pH 8, 9 and 10.

Figure 7. SEM image.

oxyhydroxide particles were coated with a layer of copper.

6.2. X-Ray Diffraction

Diffractograms were obtained with a Bruker AXS D4 Endeavor diffractometer operating with a Cu-K radiation source filtered with a graphite monochromator (λ = 1.5406 Å). The samples were wet ground to a fine powder (isopropyl alcohol from Sigma-Aldrich) and pressed into a sample holder. The XRD scans were recorded from 20° to 80° 2θ, with 0.02° step-width and with a 10 s counting time for every step-width (increment). Experiments were run at 40 kV and 40 mA power. **Figure 9**, shows a diffractogram of the filtered solid products (the feed solution contained 700 ppm of copper and the pH of the solution after EC was ~10).

This figure indicates the presence of magnetite, geothite, iron hydroxide oxide, and lepidocrocite in the solid products.

7. Conclusions

Electrocoagulation process was carried out to obtain a 99% of copper removal. The optimal operations conditions were: pH = 8, residence time: 20 minutes and 4 g/l of sodium chloride as conductivity modifier. The solid product obtained from EC process was 13% copper and 24% iron. The EC process does not generate any smell in the process of elimination of copper from the barren solution. During the EC process is not necessary to add any reagent (except Nacl), since the coagulating agent is generated in situ.

We found that it is possible to reduce the copper cyanide complex from 720 $mg \cdot l^{-1}$ to below 10 $mg \cdot l^{-1}$ within 20 minutes.

Also, the results of this study suggest that EC produces magnetic particles of magnetite. lepidocrocita and amorphous iron oxyhydroxide species that can be used to removal copper. The Scanning Electronic Microscopy, techniques demonstrate that the formed species are of magnetic type, like lepidocrocite and magnetite which adsorbed the copper particles on his surface due to the electrostatic attraction between both metals.

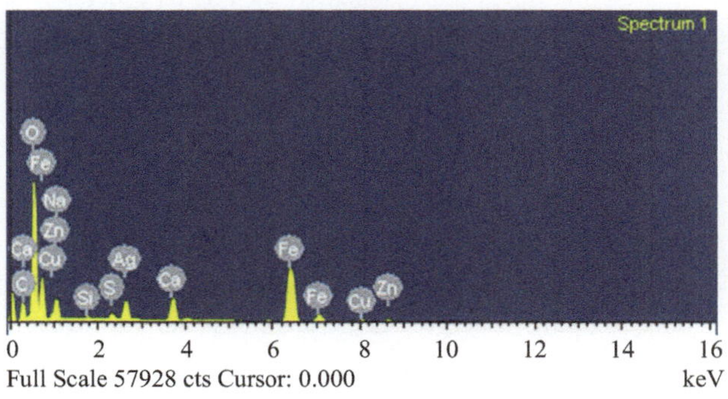

Figure 8. EDAX Analysis.

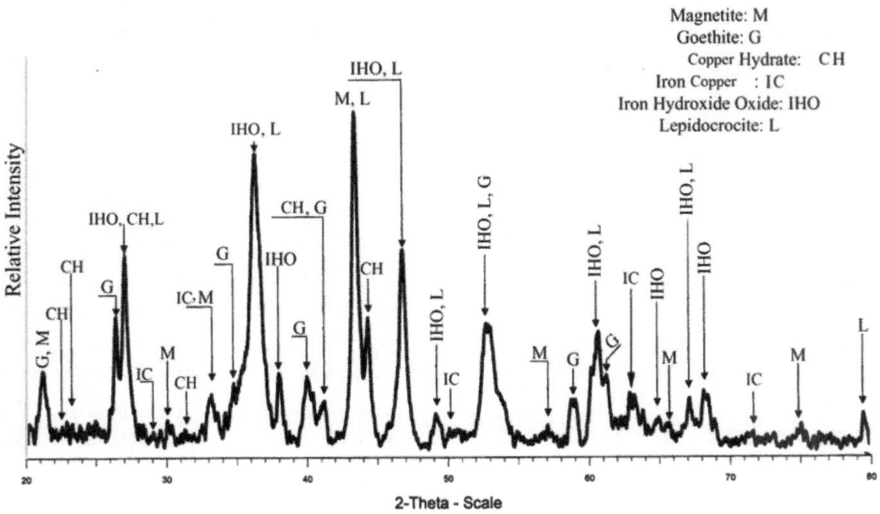

Figure 9. X-ray diffraction spectrum obtained from the EC product at pH = 10.

8. Acknowledgements

The authors wish to acknowledge support of this project to the National Council of Science and Technology (CONACYT) and to Dirección General de Educación Superior Tecnológica (DGEST).

REFERENCES

[1] D. M. Muir, S. R. LaBrooy and C. Cao, "Recovery of Gold Fromcopper-Bearing Ores," In: R. J. Harden, Ed., *Gold Forum Ontechnology and Practic*, World Gold, 1989.

[2] J. O. Marsden abd C. I. House, "The Chemistry of Gold Extraction," 2nd Edition, Society for Mining, Metallurgy and Exploration, Inc., Littleton, 2006.

[3] J. R. Parga, S. S. Shukla and D. L. Cocke, "Photocatalytic Detoxification of Cyanide and Recovery of Titanium Dioxide by Electrocoagulation," *Research Journal of Chemistry and Environment*, Vol. 9, No. 1, 2005, pp. 60-63.

[4] S. T. Mudder, "The Chemistry and Treatment of Cyanidation Wastes," Mining Journal Books Limited, London, 1991, pp. 277-278.

[5] M. D. Adams and R. Lawrance, "Biogenic Sulphide for Cyanide Recycle and Copper Recovery in Gold-Copper Ore Processing," International Workshop on Process Hydrometallurgy, Hydroprocess Brisbone, 2008, pp. 14-16.

[6] O. Asare, K. Xue and T. Ciminelli, "Solution Chemistry of Cyanide Leaching Systems. In Precious Metals Mineralogy," *Extraction and Processing-Proceedings of and International Symposium*, 1984, pp. 173-197.

[7] N. Mameri, A. R. Yeddou, H. Lounici, D. Belhocine, H. Grib and B. Bariou, "Defluoridation of Septentrional Sahara Water of North Africa by Electrocoagulation Process Using Bipolar Aluminum Electrodes," *Water Research*, Vol. 32, No. 5, 1998, pp. 1604-1612.

[8] W. A. Pretorius, W. G. Johannes and G. G. Lempert, "Electrolytic Iron Flocculant Production with a Bipolar Electrode in Series Arrangement," *Water South Africa*, Vol. 17, No. 2, 1991, pp. 133-138.

[9] G. Pavas, M. P. Camargo, C. Jones and V. T. Pineda, "Oxidación Fotocatalítica de Cianuro," Universidad EAFIT, Medellín, 2005, pp. 56-58.

[10] J. R. Parga, H. M. Casillas, V. Vazquez and J. L. Valenzuela, "Cyanide Detoxification of Mining Wastewaters with TiO_2 Nanoparticles and Its Recovery by Electrocoagulation," *Chemical Engineering and Technology*, Vol. 32, No. 12, 2009, pp. 1901-1908.

[11] G. Vicuña and I. Tuñon, "Apuntes de Química Avanzada," Departamento de Química-Física. Universidad de Valencia, 2006, pp. 103-115.

[12] A. G. Gupta and S. Kundu, "Adsorptive Removal of As(III) from Aqueous Solution Using Iron Oxide Coated Cement (IOCC): Evaluation of Kinetic Equilibrium and Thermodynamic Models," *Separation and Purification Technology*, Vol. 51, No. 2, 2006, pp. 165-172.

A Review of an Expert System Design for Crude Oil Distillation Column Using the Neural Networks Model and Process Optimization and Control Using Genetic Algorithm Framework

Lekan Taofeek Popoola[1], Gutti Babagana[2], Alfred Akpoveta Susu[3]
[1]Department of Petroleum and Chemical Engineering, Afe Babalola University, Ado-Ekiti, Nigeria
[2]Department of Chemical Engineering, University of Maiduguri, Maiduguri, Nigeria
[3]Department of Chemical Engineering, University of Lagos, Lagos, Nigeria

ABSTRACT

This paper presents a comprehensive review of various traditional systems of crude oil distillation column design, modeling, simulation, optimization and control methods. Artificial neural network (ANN), fuzzy logic (FL) and genetic algorithm (GA) framework were chosen as the best methodologies for design, optimization and control of crude oil distillation column. It was discovered that many past researchers used rigorous simulations which led to convergence problems that were time consuming. The use of dynamic mathematical models was also challenging as these models were also time dependent. The proposed methodologies use back-propagation algorithm to replace the convergence problem using error minimal method.

Keywords: Artificial Neural Network; Crude Oil Distillation Column; Genetic Algorithm Framework; Sigmoidal Transfer Function; Back-Propagation Algorithm

1. Introduction

Crude oil distillation is the separation of hydrocarbons in crude oil into fractions based on their boiling points which lie within a specified range [1]. The separation is done in a large tower that is operated at atmospheric pressure. The tower contains a number of trays where hydrocarbon gases and liquids interact. The liquids flow down the tower and the gases up [2]. The fractions that rise highest in the column before condensing are called light fractions, and those that condense on the lowest trays are called heavy fractions [3]. **Figure 1** gives a schematic representation of the process overview of a crude oil distillation column. This crude oil distillation column can be designed using artificial neural network. An artificial neural network (ANN) is an information processing paradigm that is inspired by the way biological nervous systems (such as the brain) process information. It is composed of a large number of highly interconnected processing elements (neurons) working in unison to solve specific problems [4] (**Figure 2**).

Artificial neural networks initially grew from the full understanding of some ideas and aspects about how biological systems work, especially the human brain. In biology, the cell body of neuron is called the soma. The spine-like extensions of the cell body are dendrites. They usually branch repeatedly and form a bushy tree around the cell body and provide connections to receive income-ing signals from other neurons. The axon extends away from the cell body to provide a pathway for outgoing signals. Signals are transferred from one neuron to another through a contact point called a synapse [3]. Like biological neuron networks, ANN is made of neuron and synaptic connections which are highly simplified abstracts of their counterparts in real neural networks (**Figure 3**). The neurons in the ANN are usually called units, node, or processing elements, and the efficacy of the synaptic connection, which is a measurement of excitability and inhabitability, are usually called weights. Neural networks systems (NNS) are typically organized in layers. Layers are made up of a number of interconnected nodes (artificial neurons or processing element),

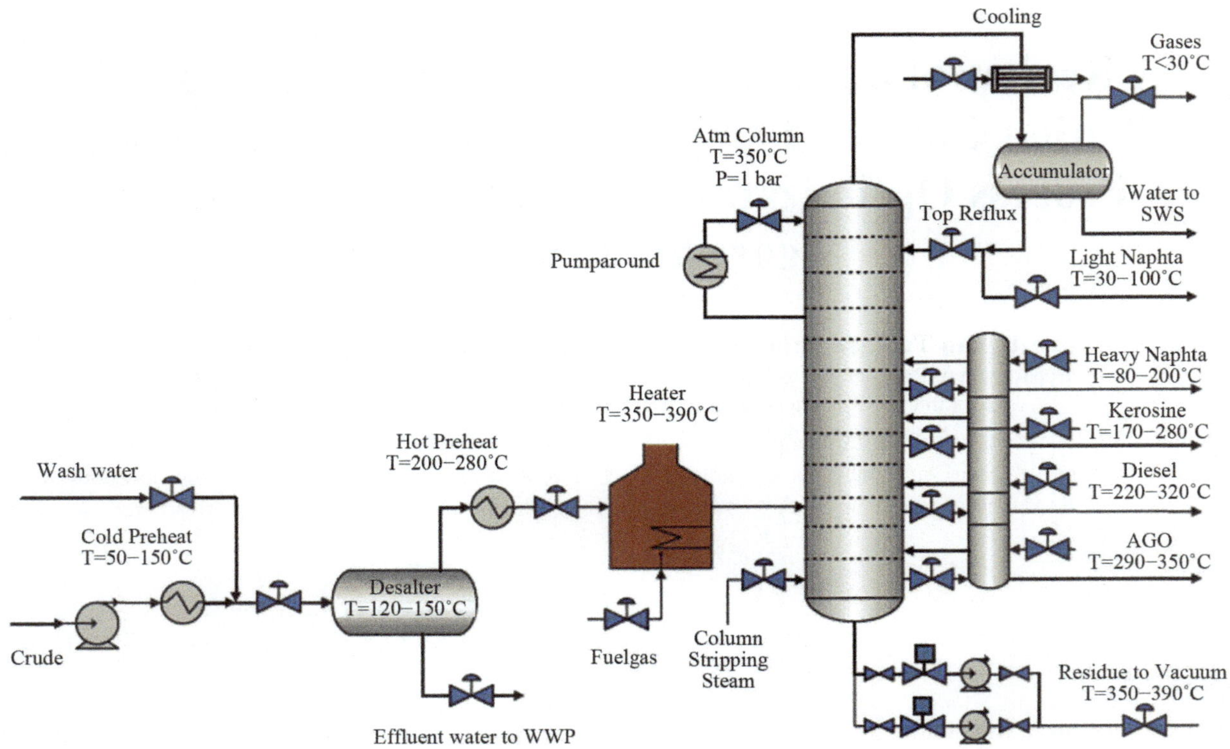

Figure 1. Process overview of crude oil distillation column.

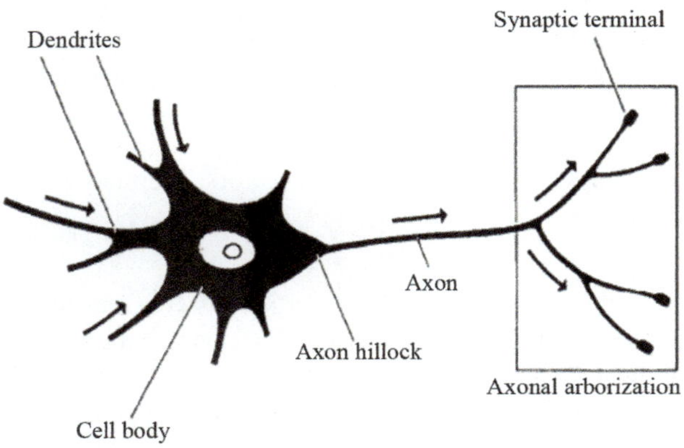

Figure 2. Basic neuron.

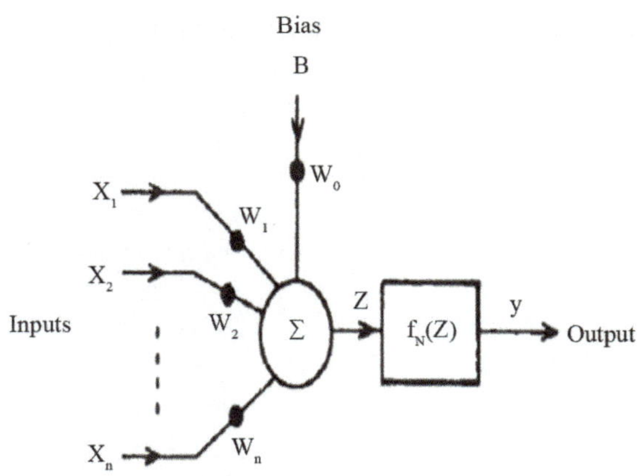

Figure 3. A biased neuron.

which contains an activation function. The data are presented to the networks via the input layer, which communicates to one or more hidden layers where the actual processing is done through a system of weighted connections. The hidden layers are then linked to an output layer, which generates the output [5,6].

The objective of crude oil distillation unit is to perform a process optimization including high production rate with required product quality and low operating costs by searching an optimal operating condition of the operating variables [7]. Genetic algorithm and fuzzy logic are adaptive methods which can be used to solve optimization problems. Genetic Algorithms are based on the principle of genetic processes of biological organisms while Fuzzy logic utilizes human knowledge by giving the fuzzy or linguistic descriptions a definite structure. An expert system design of this unit is necessary in order to achieve the earlier stated goals. An expert system is a computer system employing expert knowledge to attain high levels of performance in solving the problems within a specific domain area [8].

2. Literature Review

For effective performance and operation of the crude oil distillation column, competent design of the CDU that improve fuel properties, increase the yields of the distillate products and meet environmental specifications is very important. Traditional systems of crude oil distillation column design, optimization and control by previous researchers are discussed in this section. Manley [9] designed crude oil distillation column through the use of improved process thermodynamics which was also used by Santana *et al*. [10] to assess and improve the energy efficiency of a crude-oil tower. They were limited to the fact that the improved process designs were unproven, projected capital cost reductions were uncertain and energy savings at current prices would not justify the investment needed for process development and commercialization. The design of crude oil distillation column through the use of on-line soft-sensor for its control and optimization was discussed by Yu *et al.* [11] but they did not consider the design for offline application.

The use of neural network architectures to design refinery crude distillation column for the prediction of product quality was proposed by Bawazir *et al.* [12]. However, the design of crude oil distillation column for optimization using the neural network was left out. Liau *et al.* [13] picked up the challenges by using neural network for the expert system design of a crude oil distillation unit for process optimization. Torgashov [14] designed self-optimizing control of complex crude distillation column control using non-linear process model-based method. Khairiyah *et al.* [15] used the method proposed by Bawazir *et al.* [12] for design of crude oil distillation column for real time optimization. Gadalla *et al.* [16] used short-cut models for retrofit design of distillation columns and heat exchanger networks but were limited to the fact that short-cut models were not detailed enough and require longer time for convergence. The stage-wise modelling procedure approach was used for the design and optimization of a refinery crude distillation unit in the context of total energy requirement by Okeke *et al.* [17]. However, they failed to give full practical validation of the design method. Domijan *et al.* [18] presented a model that used the crude true boiling point curve and other routinely made laboratory measurements on crude distillation unit (CDU) for product properties prediction.

Another system of crude oil distillation column design was proposed by Macías-Hernández *et al.* [19] using evolving Takagi-Sugeno fuzzy models for the design of soft sensor for predicting crude oil distillation side streams. However, future research need to be directed at practical implementation of this technique in an online estimator in closed loop. Zalizawati [20] designed a continuous distillation column through development of multiple-input multiple-output (MIMO) and multiple-input single-output (MISO) neural network models. Haydary *et al.* [21] proposed steady-state and dynamic simulation of crude oil distillation using ASPEN PLUS and ASPEN DYNAMICS. Kanthasamy [22] used Hammerstein model and nonlinear autoregressive model with exogenous input (NARX) in designing nonlinear model predictive control (NMPC) of a distillation column. He included the activity and fugacity calculations in the model in order to account for the non-ideality of the system. Smith *et al.* [23] proposed design of heat integrated distillation system. However, their work was not detailed enough as the real practical analysis of the system was not stated. Tonnang *et al.* [3] developed a powerful controller that allows perfect control of a complex crude oil distillation column using neural network. Kansha *et al.* [24] investigated the crude oil distillation design with exergy analysis and the feasibility for application of the self-heat recuperation technology. The simulation was conducted by PRO/II Ver.8.1 (Invensys, Simsci) to calculate the energy required. However, this design required modification such as a self-heat exchange and an additional furnace heater for the first distillation column and no integration for hot charge method.

3. Research Methodologies

3.1. Design and Control of CDU Using ANN

Artificial neural network (ANN) models are black box models, consisting of layers of nodes with nonlinear basis functions and weighted connections that link the nodes. The inputs to the model are mapped to the outputs after being trained with a set of training or learning data to optimise the weights and biases of the nodes. Multilayer feedforward ANN (**Figure 4**) were mathematically proven to be a universal approximator [25]. The back propagation algorithm proposed by Rumelhart *et al.* [26] is the effective means for training or learning the set of data in order to optimize the weights. It is an approximate steepest descent algorithm in which performance index is mean square error [27]. **Figure 5** gives the implementation procedure for the ANN. The error function of the output is given by:

$$\text{error} = \frac{1}{2}\sum_{j=1}^{m}\left[d_j(n) - y_j(n)\right] \tag{1}$$

where d_j = Practical data of jth output neuron;
y_j = Computed data of jth ouput neuron;
m = Neuron number;
n = Training step.

$$y_k = F_N(Z_k) = \frac{1}{1+\exp(-Z_k)} \tag{2}$$

where $y_k = F_N(Z_k)$ = sigmoidal transfer function, Z_k = sum of the jth input to the neuron multiplied by their respective weights

$$Z_k = \sum_j w_{kj}x_j \tag{3}$$

where w_{kj} = weight of the jth input to the kth neuron of the output layer;

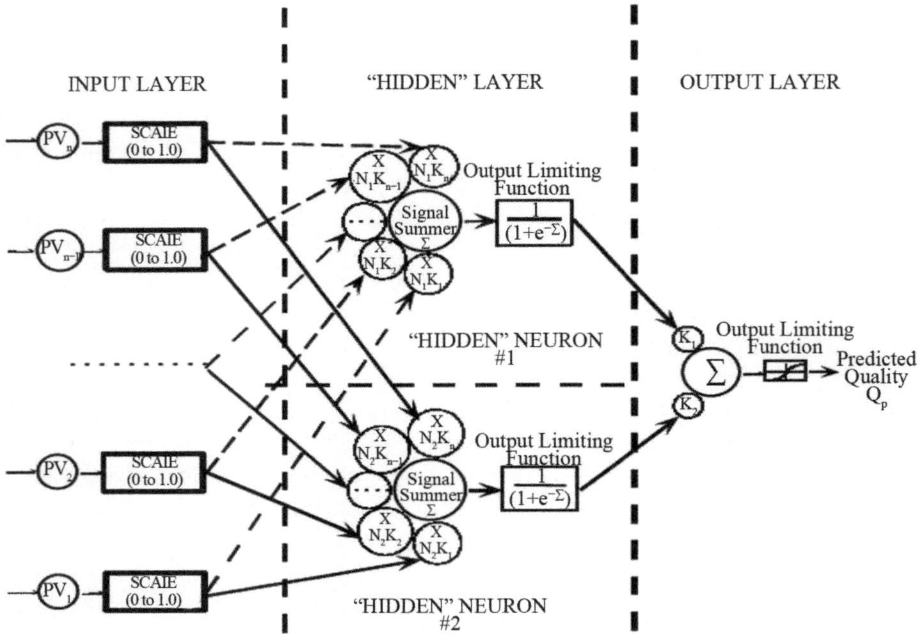

Figure 4. Multilayer feedforward ANN.

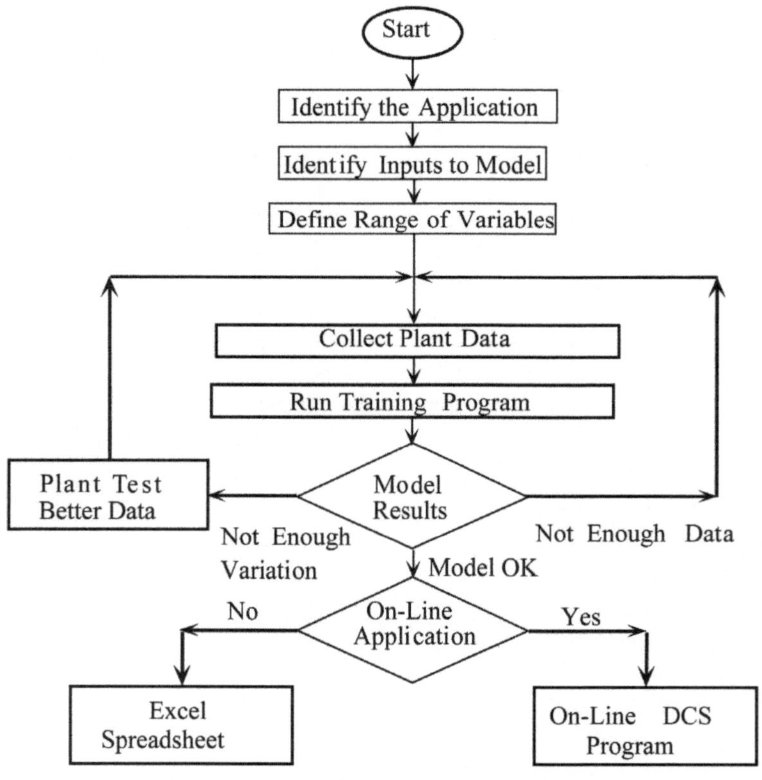

Figure 5. ANN implementation flow chart.

x_j = jth input to the neuron.

$$w_{ij}^{k+1} = w_{ij}^{k} + \eta \delta_j^k I_i f'(s) \quad (4)$$

where w_{ij}^{k+1} = Weigths of the connection from unit i in layer $k+1$ to unit j in layer $k+2$;

w_{ij}^{k} = Weigths of the connection from unit i in layer k to unit j in layer $k+1$;

η = Learning rate constant;

δ_j^k = Signal error;

I_i = Input vector to the networks;

$f'(s)$ = Derivative of the networks sigmoidal transfer function;

s = Sum of all the weigths.

The development of neural network that could be used for the control of the crude oil distillation column is also discussed in this section. The tower or column receives crude oil and steam flow as inputs. Naphthalene, Kerosene, Light Diesel Oil and Heavy Diesel Oil are its outputs. Stripping (distillate flows) is sent to the storage tank, while some quantity of Naphthalene, Kerosene and Light Diesel Oil (reflux flows) are returned into the column. The input values to the neural network controller (NNC) are: distillate flows, feed flow, feed temperature, top temperature, bottom temperature, bottom composition, reflux temperature, and the tower pressure. Its output values are used to adjust the reflux flows and steam flow. The neural network controller flow chart is presented in **Figure 6**.

3.2. Optimization of CDU Using Genetic Algorithm and Fuzzy Logic

Genetic Algorithm is a powerful optimization technique based on the principles of natural evolution and selection. In the specific case of selecting the optimum set of inputs from a larger set, GA can be used to search through a large number of input combinations with interdependent variables in the artificial neural network to be designed for the crude oil distillation column. **Figure 7** gives the standard structure of genetic algorithm.

Fuzzy theory is another powerful tool in the exploration of complex problems because of its ability to determine outputs for a given set of inputs without using a conventional, mathematical model. The development of fuzzy theory came from the inability to describe some physical phenomena with the exact mathematical models dictated by more conventional Boolean models. Fuzziness describes event ambiguity. It measures the degree to which an event occurs, not whether it occurs. In its simplest form a fuzzy logic is simply a set of rules describing a set of actions to be taken for a given set of inputs. It is easiest to think of these rules as if then statements of the form if {*set of inputs*} then {*outputs*}.

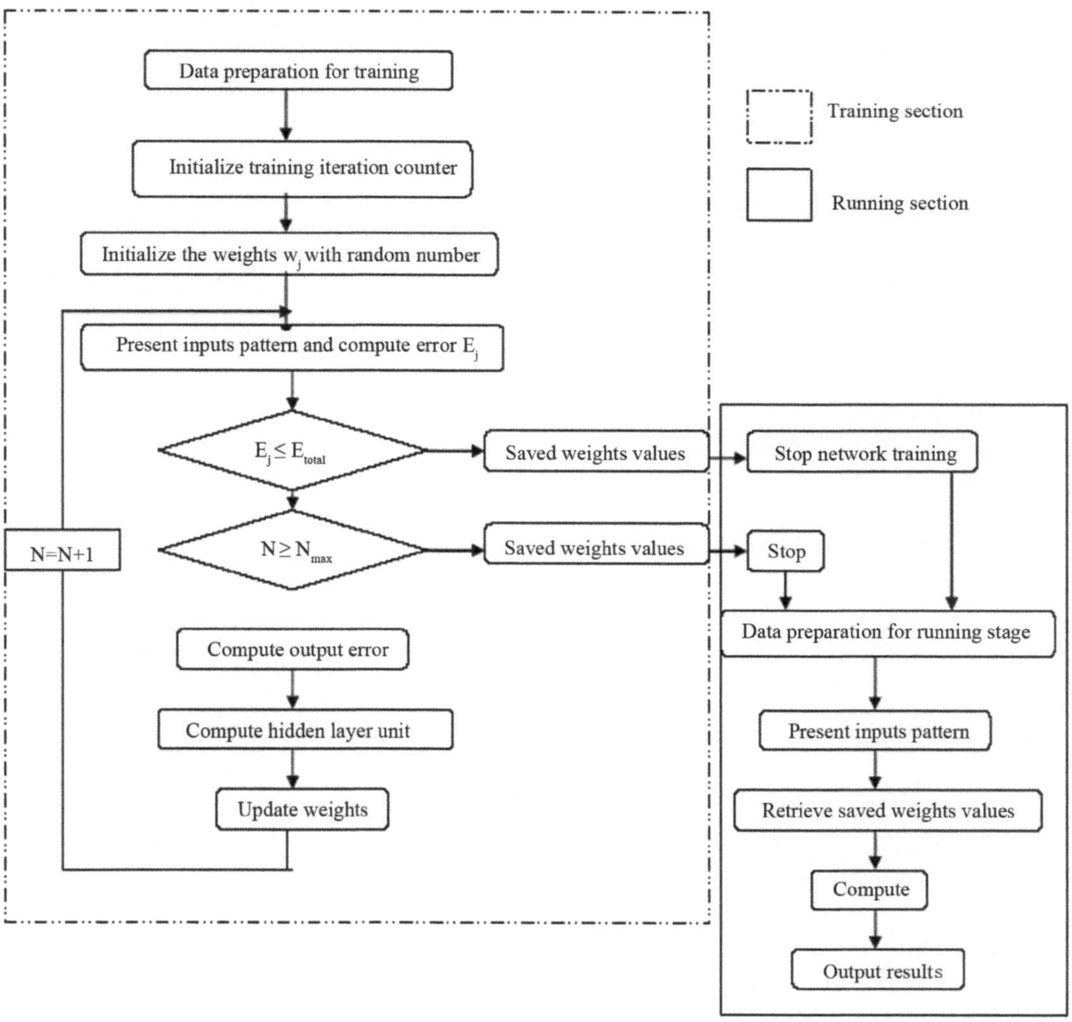

Figure 6. Neural network controller flow chart.

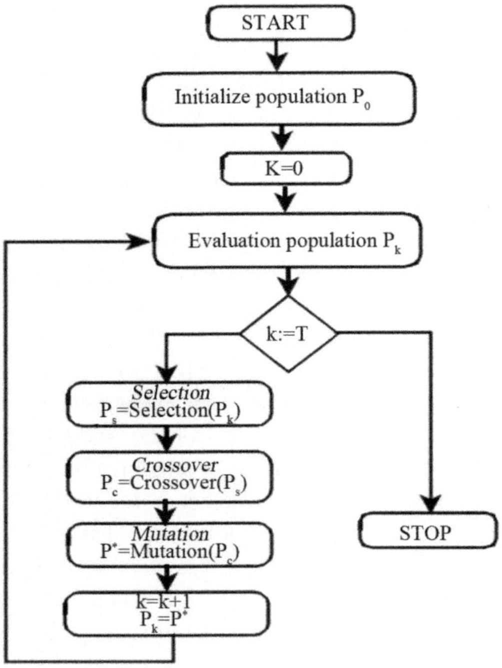

Figure 7. Standard genetic algorithm.

4. Discussion

After adequate scrutiny of various traditional systems of crude oil distillation column design, optimization and control, artificial neural network, fuzzy logic and genetic algorithm had been found as the best methodologies based on some established facts. They serve as substitutes for dynamic mathematical models as they are time independent. Many researchers used rigorous simulations which led to convergence problems and were also time consuming. These soft computing methodologies use error minimal method to replace the convergence problem. Also, artificial neural network models, fuzzy logic and genetic algorithm approach had been found as the effective ways to model complex processes due to their non-linear characteristic structures. Lastly, the proposed methodologies can remarkably enhance the regulatory and advanced control capabilities of various industrial processes such as crude oil distillation columns in refineries.

5. Conclusion

An expert system design of crude oil distillation column can be done using the artificial neural networks. The product quality specification and the optimal operation can be reached through the use of artificial neural network. Also, the crude oil distillation column optimization can be achieved using both the fuzzy logic and genetic algorithm frame work. The continuous evaluation and adjustment of process operating conditions to optimise economic productivity can be reached by these methodologies. A neural network controller can be designed for crude oil distillation column. The developed neural network controller is capable of mapping the interactions and nonlinear dynamics of the process. Artificial neural networks, fuzzy logic and genetic algorithm framework are the best soft computing methodologies for both the expert design and optimization of crude oil distillation column. The design of neural network controller for the crude oil distillation column is also recommended in order to meet the requirements with respect to environment, health and safety of the plant personnel and the quality of the finished products.

REFERENCES

[1] K. H. Bawazeer, "Prediction of Crude Oil Product Quality Parameters Using Neural Networks," MS Thesis, Florida Atlantic University, Boca Raton, 1996.

[2] W. Heather, "Refining Crude Oil," The New Zealand Refining Company Ltd., Ruakaka, 2003.

[3] Z. E. H. Tonnang, "Distillation Column Control Using Artificial Neural Networks," M.Sc Thesis, Microprocessors and Control Engineering, Department of Electrical and Electronics Engineering, Faculty of Technology, University of Ibadan, Ibadan, 2010.

[4] S. Christos and S. Dimitrios, "Introduction to Artificial Neural Network," Research Report, 2001.

[5] H. L. Hoffman, D. E. Lupfer, L. A. Kane and B. A. Jensen, "Distillation Column, Basic and Advance Controls Process Control, Instrument Engineer Handbook," 3rd Edition, Butherworth Heinemann, Oxford, 1995.

[6] T. M. Gowrie and V. V. C. Reddy, "Load Forecasting by a Novel Technique Using ANN," *ARPN Journal of Engineering and Applied Sciences*, Vol. 3, No. 1, 2008, pp. 19-25.

[7] J. W. Sea, M. Oh and T. H. Lee, "Design Optimization of Crude Oil Distillation," *Chemical Engineering Technology*, Vol. 23, No. 2, 2000, pp. 157-164.

[8] J. McCarthy, "Some Expert System Need Common Sense," Stanford University, Stanford, 1984.

[9] D. B. Manley, "Waste Minimization through Improved Process Thermodynamics: Crude Oil Fractionation," The University of Missouri, Rolla, 1993.

[10] E. I. Santana and R. J. Zemp, "Thermodynamic Analysis of a Crude-Oil Fractionating Process," *4th Mercosur Congress on Process Systems Engin eering*, Vol. 21S, 2001, pp. 523-528.

[11] J. J. Yu, C. H. Zhou, S. Tan and C. C. Hang, "An On-line Soft-Sensor for Control and Optimization of Crude Distillation Column," Research Institute of Industrial Process Control, Zhejiang University, Hangzhou, 1997.

[12] K. H. Bawazir and A. Zilouchian, "Application of Neural Networks in Oil Refineries," *Proceedings of* 1996 *IEEE International Conference on Neural Networks*, New Orleans, 11-13 May 1999, pp. 126-135.

[13] L. C.-K. Liau, T. C.-K. Yangb and M. T. Tsaib, "Expert System of a Crude Oil Distillation Unit for Process Optimization Using Neural Networks," *Expert Sy stems with Applications*, Vol. 26, No. 2, 2004, pp. 247-255.

[14] A. Torgashov, "Nonlinear Process Model-Based Self-Optimizing Control of Complex Crude Distillation Column," *European Symp osium on Computer Aid ed Pro cess Engineering*-11, Vol. 9, 2001, pp. 793-798.

[15] M. F. Khairiyah, K. Fakhri and L. D. Peter, "Connectionist Models of a Crude Oil Distillation Column for Real Time Optimisation," Regional Symposium on Chemical Engineering 2002, Songkla, 2002.

[16] M. Gadalla, M. Jobson and M. Smith, "Optimisation of Existing Heat—Integrated Refinery Distillation Systems," Ph.D. Thesis, UMIST, Manchester, 2002.

[17] E. O. Okeke and A. A. Osakwe-Akofe, "Optimization of a Refinery Crude Distillation Unit in the Context of Total Energy Requirement," APACT03, 28-30 April 2003, NNPC R&D Division, Port Harcourt, 2003.

[18] P. Domijan and D. Kalpić, "Off-Line Energy Optimization Model for Crude Distillation Unit," Ph.D. Thesis, Department of Applied Mathematics, Faculty of Electrical Engineering and Computing, University of Zagreb, Zagreb, 2004.

[19] J. J. Macías-Hernández, P. Angelov and X. Zhou, "Soft Sensor for Predicting Crude Oil Distillation Side Streams Using Evolving Takagi-Sugeno Fuzzy Models. Results Outlined," *Proceedings of 2nd International Symposium on Evolving Fuzzy Systems*, 7-9 September 2008, Lake District, IEEE Press, 2008, pp. 214-220.

[20] B. A. Zalizawati, "Development of Multiple-Input Multiple-Output and Multiple-Input Single-Output Neural Network Models for Continuous Distillation Column," M.Sc Thesis, School of Chemical Engineering, Malaysia University, 2008.

[21] R. Kanthasamy, "Nonlinear Model Predictive Control of a Distillation Column Using Hammerstein Model and Nonlinear Autoregressive Model with Exogenous Input," Ph.D. Thesis, Universiti Sains Malaysia, 2009.

[22] J. Haydary and T. Pavlik, "Steady-State and Dynamic Simulation of Crude Oil Distillation Using ASPEN Plus and ASPEN Dynamics," *Petroleum and Coal* , Vol. 51, No. 2, 2009, p. 100.

[23] R. Smith, M. Jobson, L. Chen and S. Farrokhpanah, "Heat Integrated Distillation System Design," Centre for Process Integration, School of Chemical Engineering and Analytical Science, The University of Manchester, 2010.

[24] Y. Kansha, A. Kishimoto and A. Tsutsumi, "Application of the Self-Heat Recuperation Technology to Crude Oil Distillation," Collaborative Research Centre for Energy Engineering, Institute of Industrial Science, The University of Tokyo, Tokyo, 2011.

[25] K. Hornik, M. Stinchcombe and H. White, "Multilayer Feedforward Neural Networks Are Universal Approximators," *Neural Networks*, Vol. 2, No. 5, 1989, pp. 359-366.

[26] D. Rumelhart and J. McClelland, "Parallel Distributed Processing," MIT Press, Cambridge, 1986.

[27] G. Daniel, "Principles of Artificial Neural Networks," *Advanced Series in Circuits and Systems*, Vol. 51, No. 2, 2007, pp. 100-109.

8

Polydopamine Particles Effect on Melanoma Cells Proliferation and Melanin Secretion

Sandy Eap[1*], Alice Ferrand[1*], Valérie Machi[1,2], Vincent Ball[2], Olivier Huck[1,2], Nadia Benkirane-Jessel[1,2#]

[1]INSERM, Institut National de la Santé et de la Recherche Médicale, "Osteoarticular and Dental Regenerative Nanomedicine" Team, Strasbourg, France

[2]Université de Strasbourg, Faculté de Chirurgie Dentaire, Strasbourg, France

ABSTRACT

Melanin is a biopolymer implicated in the protection of cellular membranes and DNA produced by melanocytes. This pigment has a dual role and should be considered as a photo-protector and as a photosensitizer due to its interaction with UV. The design of multifunctional and biologically responsive coatings is of major interest in modern biomaterials science. The aim of this study is not only to characterize the deposition of multilayered polyelectrolytes films made from polydopamine particles and polyamines like poly-(L-lysine hydrobromide) (PLL), but also to evaluate melanoma cells activity in terms of proliferation and their capacity to stimulate melanin secretion. One could expect that the presence of a melanin like material in the film may have a positive or a negative feedback on the melanin biosynthesis and consequently on melanoma development. Some comparisons are also done with pure polydopamine grains in suspension in the cell culture medium, to investigate if the immobilization of the polydopamine grains has an influence on their bioactivity.

Keywords: Polydopamine Particles; Polyelectrolyte Multilayers Films; Melanin; Melanoma Cells

1. Introduction

The design of multifunctional and biologically responseve coatings is of major interest in modern biomaterials science [1]. Such coatings should be biodegradable and porous [2] and should display adequate mechanical properties since adherent cells are sensitive to the stiffness of their adhesion substrate [3]. Additionally, coatings useful for bio-applications should contain molecules able to be released in a progressive manner or some growth factors able to interact with the cells in contact with them [4]. Three of the most popular surface functionalization methods consist in the deposition of hydrogels [5], of polyelectrolyte multilayer (PEM) films [6] or thin polydopamine films [7-10] which are chemically and structurally close to natural melanin [11], the natural pigment of skin.

Melanocytes have the ability to produce melanin. Melanin is a biopolymer implicated in the protection of cellular membranes and DNA. This pigment has a dual role and should be considered as a photo-protector and as a photosensitizer due to its interaction with UV [12] but has also been linked to Parkinson's disease characterized by loss of pigmented dopamine neurons in substantia nigra [13].

Films made through spin coating of melanin particles are adequate for adhesions of neurons [14]. However polydopamine films, even if they are advantageous for secondary biofunctionalisation with molecules containing thiol or amino functionalities [15,16], are highly impermeable to ions [17] and molecules of higher molecular mass. Hence it would be advantageous to combine the intrinsic properties of polydopamine with those of hydrogels or PEM films. It has already been demonstrated that polydopamine particles produced through oxidation of dopamine in the presence of dioxygen as an oxidant can be deposited in alternation with a quaternary polyamine, namely poly(dimethyldiallyl ammonium chloride) (PDADMAC) [18]. Those films are permeable to small ions and they could be an ideal platform for bio-applications. The aim of this study is not only to characterize the deposition of PEM film made from polydopamine parti-

*These authors contributed equally to this work.

#Corresponding author.

cles and polyamines like poly-(L-lysine hydrobromide) (PLL), but also to evaluate melanoma cells activity in terms of proliferation and their capacity to stimulate melanin secretion. One could expect that the presence of a melanin like material in the film may have a positive or a negative feedback on the melanin biosynthesis and consequently on melanoma development. Some comparisons were also done with pure polydopamine grains in suspension in the cell culture medium, to investigate if the immobilization of the polydopamine grains has an influence on their bioactivity. The biological activity of the $(PLL\text{-}dopamine)_n$ PEM films was also compared with that of films not containing polydopamine but a negatively charged counterpart, poly(-L-glutamic) acid.

In addition to the interaction of $(PLL\text{-}polydopamine)_n$ films with melanoma, those films could find interesting use as water resistant adhesives because lysine and L-DOPA are the two major amino acids of the Mytilus Edulis Foot Proteins, Mefp 3 and Mefp 5 which are abundant at the extremity of the mussel byssus plaque [19].

2. Materials and Methods

2.1. Materials

Poly(L-lysine)-hydrobromide (PLL, 30 kDa) and poly-L-glutamic acid (PGA, 54 kDa) were purchased from Sigma. Dopamine in the hydrochloride form was also from Sigma (ref H-8502) and was used without further purification. Dopamine was solubilized at 2 $g \cdot L^{-1}$ in 50 mM Tris buffer (Prolabo, France) at pH 8.5 and this solution was vigorously shaken (300 rpm with a magnetic stirrer) in an open air reactor during 24 h. This allowed to produce a suspension of polydopamine grains. These grains have been characterized by means of Transmission Electron Microscopy in a previous paper [20]. The obtained grains were highly anisotropic in their shape, displaying a disk like morphology with a lateral dimension of about 100 - 300 nm and a thickness of only a few nm. To increase its colloidal stability, the suspension was titrated to pH 12 with a concentrated NaOH solution. Before the beginning of the deposition of each PEM film, the polydopamine suspension was diluted to a final concentration of 0.2 $g \cdot L^{-1}$ or too lower concentrations with 10 mM Tris buffer at pH 7.5 and complemented with 150 mM NaCl. No visible sedimentation of melanin grains occurred in these conditions. Unfortunately, these suspensions were too diluted in polydopamine grains to allow for a characterization by dynamic light scattering.

2.2. Polyelectrolyte Multilayer Film Preparation

For all in vitro experiments, PEM films were prepared on glass coverslips (CML, France) cleaned with 10 - 2 M sodium dodecyl sulfate and subsequently with 0.12 N HCl for 15 min at 100°C, and then extensively rinsed with deionized water. Glass coverslips were deposited in 24-well plates (Nunc, Danemark). Films having the nominal dipping sequence of $(PLL\text{-}polydopamine)_5$ or $(PLL\text{-}PGA)_5$ were prepared by alternated immersion of the pre-treated coverslips during 5 min in polyelectrolyte solutions (300 μL) at the concentration of 50 $\mu g \cdot mL^{-1}$ for PLL or PGA, and 200 $\mu g \cdot mL^{-1}$ for polydopamine in the presence of 10 mM Tris buffer at pH 7.5, complemented with 150 mM of NaCl. After each deposition step, the coverslips were rinsed one times for 5 min with 0.15 M NaCl 10 mM Tris at pH 7.5. All the films were sterilized for 30 min using UV light (254 nm). The adsorption duration of 5 min per dipping step was shown to be sufficient to reach a steady state surface coverage by means of Crystal Microbalance with Dissipation monitoring (QCM-D).

2.3. QCM-D

The deposition of the PEM films was monitored in situ with a quartz crystal microbalance using an axial flow chamber QAFC 302 (QCM-D, D300, Q-Sense, Göteborg, Sweden). QCM works by measuring the resonance frequency shift (Δf) of a quartz crystal induced by polyelectrolyte or protein adsorption on to the crystal in comparison to the crystal in contact with buffer. Changes in the resonance frequencies were measured at the third ($\nu = 3$), the fifth ($\nu = 5$) and the seventh ($\nu = 7$) overtones of the quartz crystal, corresponding to frequencies close to 15 MHz, 25 and 35 MHz respectively. As a first approximation, a shift in $\Delta f/\nu$ can be related to a variation of the mass adsorbed to the crystal. This mass is made of the polyelectrolytes and the polydopamine, the ions eventually bound to the film and the water of hydration that is moved during the shearing motion of the piezolelectric quartz crystal. The Sauerbrey relation [20]:

$$\Gamma = -C.\Delta f/\nu \qquad (1)$$

relates the reduced frequency changes, $\Delta f/\nu$ to the total mass bound to the crystal per unit area, Γ. C is a proportionality constant characteristic of the crystal used (in our case: $C = 17.7\ ng \cdot cm^{-2}$ Hz-1 as provided by the furnisher). Note that the Sauerbrey equation can only be used for films strongly adhering to the substrate. A good qualitative criterion for the validity of the use of Equation (1) is the superposition of the $\Delta f/\nu$ signals for all the considered overtones. The details of the methodology applied in the present work were described elsewhere [21].

The film thickness was calculated from the surface concentration calculated in Equation (1) and from the density of the multilayer film, ρ, according to:

$$d = \Gamma/\rho \qquad (2)$$

we assumed a density of 1.2 $g \cdot cm^{-3}$ for the hydrated multilayer film.

2.4. Atomic Force Microscopy

The AFM topographical images were acquired using a Pico-SPM device (Molecular Imaging) at a constant resolution of 512 × 512 pixels over an area of 10 μm × 10 μm and with a frequency of 1 Hz. The film thickness was evaluated after needle scratching of the (PLL-polydopamine)$_5$ films and imaging the surface by scanning the cantilever in the direction perpendicular to the needle scratch.

2.5. Cell Culture

Murine melanoma cells (B16-F1) were maintained at 37°C in a humidified air-CO_2 (5%) atmosphere in Dulbecco's modified eagle medium (D-MEM) with 4.5 $g \cdot L^{-1}$ D-glucose, L-Glutamine, and Hepes (Invitrogen) supplemented with 10% heat-inactivated fetal calf serum (FCS, Hyclone), 1 mM Sodium pyruvate (Invitrogen), 100 $Um \cdot L^{-1}$ of penicillin and 100 $\mu g \cdot mL^{-1}$ streptomycin (Hyclone). Cells were maintained at 37°C in 5% CO_2 humidified atmosphere. Cells were treated with polydopamine particles in solution at different concentrations: 1, 5, 10, 25, 50, 100 or 200 $\mu g \cdot mL^{-1}$ and their proliferation and melanin secretion was compared to that of cells adhering to (PLL-PGA)$_5$ and to (PLL-polydopamine)$_5$ polyelectrolyte multilayer films.

2.6. Cellular Proliferation

AlamarBlue® (Serotec) was used to assess cellular proli- feration. The AlamarBlue® test is a nontoxic, water-soluble, colorimetric redox indicator that changes color in response to cell metabolism. In this study, 10,000 cells/well were seeded in 24 well plates in cell culture medium and incubated at 37°C in a humidified air-CO_2 (5%) on the top of PEM-coated 14 mm diameter coverslips. The coverslips coated with either the (PLL-PGA)$_5$ or the (PLL-polydopamine)$_5$ films placed in 24-well plates. After 3 and 7 days of culture, the cells were incubated in 10% AlamarBlue®/DMEM solution in a humidified atmosphere at 37°C and 5% CO_2. After 4 h, 200 μL of incubation media was transferred to 96-well plates and measured at 570 and 630 nm in order to determine the percentage of AlamarBlue® reduction. After the lecture, the 10% AlamarBlue®/D-MEM solution was removed and replaced with classical medium until the next measurement.

2.7. Melanin Assay

The melanin cells content was quantified after 3 or 7 days of incubation of cells with different PEM film architectures or after contact with polydopamine in solution. 10^4 cells/well were seeded in 24 well plates in cell culture medium and incubated at 37°C in a humidified air-CO_2 (5%). The melanin supernatant was quantified after 3 and 7 days of incubation of cells. The absorbance of the supernatant was measured at a wavelength of 410 nm using an Elisa reader. Cell pellets were dissolved in 0.85 M KOH, 10% DMSO at 80°C for 90 min. Optical absorbance was measured at 410 nm using an Elisa reader. Before each measurement the number of cells was evaluated by numeration in a Neubauer counting chamber. The absorbance was then normalized for each measurement to allow comparison between different culture conditions. The melanin content was evaluated according to a standard curve of melanin concentration. All experiments were performed at least three times, each yielding at least triplicate values. When the cells were put in presence of a solution containing polydopamine, the absorbance measurement corresponds to the absorbance of polydopamine and that of the melanin secreted by the cells. Indeed, polydopamine presents an absorption spectrum without peaks and very similar to that of melanin which is a black pigment [22]. This means that the absorbance red from the cells put in contact with the polydopamine solution corresponds to an overestimation of the amount of melanin secreted by the cells.

2.8. Immunoflorescence

B16-F1 cells were rinsed with PBS and fixed with 4% paraformaldehyde in PBS for 1 hour at 37°C. The cells were then washed three fold during 10 min with PBS. After PBS washing, the cells were permeabilized, during one hour at room temperature, with PBS containing 1% BSA and 0, 1% Triton X100 (PBS-Tx). The cells were then incubated overnight at 4°C with primary antibody diluted at 1/200 in PBS-Tx. After overnight incubation at 4°C, the cells were washed with PBS three fold and incubated with a fluorochrome-conjugated secondary antibody diluted at 1/500 in PBS-Tx for 1 h at room temperature. AlexaFluor 488 goat anti-mouse (Invitrogen) was used as the secondary antibody. The cells were incubated with AlexaFluor 546-Phalloidin (Invitrogen) diluted at 1/100 for 20 min at room temperature, and counterstained with DAPI at 200 $ng \cdot mL^{-1}$ for 5 min at room temperature. They were then washed with PBS three fold during 10 min. The cells were covered with fluorescent mounting medium (Dako) and analyzed by fluorescence microscopy (Leica DM 4000B).

2.9. Statistical Analysis

All values are expressed as mean ± SEM, and all experiments were repeated at least three times. Statistical analysis was performed using the Mann-Whitney U test. A probability p value < 0.05 was considered significant to reject the null hypothesis.

3. Results and Discussion

3.1. Deposition of Pem Films

The deposition of the (PLL-polydopamine)$_n$ PEM films was recorded in situ by means of QCM-D (**Figure 1**). Within the experimental error, the normalized frequency changes, $\Delta f/v$, overlap for the different harmonics ($v = 3, 5, 7$) of the shear waves propagating along the piezoelectric crystal (**Figure 1(a)**). This means that the (PLL-polydopamine)$_n$ coatings behave as rigid films and that the Sauerbrey Equation (1) can be used to calculated the hydrated mass of the film as well as the film thickness. The thickness of the (PLL-polydopamine)$_n$ coatings increases almost linearly with the number of layer pairs (*i.e.* one deposition of PLL followed by the deposition of one polydopamine layer) with an average thickness increment of 3.2 nm per layer pair. However a closer inspection of **Figure 1(b)** shows that polydopamine contributes more than PLL in the thickness increment of each deposited layer pair.

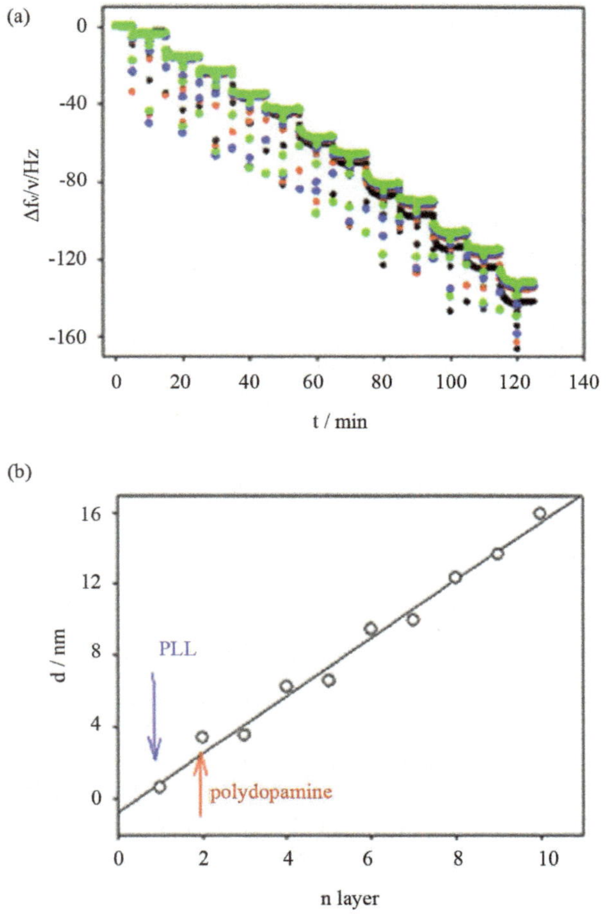

Figure 1. Quartz crystal microbalance measurement to follow in situ the deposition of a (PLL-polydopamine)n film. (a) Reduced frequency shifts at 5 (●), 15 (●), 25 (●) and 35 MHz (●) are shown as a function of sequential layer deposition of PLL and polydopamine. (b) The thickness of the coating, calculated according to Equation (2) is plotted as a function of the number of the deposited layer. Odd integers correspond to the deposition of PLL whereas even integers correspond to the deposition of polydopamine particles. The thickness increment corresponding to the first PLL deposition and the first polydopamine deposition are marked with an arrow. The full line corresponds to a linear regression to the experimental data.

In addition, the (PLL-polydopamine)$_n$ films appear homogeneous and are made from distinct particles about 100 nm in diameter (**Figure 2**). This size is in agreement with the size of polydopamine particles produced in solution upon oxygenation of slightly basic dopamine solutions and determined by means of transmission electron microscopy [20]. The thickness of the (PLL-polydoapmine)$_n$ films increases proportionally with the number of polycation-polyanion deposition cycles, n. This behavior is similar to that found previously for PEM films made from poly(diallyldimethyl ammonium chloride) and polydopamine [18].

The thickness evolution and surface morphology of the (PLL-PGA)$_5$ films, used as a reference for the behavior of melanoma, has been published elsewhere [21].

3.2. Cellular Proliferation

The proliferation of melanoma in solution decreases when the concentration of polydopamine in the cell culture medium increases (**Figure 3**). The polydopamine used in this study is in the form of platelet like particles of about 100 - 200 nm in extension and a few nm in thickness. The effect is more pronounced when the incubation time of the cells with the culture medium containing the polydopamine increases from 3 to 7 days. These data show that increasing dopamine concentration in the cell medium induces a pronounced reduction of their proliferation.

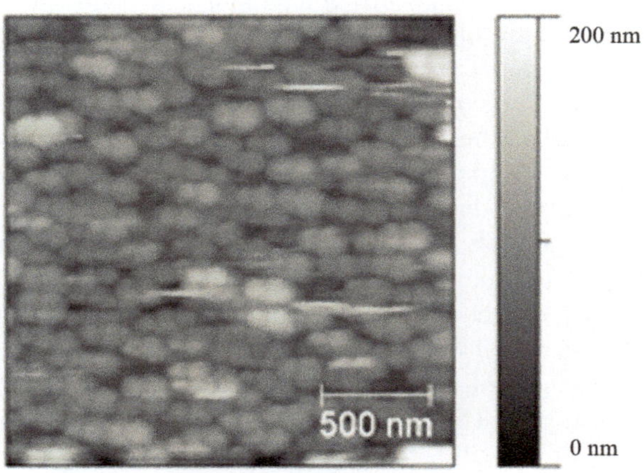

Figure 2. AFM topographical image (2 µm × 2 µm) of a (PLL-polydopamine)$_5$ film deposited on a quartz crystal used for the QCM-D experiments displayed in Figure 1.

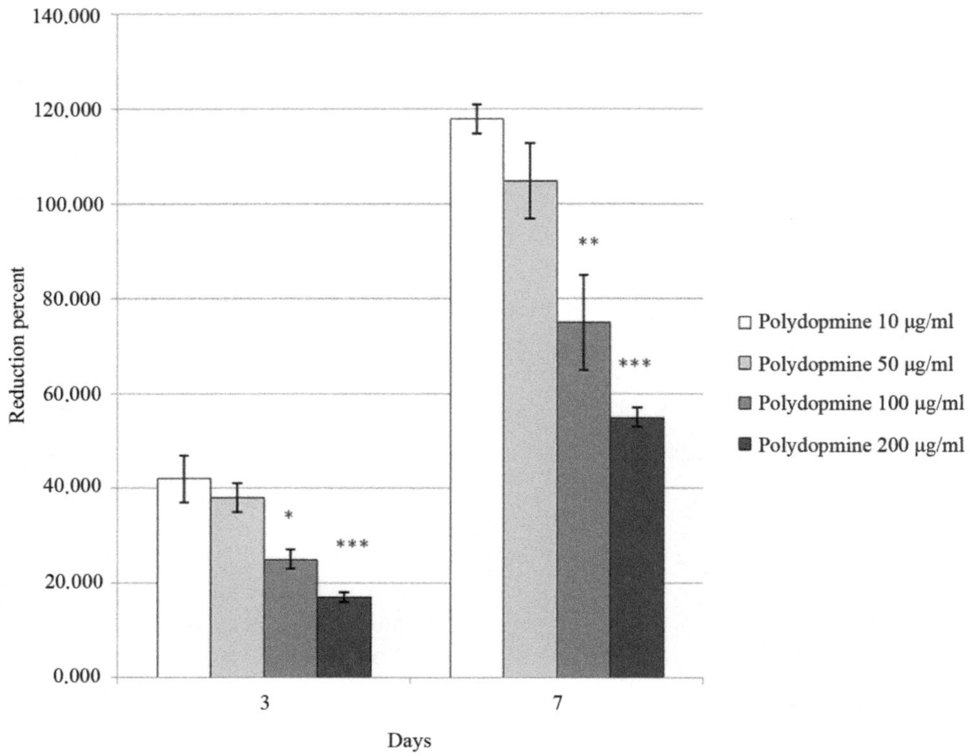

Figure 3. Proliferation of B16F1 murine melanoma cells growing in contact with melanin in solution at various concentrations in the culture medium (10, 50, 100 or 200 µg·mL^{-1}). Cells were deposited on 24 well plates at a density of 1 × 10^4 cells/well. Proliferation was measured after 3 days or 7 days of culture with polydopamine in solution and was quantified with the Alamar® Blue test.

On the other hand when polydopamine is present in the PEM films as the last deposited layer, hence with the possibility of the melanine-like material to be in direct contact with the cells, their proliferation is significantly enhanced with respect to the same cells put in contact with (PLL-PGA)$_5$ films (**Figure 4**) which are made from the same polycation, PLL, but differ from the polydopamine containing films with respect to the incorporated polyanion. This result means that contrarily to polydopamine particles present in solution, their immobilization on the surface allows for the melanoma to have a better proliferative behavior. It cannot be excluded that the positive effect of polydopamine is due to some enhancement of the mechanical properties of the (PLL-polydopamine)$_5$ film with respect to the soft (PLL-PGA)$_5$ film. It is well known that the behavior of cells on surfaces is not only dependent on the composition of that surface but also on its mechanical properties [3]. Indeed, one may expect the (PLL-polydopamine)$_n$ films to be stiffer than their (PLL-PGA)n counterparts owing to the establishment of covalent bonds between polydopamine and PLL [15]. The quantification of the mechanical properties of the (PLL-polydoamine)$_n$ films with colloidal probe AFM requires to produce thick coatings in order to avoid a strong influence of the rigid silica substrate. The mechanical properties of these coatings will be investigated in a forthcoming investigation.

3.3. Melanine Secretion

In the presence of polydopamine in solution, the secretion of melanin by the melanocytic cells is considerably amplified when the polydopamine concentration increases particularly in the supernatant and the more so for shorter incubation times of 3 days (**Figure 5**).

However, after 3 days of culture, the secretion of melanin by melanoma cells adhering on (PLL-polydopamine)$_5$ films is significantly lower than on (PLL-PGA)$_5$ films (**Figure 6**) even if the latter are less favorable than the former ones concerning the cellular proliferation (**Figure 4**). This result shows that the presence of polydopamine, chemically and structurally similar to melanin, on the PEM films significantly hinders the secretion of melanin by the melanoma cells in presence of the films. This result holds true not only in the cell lysate but also in the supernatant. But after 7 day of culture the difference in melanin secretion by the cells in contact with the (PLL-polydopamine)$_5$ and the (PLL-PGA)$_5$ films decreases, even if it remains significant. These results could be explained by the limited size of our active nanoreservoir here (n = 5). In our group, we have shown that the mechanism by which the cells interact with our

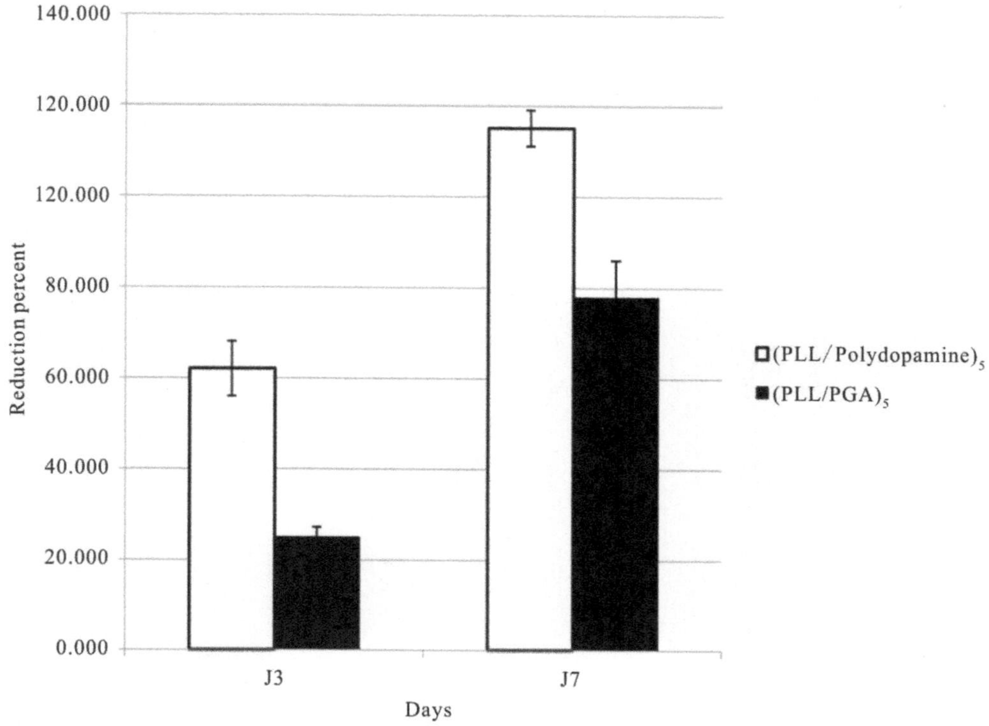

Figure 4. Proliferation of B16F1 murine melanoma cells growing on the surface of (PLL-polydopamine)$_5$ or (PLL-PGA)$_5$ films. Cells were deposited on 24 well plates at a density of 1 × 10^4 cells/well. Proliferation was measured after 3 days or 7 days of culture in contact with multilayered films and was quantified by the Alamar® Blue test.

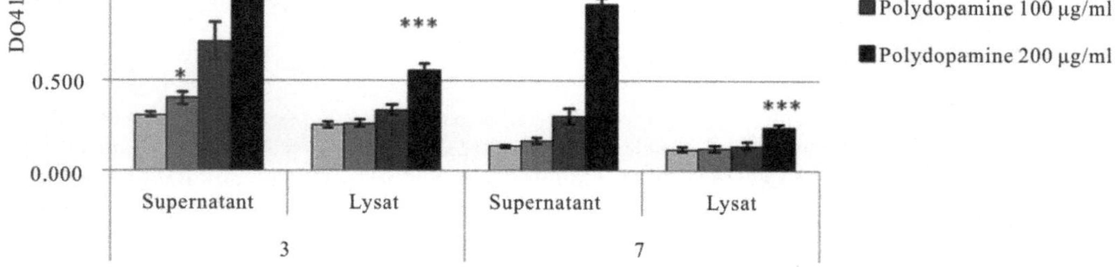

Figure 5. Melanin production by B16F1 Murine melanoma cells growing in contact with polydopamine in solution at different concentration in the culture medium (10, 50, 100 or 200 µg·mL^{-1}). Cells were deposited on 24 well plates at a density of 1 × 10^4 cells/well. The melanin production was measured after 3 or 7 days of culture.

active films is the local degradation by cells. We have also reported that by increasing the size of our reservoir (n), we increase the activity [23-30].

Finally, the morphology and organization of the melanoma cells adhering on both kinds of multilayer films was investigated with staining methods after 3 days (**Figure 7**) and 7 days (**Figure 8**) of culture.

Our results indicated clearly here that after 3 days of culture, melanocytics cells adhesion is clearly better onto (PLL-polydopamine)$_5$ film than onto (PLL/PGA)$_5$ film. The cells are widespread onto the (PLL-polydopamine)$_5$ film, their cytoskeleton shown by actin fibers are making as much filopods as possible in comparison to the round and non-spread cells onto the (PLL/PGA)$_5$ film. After 7

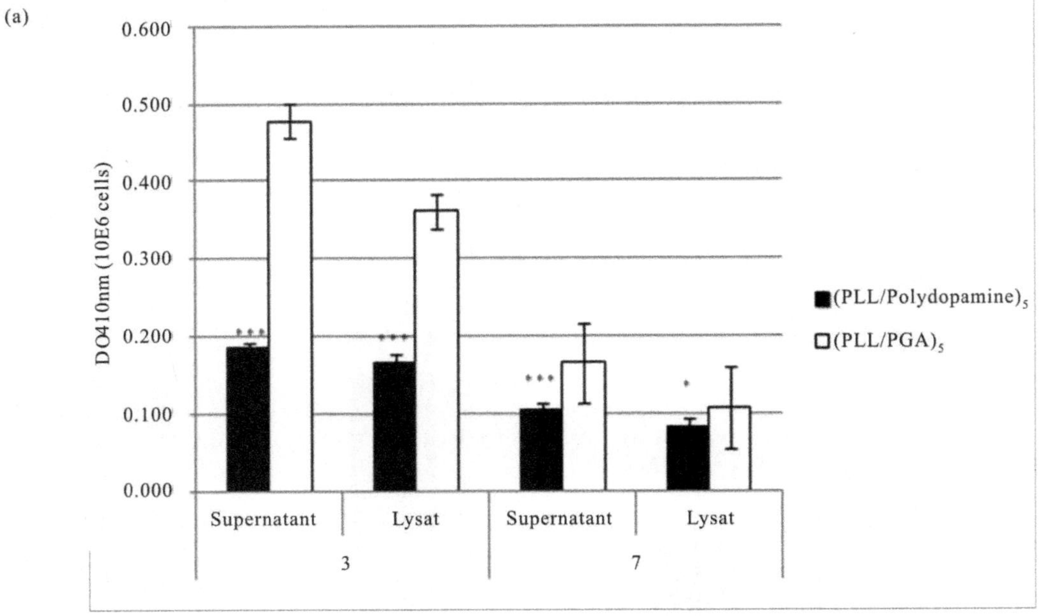

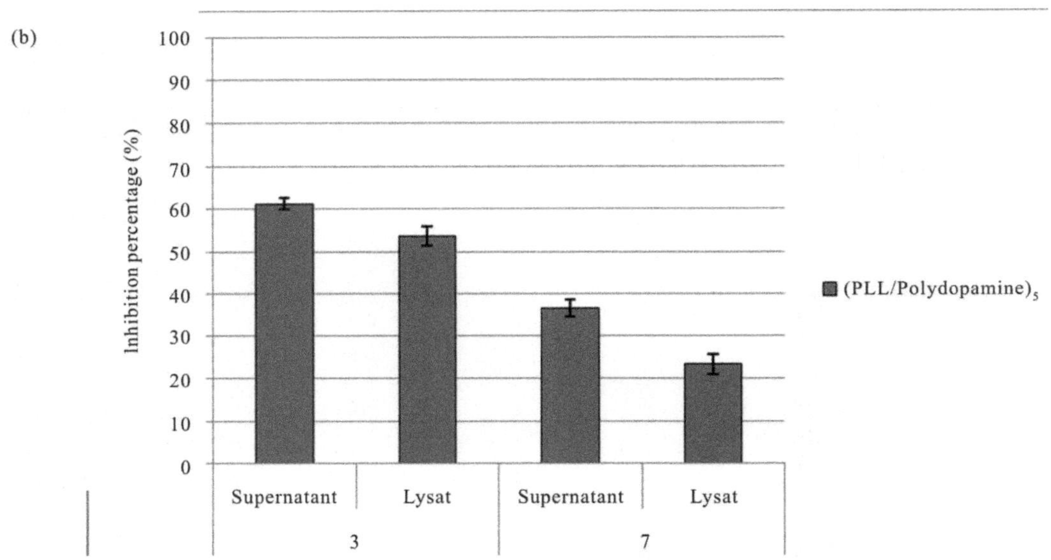

Figure 6. (a) Melanin production by B16F1 murine melanoma cells growing on the surface of the (PLL-polydopamine)$_5$ or (PLL-PGA)$_5$ films. The cells were deposited on 24 well plates at a density of 1 × 10^4 cells/well. The melanin production was measured after 3 or 7 days of culture. (b) Percentage of inhibition of melanin productioon by B16F1 murine melanoma cells growing on the surface of the multilayered films (PLL-polydopamine)$_5$ in comparison with (PLL/PGA)$_5$ films. Cells were deposited on 24 well plates at a density of 1 × 10^4 cells/well. The melanin production was measured after 3 or 7 days of culture.

days (**Figure 8**), we have shown that the multiplication and morphology of these cells confirmed this tendency, which clearly means that (PLL-polydopamine)$_5$ film is greatly favorable to our murine melanoma cells adhesion and proliferation as we have observed earlier.

Clearly, our results show that dopamine is more favorable than glutamic acid and promotes good proliferation of these melanocytics cells.

To understand these results, we have also shown that in term of melanin secretion, increasing polydopamine concentration in the medium greatly increases the production of melanin by melanoma cells. However, on the contrary, incorporation of polydopamine in LBL film seems to significantly reduce melanin secretion in comparison to the control film without polydopamine, which means that the inhibition effect of polydopamine on the melanin secretion is not only due to polydopamine sole presence in solution but also to its incorporated and immobilized form into the multilayered polyelectrolyte film. Indeed, incorporated into the LBL film, polydopamine

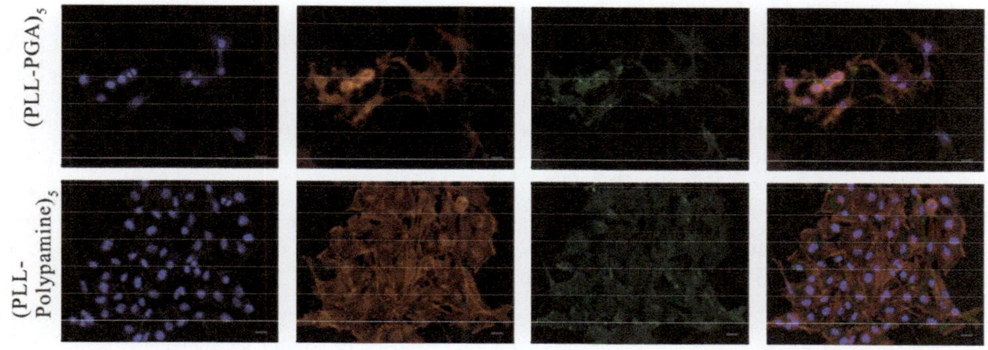

Figure 7. Morphology and organization of B16F1 murine melanoma cells growing on the surface of the multilayered films. Proliferation of B16F1 cells after 3 days of culture on the surface of the (PLL-PGA)$_5$ or (PLL-polydopamine)$_5$ films. Morphology and actin organization of cells observed by fluorescent microscopy: Integrin *αv* (green), Alexa Fluor 546-labeled phalloidin (red) and DAPI (blue), scale bar = 20 μm.

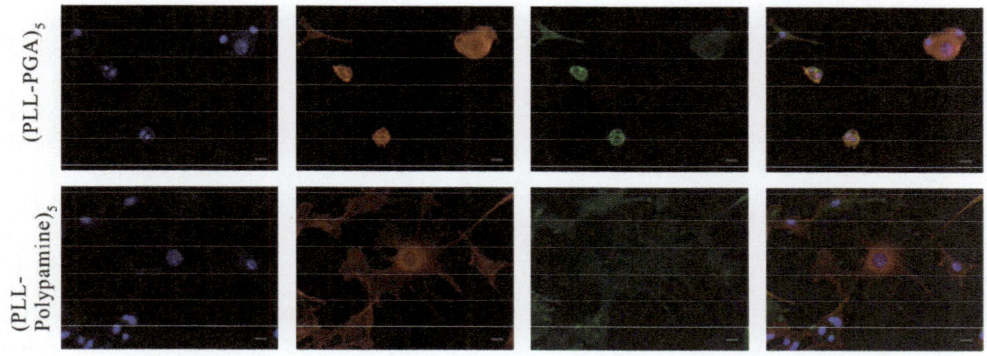

Figure 8. Morphology and organization of B16F1 Murine melanoma cells growing on the surface of the multilayered films. Proliferation of B16F1 cells after 7 days of culture on the surface of (PLL-PGA)$_5$ or (PLL-Melanin)$_5$ films. Morphology and actin organization of cells observed by fluorescent microscopy: Integrin *αv* (green), Alexa Fluor 546-labeled phalloidin (red) and DAPI (blue), scale bar = 20 μm.

would be directly internalized by melanoma cells throughout the different layers whereas contact between cells and polydopamine in solution is highly limited in comparison.

4. Conclusion

As we have shown, even if increasing concentration of polydopamine in solution leads to a decrease in melanoma cells proliferation, it is not the case when polydopamine is incorporated into the LBL film; in contrast, cell proliferation is even higher than the control without polydopamine. However, in terms of melanin secretion, the polydopamine incorporation into the LBL film seems to significantly reduce the production of melanin by melanoma cells in comparison to the effect of polydopamine in solution.

5. Acknowledgements

This work was supported by the project Angiomatrix from the "Agence Nationale de la Recherche, ANR" and "Alsace contre le Cancer", C.M. thanks the Faculté de Chirurgie Dentaire of Strasbourg for financial support. N.J. is indebted to CHU de Nancy, Hôpital Central, "Chirurgie Orthopédique et Traumatologie" (Contrat d'interface INSERM vers l'hôpital).

REFERENCES

[1] N. A. Peppas and R. Langer, "New Challenges in Biomaterials," *Science*, Vol. 263, No. 5154, 1994, pp. 1715-1720.

[2] R. Langer and D. A. Tirrell, "Designing Materials for Biology and Medicine," *Nature*, Vol. 428, No. 6982, 2004, pp. 487-492.

[3] D. E. Discher, P. Janmey and Y.-L. Wang, "Tissue Cells Feel and Respond to the Stiffness of Their Substrate," *Science*, Vol. 310, No. 5751, 2005, pp. 1139-1143.

[4] Y. Hirano and D. J. Mooney, "Peptide and Protein Presenting Materials for Tissue Engineering," *Advanced Materials*, Vol. 16, No. 1, 2004, pp. 17-25.

[5] K. Y. Lee and D. J. Mooney, "Hydrogels for Tissue Engineering," *Chemical Review*, Vol. 101, No. 7, 2001, pp. 1869-1879.

[6] Z. Tang, Y. Wang, P. Podsiadlo and N. A. Kotov, "Bio-

medical Applications of Layer-by-Layer Assembly: From Biomimetics to Tissue Engineering," *Advanced Materials*, Vol. 18, No. 24, 2006, pp. 3203-3224.

[7] H. Lee, S. M. Dellatore, W. M. Miller and P. B. Messersmith, "Mussel-Inspired Surface Chemistry for Multifunctional Coatings," *Science*, Vol. 318, No. 5849, 2007, pp. 426-430.

[8] B. Zhu and S. Edmondson, "Polydopamine-Melanin Initiators for Surface-Initiated ATRP," *Polymer*, Vol. 52, No. 10, 2011, pp. 2141-2149.

[9] K. C. Black, J. Yi, J. G. Rivera, D. C. Zelasko-Leon and P. B. Messersmith, "Polydopamine-Enabled Surface Functionalization of Gold Nanorods for Cancer Cell-Targeted Imaging and Photothermal Therapy," *Nanomedicine*, Vol. 8, No. 1, 2013, pp. 17-28.

[10] M. E. Lynge, R. van der Westen, A. Postma and B. Stadler, "Polydopamine—A Nature-Inspired Polymer Coating for Biomedical Science," *Nanoscale*, Vol. 3, No. 12, 2011, pp. 4916-4928.

[11] M. d'Ischia, A. Napolitano, A. Pezzella, P. Meredith and T. Sarna, "Chemical and Structural Diversity in Eumelanins: Unexplored Bio-Optoelectronic Materials," *Angewandte Chemie International Edition in English*, Vol. 48, No. 22, 2009, pp. 3914-3921.

[12] H. Swalwell, J. Latimer, R. M. Haywood and M. A. Birch-Machin, "Investigating the Role of Melanin in UVA/UVB- and Hydrogen Peroxide-Induced Cellular and Mitochondrial ROS Production and Mitochondrial DNA Damage in Human Melanoma Cells," *Free Radical Biology and Medicine*, Vol. 52, No. 3, 2012, pp. 626-634.

[13] T. Pan, J. Zhu, W.-J. Hwu and J. Jankovic, "The Role of Alpha-Synuclein in Melanin Synthesis in Melanoma and Dopaminergic Neuronal Cells," *PLoS ONE*, Vol. 7, No. 9, 2012, Artciel ID: e45183.

[14] C. J. Bettinger, R. Langer and J. T. Borenstein, "Engineering Substrate Topography at the Micro- and Nanoscale to Control Cell Function," *Angewandte Chemie—International Edition*, Vol. 48, No. 30, 2009, pp. 5406-5415.

[15] H. Lee, J. Rho and P. B. Messersmith, "Facile Conjugation of Biomolecules onto Surfaces via Mussel Adhesive Protein Inspired Coatings," *Advanced Materials*, Vol. 21, No. 4, 2009, pp. 431-434.

[16] F. Bernsmann, B. Frisch, C. Ringwald and V. Ball, "Protein Adsorption on Dopamine-Melanin Films: Role of Electrostatic Interactions Inferred from Zeta-Potential Measurements versus Chemisorption," *Journal of Colloid Interface Science*, Vol. 344, No. 1, 2010, pp. 54-60.

[17] F. Bernsmann, J.-C. Voegel and V. Ball, "Different Synthesis Methods Allow to Tune the Permeability and Permselectivity of Dopamine—Melanin Films to Electrochemical Probes," *Electrochimica Acta*, Vol. 56, No. 11, 2011, pp. 3914-3919.

[18] F. Bernsmann, O. Ersen, J. C. Voegel, E. Jan, N. A. Kotov and V. Ball, "Melanin-Containing Films: Growth from Dopamine Solutions versus Layer-by-Layer Deposition," *Chemphyschem*, Vol. 11, No. 15, 1002, pp. 3299-3305.

[19] B. P. Lee, P. B. Messersmith, J. N. Israelachvili and J. H. Waite, "Mussel-Inspired Adhesives and Coatings," *Annual Review of Materials Research*, Vol. 41, 2011, pp. 99-132.

[20] G. Sauerbrey, "Verwendung von Schwingquarzen zur Wägung dünner Schichten und zur Mikrowägung," *Zeitschrift für Physik*, Vol. 155, No. 2, 1959, pp. 206-222.

[21] V. Ball, M. Michel, F. Boulmedais, J. Hemmerle, Y. Haikel, P. Schaaf and J. C. Voegel, "Nucleation Kinetics of Calcium Phosphates on Polyelectrolyte Multilayers Displaying Internal Secondary Structure," *Crystal Growth & Design*, Vol. 6, No. 1, 2005, pp. 327-334.

[22] P. Meredith, B. J. Powell, J. Riesz, S. P. Nighswander-Rempel, M. R. Pederson and E. G. Moore, "Towards Structure-Property-Function Relationships for Eumelanin," *Soft Matter*, Vol. 2, No. 1, 2006, pp. 37-44.

[23] N. Jessel, F. Atalar, P. Lavalle, J. Mutterer, G. Decher, P. Schaaf, J. C. Voegel and J. Ogier, "Bioactive Coatings Based on a Polyelectrolyte Multilayer Architecture Functionalized by Embedded Proteins," *Advanced Materials*, Vol. 15, No. 9, 2003, pp. 692-695.

[24] N. Benkirane-Jessel, P. Lavalle, F. Meyer, F. Audouin, B. Frisch, P. Schaaf, J. Ogier, G. Decher and J. C. Voegel, "Control of Monocyte Morphology on and Response to Model Surfaces for Implants Equipped with Anti-Inflammatory Agents," *Advanced Materials*, Vol. 16, No. 17, 2004, p. 1507.

[25] N. Benkirane-Jessel, P. Schwinte, P. Falvey, R. Darcy, Y. Haikel, P. Schaaf, J. C. Voegel and J. Ogier, "Build-Up of Polypeptide Multilayer Coatings with Anti-Inflammatory Properties Based on the Embedding of Piroxicam-Cyclodextrin Complexes," *Advanced Functional Materials*, Vol. 14, No. 2, 2004, pp. 174-182.

[26] N. Jessel, M. Oulad-Abdelghani, F. Meyer, P. Lavalle, Y. Haikel, P. Schaaf and J. C. Voegel, "Multiple and Time-Scheduled *in Situ* DNA Delivery Mediated by Beta-Cyclodextrin Embedded in a Polyelectrolyte Multilayer," *Proceedings of the National Academy of Sciences*, Vol. 103, No. 23, 2006, pp. 8618-8621.

[27] B. S. Kim, S. W. Park and P. T. Hammond, "Hydrogen-Bonding Layer-by-Layer-Assembled Biodegradable Polymeric Micelles as Drug Delivery Vehicles from Surfaces," *ACS Nano*, Vol. 2, No. 2, 2008, pp. 386-392.

[28] N. Benkirane-Jessel, P. Lavalle, E. Hubsch, V. Holl, B. Senger, Y. Haikel, J. C. Voegel, J. Ogier and P. Schaaf, "Short-Time Tuning of the Biological Activity of Func-

tionalized Polyelectrolyte Multilayers," *Advanced Functional Materials*, Vol. 15, No. 4, 2005, pp. 648-654.

[29] A. Dierich, E. Le Guen, N. Messaddeq, J.-F. Stoltz, P. Netter, P. Schaaf, J.-C. Voegel and N. Benkirane-Jessel, "Bone Formation Mediated by Synergy-Acting Growth Factors Embedded in a Polyelectrolyte Multilayer Film," *Advanced Materials*, Vol. 19, No. 5, 2007, pp. 693-697.

[30] S. Facca, C. Cortez, C. Mendoza-Palomares, N. Messadeq, A. Dierich, A. P. R. Johnston, D. Mainard, J. C. Voegel, F. Caruso and N. Benkirane-Jessel, "Active Multilayered Capsules for *in Vivo* Bone Formation," *Proceedings of the National Academy of Sciences of the United States of America*, Vol. 107, No. 8, 2010, pp. 3406-3411.

Investigation of Inlet Gas Streams Effect on the Modified Claus Reaction Furnace

Reza Rezazadeh[1], Sima Rezvantalab[2*]
[1]Shahid Hasheminejad Gas Processing Company (S.G.P.C), Khangiran, Iran
[2]Department of Chemical Engineering, Urmia University of Technology, Urmia, Iran

ABSTRACT

The objective of this paper is to model the main reactions that take place in the Claus reactor furnace and compare it with actual data and simulated process. Since the most important point is the selection of suitable reaction conditions to increase the reactor performance, the model is formulated to predict the performance of the Claus plant. To substantiate the theoretical model, we used actual process condition and feed composition in Shahid Hasheminejad Gas Refinery. Model equations have been solved by using MATLAB program. Results from MATLAB are compared with those from SULSIM® simulator and with actual plant data. The AAD (Average Absolute Deviation) of modeling results with actual data is 2.07% and AAD of simulation results with real data is 4.77%. Error values are very little and show accuracy and precision of modeling and simulation. The predicting curve for different parameters of the reactor furnace according to variable conditions and specifications are given.

Keywords: Claus Process; Hydrogen Sulfide; Modeling; Reactor Furnace

1. Introduction

Hydrogen sulfide (H_2S) is produced from sulfur compounds in fossil fuels such as natural gas or oil. Sour gases (H_2S and CO_2), are removed from the natural gas or refinery gas by means of one of the gas treating processes. Due to global environmental rules, refineries have to recover sulfur from nature. H_2S containing acid gas stream is flared, incinerated, or fed to a sulfur recovery unit. The Claus process is commonly used to reduce the emission of sulfur compounds into the atmosphere. Recently recovery of sulfur is done by means of the modified Claus tail gas clean-up processes. In these processes, H_2S over a catalyst converts to elemental sulfur where the reaction takes place in a high temperature furnace. The recovery process is the reaction between H_2S and air to form sufur and water. Following reaction is the main reaction in the recovery process:

$$H_2S + \frac{1}{2}O_2 \rightarrow \frac{1}{x}S_x + H_2O + \text{heat}\left(\cong 626000\frac{\text{KJ}}{\text{Kgmole}}\right) \quad (1)$$

In the original Claus process, control of this reaction was difficult and sulfur recovery efficiencies were low. In order to overcome these difficulties and also increase the efficiency of the process, several modifications of the Claus process have been developed. In modified process, free flame total oxidation of 1/3 of the H_2S to SO_2 followed by a reaction over the catalyst of SO_2 with the remaining 2/3 H_2S. According to Mohamed Sassi and Ashwani K. Gupta modified Claus process for a Sulfur Recovery Plant consists of several stages [1]:

1) Combustion (In the Reactor Furnace)

$$H_2S + 3/2\,O_2 \xrightarrow{\text{High T}} SO_2 + H_2O + \text{heat}(\cong 518000\ \text{KJ/Kgmole}) \quad (2)$$

$$2H_2S + SO_2 \xleftrightarrow{\text{High T}} 3/2\,S_2 + 2H_2O \qquad \Delta H_R \cong +47000\ \text{KJ/Kgmole} \quad (3)$$

2) Redox (Catalytic Converter)

These are simplified reactions which actually take place in a Claus unit. There are various species of gaseous sulfur S_2, S_3, S_4, S_5, S_6, S_7, and S_8. Equilibrium concentrations of these sulfur compounds are not known in the

$$2H_2S + SO_2 \xleftrightarrow{\text{Catalytic, Low T}} \frac{3}{X}S_x + 2H_2O + \text{heat} \quad (\cong 108\ \text{KJ/Kgmole}) \quad (4)$$

*Corresponding author.

$$3H_2S + 3/2O_2 \xrightarrow{\text{Overall}} \frac{3}{X}S_x + 3H_2O + \text{heat} \quad (\cong 626000\ \text{KJ/Kgmole}) \tag{5}$$

entire of process. Additionally, gas stream contains water saturated with 15 - 80 mol% H_2S, 0.5 - 1.5 mol% hydrocarbons, and CO_2 which can result in carbonyl sulfide (COS) carbon disulfide (CS_2), carbon monoxide (CO), and hydrogen [2].

Most Claus plants operate in the multistep process "straight-through" mode as shown in **Figure 1**. The combustion is carried out in a reducing atmosphere with only enough air 1) to oxidize one-third of the H_2S to SO_2, 2) to burn hydrocarbons and mercaptans, and 3) for many refinery Claus units, to oxidize ammonia and cyanides. The process includes the following operations:

- Combustion: burn hydrocarbons and other combustibles and oxidize one-third of the H_2S to provide necessary SO_2 to react with remainder H_2S for producing S_2 in the furnace.
- Waste Heat Recovery: Cool combustion products.
- Sulfur Condensing: Cool outlet streams from waste heat recovery unit and from catalytic converters.
- Reheating: Reheat process stream, after sulfur condensation and separation, to a temperature high enough to remain sufficiently above the sulfur dew point.

In order to investigate different aspects of the modified Claus process, a number of studies have been performed on main burner and sulfur recovery in this process. Monnery *et al.* modeled the modified Claus process [3]. Kelly Anne Hawboldt has studied mathematical modeling of reactions in the process [4]. Recently, S. Asadi *et al.* used TSWEET simulator to optimize the recovery of sulfur [5].

At first approach, we have used a mathematical model for the key reactions that take place in the reactor furnace. In the second approach, we have simulated the process with a commercial simulator. Finally using the model and simulation, we have compared obtained results and proposed some improvements on the base case.

2. Kinetic Studies

Claus process has been investigated via different aspects, experimental and theoretical perspectives. Paskal *et al.* gives a summary of the main reactions thought to occur within the Claus furnace [6,7]. Clark *et al.* discussed the mechanisms behind the formation of key sulfur containing species found within the furnace, and in a subsequent study outline primary reaction pathways for the principal components in the furnace [8,9]. While there have been numerous attempts to model the Claus process based on simplified kinetic expressions, the complexity of the chemistry and the number of involved reactions has precluded the accurate prediction of outlet compositions. As it mentioned before, gas stream contains different compounds such as H_2S, CO_2 and hydrocarbons. Most important compound is H_2S and several groups have studied it's decomposition under different condition. As result, it suggested that there are numerous reactions on the catalytic decomposition of H_2S in the clause process. According to the studies, gaseous H_2S exists in chemical equilibrium with elemental hydrogen and sulfur by the following equation:

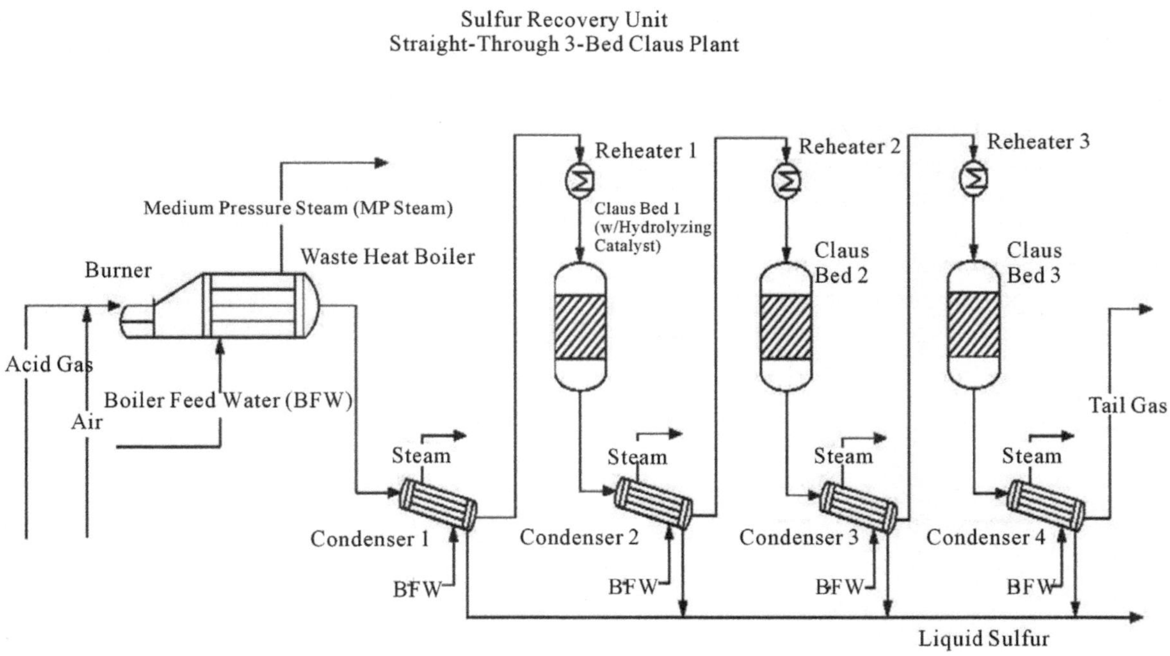

Figure1. Three-stage straight-through sulfur plant.

$$H_2S \leftrightarrow H_2 + \frac{1}{x}S_x \quad (6)$$

Oxidation includes two staged reaction, first oxidation of H_2S and followed by reaction between H_2S and SO_2 that limiting stage of the Claus reaction is the second part [10].

$$H_2S + \frac{3}{2}O_2 \leftrightarrow H_2O + SO_2 \quad (7)$$

$$2H_2S + SO_2 \leftrightarrow 2H_2O + S_2 \quad (8)$$

During the reaction in the furnace and according to the existence of ammonia in the gas stream, oxidation of NH_3 will take place. Recently, Clark has mentioned that under 1100°C ammonia oxidation is negligible. Additionally, he noted in competitive oxidation, first of all H_2S, and then methane, finally NH_3 react. On the other hand, Goar *et al.* found rate of hydrocarbon combustion is more than ammonia and NH_3 is more than H_2S [11]. Formation of COS and CS_2 in the Claus reaction furnace are also very important in the modeling. Field studies have revealed that concentrations of COS and CS_2 at the exit of the reaction furnace/ waste heat boiler typically lie between 100 ppm and 2 mol% [12]. However, these seemingly small concentrations in the furnace product stream can represent nearly half the sulfur emissions from a tail gas clean-up unit [13]. It is possible to hydrolyze COS and CS_2 back into H_2S in the Claus catalytic converters according to the following stoichiometry:

$$COS + H_2O \leftrightarrow CO_2 + H_2S \quad (9)$$

$$CS_2 + 2H_2O \leftrightarrow CO_2 + 2H_2S \quad (10)$$

As it mentioned, there many reactions which may take place in the furnace according to the conditions such as temperature and pressure. Full list of reactions that occur in the furnace is not obvious and for the known reactions, reaction rate expressions are not available. In current work we assumed that gas stream consists of CH_4, CO_2, H_2S, H_2O, O_2, N_2, CO, CS_2, COS, S_2, SO_2, H_2. Regarding gas stream composition, important reactions which take place in furnace and we use in the modeling are listed below.

$$H_2S + 3/2O_2 \rightarrow SO_2 + H_2O \quad (11)$$

$$2H_2S + SO_2 \leftrightarrow 3/2S_2 + 2H_2O \quad (12)$$

$$CH_4 + 2O_2 \rightarrow CO_2 + 2H_2O \quad (13)$$

$$CO_2 \rightarrow CO + 1/2O_2 \quad (14)$$

$$H_2S \rightarrow 1/2S_2 + H_2 \quad (15)$$

$$CH_4 + 2S_2 \rightarrow CS_2 + 2H_2S \quad (16)$$

$$CO + 1/2S_2 \rightarrow COS \quad (17)$$

3. Mathematical Model

The basic structure of the model consist of the equations of mole and energy conservative rule the furnace, which are related to each other and are function of molar conversion of H_2S in equilibrium reaction and temperature. In order to model the reactor, a steady-state simulation has been used for mole and energy balance.

Sames and Paskal presented empirical correlations to predict the fraction of CO, H_2, COS, CS_2 and sulfur (as S) in the effluent of the Claus furnace. The correlations are obtained from more than 300 tests on 100 different sulfur trains; with different flow configurations processing acid gas feed streams [12]. These empirical correlations are presented in **Appendix**. We use these equations to model the furnace and mole balance. In this work, furnace pressure is 130 kPa (absolute) and pressure drop (ΔP) is 10 kPa. Using empirical equations and applying in the mole balance for the compounds, we get the mole balance equations, for each compound.

For gas components in the outlet gas stream:

H_2S:

$$f_{i(H_2S)} + 2R(CS_2)\times\left(f_{i(CH_4)}\right) - f_{i(H_2S)} - \frac{1}{3}f_{i(H_2S)} - R(H_2)\times f_{i(H_2S)} - X \quad (18)$$

$$H_2O: f_{i(H_2O)} + 2\left(1 - R(CS_2)\right)\times f_{i(CH_4)} + \frac{1}{3}f_{i(H_2S)} + X \quad (19)$$

$$CO_2: f_{i(CO_2)} + f_{i(CH_4)}\times\left(1 - R(CS_2)\right) - \left(R(COS) + R(CO)\right)\times f_{i(CH_4)} + f_{i(CO_2)} \quad (20)$$

$$CO: R(CO)\times\left(f_{i(CH_4)} + f_{i(CO_2)}\right) \quad (21)$$

$$SO_2: \frac{1}{3}f_{i(H_2S)} - \frac{1}{2}X \quad (22)$$

$$CH_4: 0 \quad (23)$$

$$O_2: \frac{1}{2}\left(R(COS)\right) + R(CO)\times\left(f_{i(CH_4)} + f_{i(CO_2)}\right) \quad (24)$$

$$N_2: \frac{79}{21}f_{i(H_2S)} + 2f_{i(CH_4)}\left(1 - R(CS_2)\right)\times\frac{1}{2} \quad (25)$$

$$S_2: \frac{3}{4}XR(COS)\times\left(f_{i(CH_4)} + f_{i(CO_2)}\right) + \frac{1}{2}f_{i(H_2S)} - 2R(CS_2)\times f_{i(CH_4)} - \frac{1}{2}R(H_2) \quad (26)$$

$$H_2: R(H_2)\times f_{i(H_2S)} \quad (27)$$

$$CS_2: R(CS_2)\times f_{i(CH_4)} \quad (28)$$

$$COS: R(COS)\times\left(f_{i(CH_4)} + f_{i(CO_2)}\right) \quad (29)$$

Total moles:

$$f_{i(CH_4)}\left(3-2R(CS_2)+\frac{1}{2}R(CO)\right)$$
$$+f_{i(CO_2)}\left(1-R(CS_2)+\frac{1}{2}R(CO)\right) \quad (30)$$
$$+f_{i(H_2S)}\left(\frac{4}{3}+\frac{1}{2}R(H_2)\right)+f_{i(N_2)}+f_{i(H_2O)}+\frac{1}{4}X$$

X: Flow rate of H_2S conversion at the equilibrium Claus reaction

At equilibrium:

$$K_p=\frac{[H_2O]^2[S_2]^{3/2}}{[H_2S]^2[SO_2]}\left[\frac{\pi}{\text{total mols}}\right]^{3/2-1} \quad (31)$$

By interpolation in the K_p curve vs. furnace temperature (GPSA), we calculate:

$$K_p=0.6021T-420.49T(^\circ C), K_p(\text{Kpa}^{0.5}) \quad (32)$$

Replacing molar flows in Equation (18) through 30 in Equation (31) resulted into a new equation, function of temperature, in the following form (see below):

where the A, B, C, D are function of feed composition and would be calculated numerically.

In order to calculate the reactions enthalpy, we used equation 34 and data provided in **Table 1**. **Table 1** represents standard enthalpy of formation and heat capacity parameters for each component.

$$SUM(\Delta H_{rxn})=\sum_1^7 \Delta H_{rxn}\times f_{rxn(i)}\times \text{Conversion}_i \quad (34)$$

ΔH_{rxn} is standard enthalpy of reactions in 25°C, $f_{rxn(i)}$ is molar flow rate of limiting compound in each reaction, Conversion_i *is* conversion value of limiting compound in each reaction.

Replacing all mentioned equations resulted into a new Equation (35) in which conversion of H_2S is function of temperature of reaction furnace (**Table 2**).

$$X=\frac{A_0+A_4T+B_4T^2+C_4T^3+D_4T^4}{B_0+A_5T+B_5T^2+C_5T^3+D_5T^4} \quad (35)$$

A_0, A_4, A_5, B_0, B_4, B_5, C_4, C_5, D_4, D_5 are constant parameters that would calculated numerically according to the inlet gas stream condition.

In order to establish a reference point, calculations are carried out for a "base case" and the operating conditions used are given in **Table 3**. Our base case is Shahid Hasheminejad Gas Refinery. Shahid Hasheminejad (Khangiran) gas refinery is in Sarakhs, Khorasan province.

Operating conditions is composition of inlet gas stream to the Claus process in the refinery. Next step, application of the feed gas condition in the equations resulted into the parameters of mole and energy balance equations. Calculated parameters are presented in **Table 4**.

4. Process Simulation

In order to compare the implications of the reaction fur-

Table1. Compound heat capacity parameters and H_{298}^0 kJ/kg mole.

	a_i	b_i	c_i	d_i	$H^0_{(i,298)}$ at 25°C kJ/kgmole
CH_4	34.31	0.05469	3.66×10^{-6}	-11×10^{-9}	881
CO_2	19.8	0.0734	-5.6×10^{-5}	1.72×10^{-8}	912
H_2S	31.94	1.44×10^{-3}	2.43×10^{-5}	-1.18×10^{-8}	846
H_2O	32.2	1.92×10^{-3}	1.06×10^{-5}	3.6×10^{-9}	839
O_2	29.1	11.58×10^{-3}	-6.0759×10^{-6}	13.11×10^{-10}	733
N_2	21.15	13.6×10^{-4}	2.6×10^{-5}	-1.17×10^{-8}	726
SO_2	23.85	67×10^{-3}	-4.9×10^{-5}	1.33×10^{-8}	986
S_2	32.47	0.0067	0	0	805
CO	30.87	-12.9×10^{-3}	2.79×10^{-5}	-1.27×10^{-8}	729
CS_2	27.44	81.3×10^{-3}	-7.67×10^{-5}	2.87×10^{-8}	1127
COS	23.57	79.8×10^{-3}	-7.02×10^{-5}	2.45×10^{-8}	1025
H_2	27.14	9.27×10^{-3}	-1.38×10^{-5}	7.65×10^{-9}	719

Table 2. limiting specie and calculated ΔH_{rxn} [4].

Limiting specie	Reactions	ΔH_{rxn} (Kj/kgmole)
CO_2	$CO_2 \rightarrow CO + 1/2O_2$	280000
H_2S	$H_2S + 3/2O_2 \rightarrow SO_2 + H_2O$	−520000
CH_4	$CH_4 + 2O_2 \rightarrow CO_2 + 2H_2O$	−803000
H_2S	$2H_2S + SO_2 \leftrightarrow 3/2S_2 + 2H_2O$	46500
H_2S	$H_2S \rightarrow 1/2S_2 + H_2$	81000
CH_4	$CH_4 + 2S_2 \rightarrow CS_2 + 2H_2S$	−296000
CO	$CO + 1/2S_2 \rightarrow COS$	−142000

Table 3. Refinery claus process inlet gas condition.

Feed Temperature	325 K (52°C)
Inlet air temperature (average in summer and winter)	350 K (77°C)
Furnace pressure	130 kPa A
Feed molar rate	2682.9 Kgmole/h
CH_4 mole fraction	0.0105
CO_2 mole fraction	0.563
H_2S mole fraction	0.336
H_2O mole fraction	0.0905

$$T(^\circ K)=\left(\left(\frac{(A+X)^2(B+0.75X)^{1.5}\pi^{0.5}}{(C-X)^2\left(1/3 f_{i(H_2S)}-0.5X\right)(D+0.25)^{0.5}}+420.49\right)\Big/0.6021\right)+273 \quad (33)$$

Table 4. Calculated parameters for case study.

i	A_i	B_i	C_i	D_i
-	560.6297	−6.5079	589.4866	3748.3
0	2.0757×10^8	4.3965×10^4	-	-
4	-1.1318×10^5	−68.2566	0.0115	-2.3178×10^{-7}
5	12.6875	0.014	3.7×10^{-6}	3.875×10^{-10}

nace model on design, overall sulfur recovery and emissions, this modified Claus plant was simulated using SULSIM®.

SULSIM® is program for Sulfur Plant Simulation and represents simulation package for SRU and TGTU design. This software has widely accepted thermodynamic data and propriety thermodynamic properties for all of the gas components and sulfur species found in sulfur recovery processes. **Figure 2** shows the flow diagram of the simulated Claus unit of Shahid Hasheminejad Gas Refinery.

5. Results and Discussion

As described previously, we implemented real data from gas refinery in the model. In this section to verify the model, we compare the output result from reaction furnace, model values and simulation results. **Table 5** listed three set of results.

It is obvious that furnace temperature obtained using model (1098 K) is lower than actual temperature of reaction furnace (1113 K), as the 15°C temperature difference is negligible and it results into 1.35% error. Simulated results shows that predicted temperature using SULSIM® software is 1121 K and higher than Claus furnace temperature. Error occurred using software is lower (0.72%). In this case simulation is more reliable. Additionally, sulfur conversion obtained from model results (56.635%) is in good agreement with conversion of sulfur in Claus plant (54%) in gas refinery. On the other hand, results from simulation indicate that sulfur conversion is 60.28%. According to the fact that obtained conversions from model are attained in lower temperature in comparing with the plant temperature while sulfur product is higher, we can conclude that model is more efficient in sulfur conversion. It is obvious that presented model in these conditions, is effective even more than simulated process. As it is presented in the **Table 5**, in the furnace effluent there is sulfur vapor. It's due to the high temperature in the furnace; this high temperature converts sulfur to S_2 vapor. According to **Table 5**, inlet air ratio to acid gas feed in Claus plant is 0.86 and in our model this ratio is 0.83. As we used this obtained ratio in our simulation, simulation and modeling resulted in the same ratio. Therefore modeling and simulation errors are low and about 3.5%. Since sulfur production is high in comparing with actual plant; air consumption is low, CO_2 concentration difference is about 0.05 (mole %), it means error is 0.34%; we can conclude that in all cases, simulated and modeling are more efficient.

Predicted N_2 concentration using equilibrium model is 36.067 (mole %) whereas N_2 content in the plant outlet gas stream is 39 (mole %). Since In empirical equations of model, it is assumed air consumption is low, results are logic and difference between real state and model results is acceptable. Since air consumption is low, its acid gas capacity is more than actual plant and can predict better results. Simulation has the same manner in the prediction of N_2 concentration. H_2O concentration in the outlet stream from model is less than plant outlet water content. The model performance was not good in water case and error value is about 7.8%. While H_2O content in simulation is closer to the actual data and lower error has occurred. Simulation performance is better in this case.

O_2 component in Claus plant damages the equipments (catalyst exchanger) and must be minimized, in Claus plants O_2 content is zero. Predicted concentration is

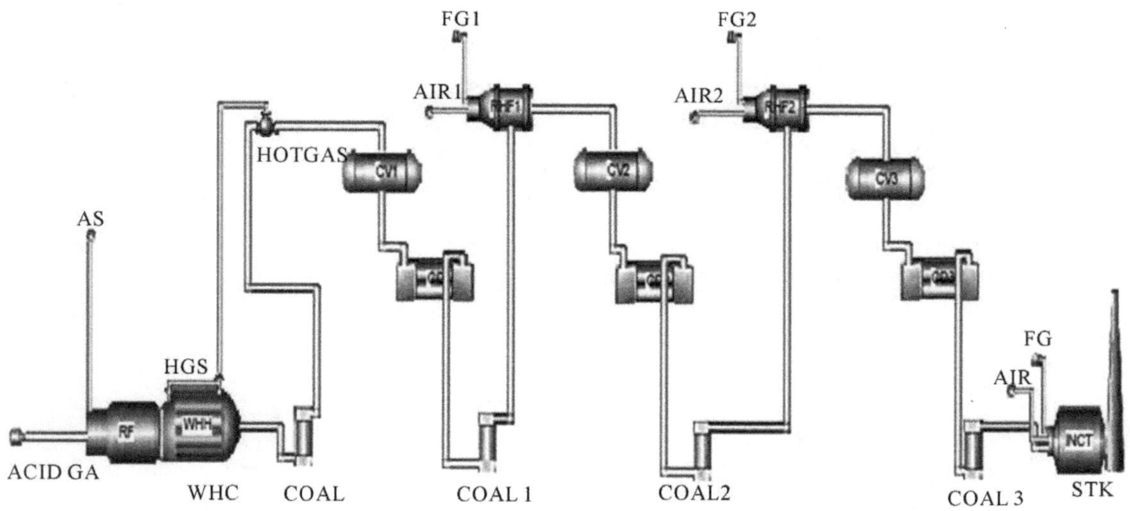

Figure 2. SULSIM® simulation used for gas for Khangiran gas refinery (S.G.P.C).

Table 5. Comparison of plant value and results from model and simulation for S.G. P.C.

Condition of Outlet Stream	Actual values for outlet concentration from W.H.B in Claus unit	Predicted Values using Model	Predicted Values using SULSIM® simulation
T (furnace temp)	1113 K (840°C)	1098 K (825°C)	1121 K (848°C)
F_{out} (kgmole/hr)	4789.7	4879.593	4882.196
X_{CH_4} (mole%)	0	0	0
X_{CO_2} (mole%)	31	31.105%	30.698%
X_{H_2S} (mole%)	4.9%	5.051%	4.952%
X_{H_2O} (mole%)	20.1%	18.519%	19.214%
X_{O_2} (mole%)	0%	0.014%	0
X_{N_2} (mole%)	39%	36.067%	36.076%
X_{SO_2} (mole%)	2.8%	2.643%	2.301%
X_{S_2} (mole%)	1%	5.139%	5.516%
X_{CO} (mole%)	0.2%	0. 325%	0.737%
X_{COS,CS_2} (mole%)	0.9%	0.102%	0.081%
X_{H_2} (mole%)	0.1%	1.035%	0.425%
Ratio (air/feed)	0.86	0.83	0.83
Sulfur conversion	54%	55.635%	60.28%

0.014% and error value is acceptable. The model predicted CH_4 concentration would drop to zero in the outlet of furnace. Checking the actual outlet concentration it is obvious that there is no methane in outlet stream, therefore no error has occurred in predicting of methane concentration. **Table 5** represents actual data for the concentration of CS_2 and COS are greater than equilibrium model, data obtained from simulation. As you know, production of CS_2 and COS in the Claus furnace reduces the efficiency of total plant and it is better to decrease these components. Therefore simulation and model outputs are more effective. According to Sames (1990), COS forms in WHB exchanger, therefore difference between predicted concentration and plant data verifies the formation reaction of COS [14]. This take place in W.H.B as following equation:

$$CO + 1/2S_2 \rightarrow COS \quad (40)$$

Predicted S_2 content in both methods is greater than plant data. There is a big difference between real and predicted values for S_2; it is due to the formation of liquid sulfur in WHB. According to **Table 5** and comparison between results and plant data, and also neglecting the error in CO and H_2 predicted concentrations, average error is about 3.5% and 5.36% for model and SULSIM® simulation; also AAD (Average Absolute Deviation) in comparing actual data with modeling and simulation results are 2.07% and 4.92%, respectively. We can conclude that our model is more efficient and applicable for other Claus plants with different inlet composition.

5.1. H_2S Concentration Effects

Figure 3 shows reaction furnace predicted temperature vs. inlet H_2S content using model and simulation. Both simulation and model have similar trend. According to the figure, model predicts that 1% increase in H_2S content will result into 7.5°C increase in furnace temperature. In order to combust hydrocarbons and aromatics, furnace temperature must be 1050°C. According to the model, if inlet gas stream contains more than 26% H_2S in current plant, temperature would increase to higher than 1050°C (1323 K). **Figures 4(a)** and **(b)** illustrate the predictions of model and simulation for H_2S and sulfur conversion in furnace vs. increase in H_2S content in feed. As model predicts, for one percent of the mole fraction of H_2S in feed stream, sulfur conversion increases by 0.54% in reaction furnace and S_2 mole fraction in outlet gas stream increases 0.12%. Simulation has similar manner in this case. Obtained results both are matched. It should be noted that sulfur conversion in **Figure 4(b)** shows model and simulation have similar trend. Therefore there is a negligible difference between the model and simulation prediction in 18% H_2S in **Figure 4(a)**. It is due to errors occurred in the simulation.

Figure 5 shows the effect of H_2S concentration in feed stream on the effluent H_2S content. As H_2S content increase in the feed, model predicts a trend for H_2S content in outlet which decreases and then with a lower slope increases and finally decreases.

The results indicate that in lower concentrations, furnace temperature is low and increase in the H_2S (to 30%) content would increase the slope of H_2S conversion line that results to decrease in H_2S content in the outlet stream of furnace.

In the higher concentrations, since furnace temperature is more than previous, H_2S cracking and conversion in Claus reaction increases and unconverted hydrogen sulfide

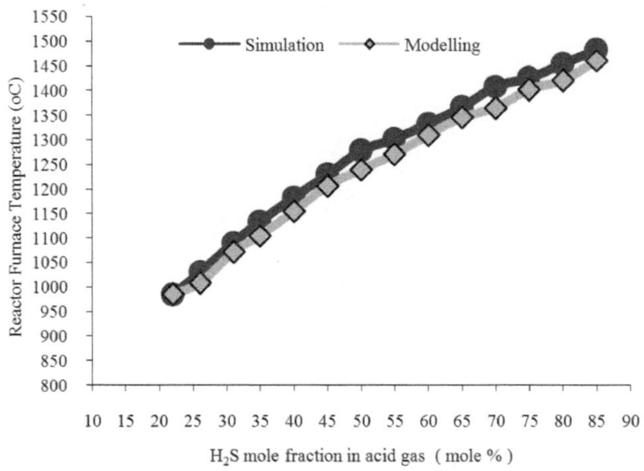

Figure 3. Simulation and model estimation for Modified Claus plant reaction furnace temperature vs. H_2S content.

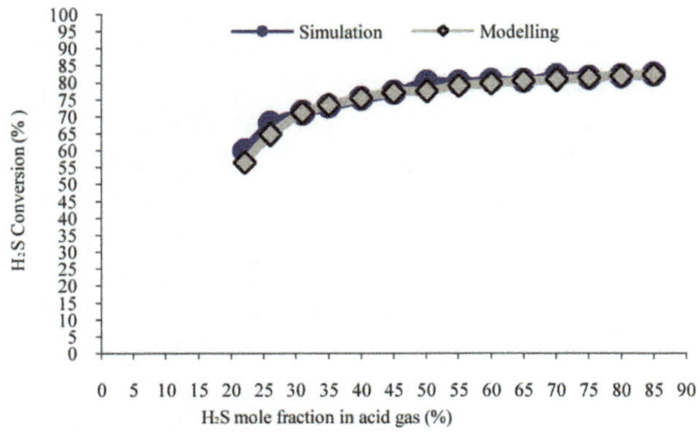

(a)

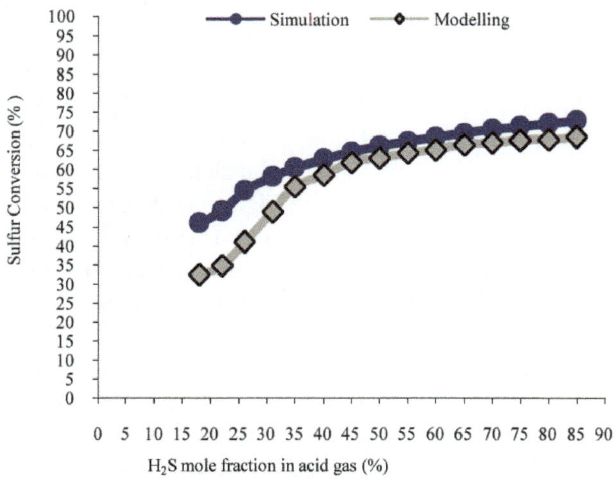

(b)

Figure 4. Comparison of predicted. (a) H_2S conversion; (b) sulfur conversion vs. H_2S mole fraction in feed stream.

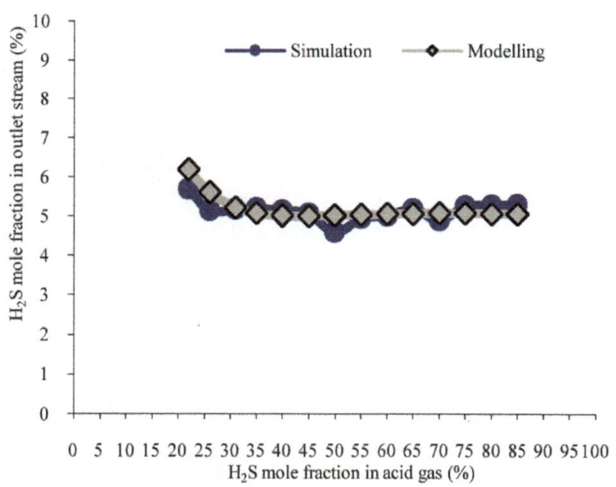

Figure 5. Estimation for H_2S content in outlet stream vs. H_2S mole fraction in feed stream.

in the outlet decreases. As can be seen in the figure, the reduction of H_2S in the effluent is agreement with increase in H_2S content in acid feed gas. For this case, model prediction is more reliable than simulation.

5.2. Effect of Inlet Temperature

In this section, temperature effects have been investigated. We can predict the effect of preheating on the furnace temperature and conversion in furnace. **Figure 6(a)** shows the variation of furnace temperature vs. inlet temperature of acid gas. According to the figure, furnace temperature increases 4.4°C by 10°C increase in inlet temperature. Therefore, if we can increase the design temperature (52°C) to 252°C, furnace reactions will take place in 914°C.

Figure 6(b) shows the H_2S conversion by increase in the temperature. As can be seen from **Figure 6**, H_2S

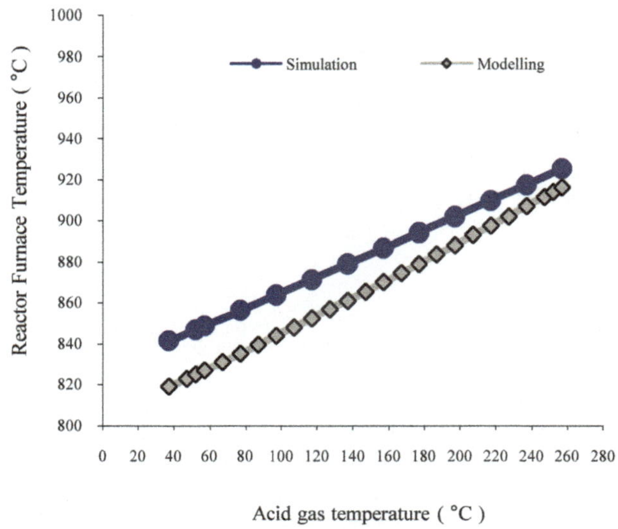

(a)

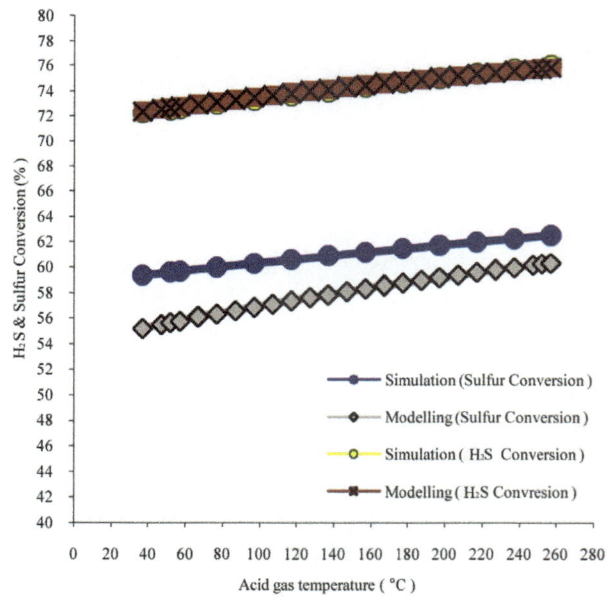

(b)

Figure 6. (a) Modified Claus furnace temperature prediction for Khangiran Gas refinery vs. inlet acid gas temperature; (b) H_2S and sulfur conversion vs. inlet acid gas temperature.

conversion increases by 0.156% when inlet temperature increases 10˚C. Also this figure demonstrates that H_2S conversion in furnace would increases from 72.66% to 75.74% in preheated feed (252˚C).

Also 10˚C increase in temperature of inlet air results into 2.52˚C increases in furnace temperature, 0.103% increase in H_2S conversion and 0.153% increase in sulfur conversion. If both acid gas feed and inlet air preheated separately and equally 10˚C, reaction furnace temperature increases 7.1˚C, H_2S conversion 0.21% and sulfur conversion 0.31%.

From a theoretical point of view, there is an optimal temperature in the furnace reactor to get the more efficient performance, maximizes sulfur production and H_2S conversion as reported in the previous sections for Claus process. A solution for this problem is fuel gas injection to the furnace in order to increase the temperature. By one percent increase fuel content in the inlet gas, furnace temperature would increase 30˚C - 50˚C. Calculation showed that 2000 Sm^3/hr fuel injections in acid gas feed (50000 Sm^3/hr) for Shahid Hasheminejad Refinery, furnace temperature would increase 130˚C. More hydrocarbon content in the feed will produces more CS_2 and COS in the furnace. Increasing flow rate causes decrease in furnace capacity. It was observed the positive effect of fuel injection by increasing the temperature, led to reduction in plant capacity.

6. Conclusion

The reactor furnace for an industrial three-stage straight-through sulfur plant with identical feed gas composition and operating conditions was molded and compared. The results of the modeling and steady state simulation have been presented in **Table 5**. The results showed that H_2S conversion could be promoted by an increase in hydrogen sulfide content in the feed gas. Therefore if we could enhance the H_2S concentration, sulfur conversion and overall efficiency of the furnace would improve. This also could lead to decomposition of aromatic compounds such as BTEX, additionally furnace temperature would increase. In order to increase furnace temperature, fuel injection is possible but, it must be optimized to prevent plant capacity reduction. On the other hand, reduction in CO_2 and N_2 inlet flow helps to reduce the volume of effluent and increases the furnace capacity in the Claus plant. Also results demonstrated that by utilization of heat input (preheated feed and air) in a furnace of a plant, the performance of the reactor would improve.

REFERENCES

[1] M. Sassi and A. K. Gupta, "Sulfur Recovery from Acid Gas Using the Claus Process and High Temperature Air Combustion (HiTAC) Technology," *American Journal of Environmental Sciences*, Vol. 4, No. 5, 2008, pp. 502-511.

[2] K. Karan, "An Experimental and Modeling Study of Homogeneous Gas Phase Reactions Occurring in the Modified Claus Process," Ph.D. Thesis, University of Calgary, Calgary, 1998.

[3] W. D. Monnery, *et al.*, "Modeling the Modified Claus Process Reaction Furnace and the Implications on Plant Design and Recovery," *The Canadian Journal of Chemical Engineering*, Vol. 71, No. 5, 1993, pp. 711-724.

[4] K. A. Hawboldt, "Kinetic Modeling of Key Reaction in the Modified Claus Plant Front End Furnace," Thesis, University of Calgary, Calgary, 1998.

[5] S. Asadi, *et al.*, "Effect of H_2S Concentration on the Reaction Furnace Temperature and Sulphur Recovery," *International Journal of Applied Engineering Research*, Vol. 1, No. 4, 2011, p. 961.

[6] H. G. Paskall, "Capability of the Modified-Claus Process: A Final Report to the Department of Energy and Natural Resources of the Province of Alberta," Western Research & Development, Cheyenne, 1979.

[7] H. G. Paskall, "Reaction Furnace Chemistry and Operational Models," Sulphur Recovery, Western Research, Calgary, 1981.

[8] P. D. Clark, *et al.*, "Mechanism of CS_2 Formation in the Claus Front-End Reaction Furnace," *Proceedings of the 47th Annual Laurance Reid Gas Conditioning Conference*, Norman, 2-5 March 1997, pp. 321-337.

[9] P. D. Clark, *et al.*, "Mechanisms of CO and COS Formation in the Claus Furnace," *Industrial Engineering and Chemistry Research*, Vol. 40, No. 2, 2001, pp. 497-508.

[10] P. A. Tesner, M. N. Nemirovskii and D. N. Motyl, "Kinetics of the Thermal Decomposition of Hydrogen Sulphide at 600 - 1200˚C," *Kinetics and Catalysis*, Vol. 31, 1990, p. 1232.

[11] B. Goar, *et al.* "Sulfur Recovery Technology," *Energy Progress*, Vol. 6, No. 2, 1986, pp. 71-75.

[12] J. A. Sames and H. G. Paskall, "Simulation of Reaction Furnace Kinetics for Split-Flow Sulphur Plants," Sulphur Recovery, Western Research, Calgary, 1990.

[13] E. A. Luinstra and P. E. Haene, "Catalyst Added to Claus Furnace Reduces Sulfur Losses," *Hydrocarbon Process*, Vol. 68, No. 7, 1989, pp. 53-60.

[14] J. A. Sames, *et al.*, "Evaluation of Reaction Furnace Variables in Modified Claus Plants," Sulphur Recovery, Western Research, Calgary, 1990.

Appendix

The Western Research (Sames and Paskall) correlations are given below:

$R(CO) = 0.002A^{0.0345} \exp(4.534\ A)$

$R(H_2)$ = fraction of furnace inlet H_2S that cracks to H_2 and S

$R(H_2) = 0.056\ (\pm 0.024)$

R(COS) = fraction of furnace inlet carbon that forms COS

$R(COS) = 0.01$ tangent (100A) for $0 \leq A \leq 0.86$

$R(COS) = 0.143$ for $A > 0.86$

$R(CS_2)$ = fraction of furnace inlet hydrocarbon that forms CS_2

$R(CS_2) = 2.6\ A^{0.971} \exp(-0.965\ A)$

R(S) = fraction of furnace inlet H_2S that forms elemental S

$R(S) = 1.58A^{1.099} \exp(-0.73A)$

where A = mole fraction of H_2S in the acid gas feed on a dry basis.

Nomenclature

f_i: Molar flow rate of ith compound
X: Mole fraction of compound
$R(i)$: fraction of furnace inlet of ith compound
K_p: Equilibrium constant
T: temperature
π: furnace total pressure
$C_{p,i}$: Heap capacity of ith compound
ΔH_{rxn}: Reaction enthalpy
$H_{i,298}{}^{0}$: Standard enthalpy of ith compound formation
W.H.B: Waste Heat Boiler

10

Theoretical Evaluation of Two-Phase Flow in a Horizontal Duct with Leaks

Morgana de Vasconcellos Araújo[1], Severino Rodrigues de Farias Neto[1], Antonio Gilson Barbosa de Lima[2]

[1]Department of Chemical Engineering, Center of Science and Technology, Federal University of Campina Grande (UFCG), Campina Grande, Brazil

[2]Department of Mechanical Engineering, Center of Science and Technology, Federal University of Campina Grande (UFCG), Campina Grande, Brazil

ABSTRACT

The transport of oil and its derivates are done, mostly, by pipeline. The time to detect leaks has to be short for preventing big disasters in the nature and decreasing losses for industries. The techniques available for leak detection vary from visual inspection to the use of computational techniques such as mathematical modeling. This paper aims to study the fluid dynamics of two-phase flow (water-oil) in the pipe with leakage. The equations of the mass and momentum conservation are numerically solved by using the ANSYS® CFX commercial code with the aid of a structured mesh of a horizontal pipe with three holes of leaks. The Eulerian-Eulerian model was adopted considering the oil as continuous phase and water as dispersed phase, and constant fluid properties. With profiles of pressure and volume fraction along the time in the pipe, the influence of leakage on the single-phase (oil) and two-phase (water-oil) was evaluated.

Keywords: Two-Phase Flow; Numerical Simulation; Leakage; Pipeline

1. Introduction

The transportation of oil and its derivatives are mostly conducted through pipelines that connect production facilities, refineries and, in some situations, the consumer centers.

The materials used for making pipes come through technological improvements, where there is a significant use of materials based on special steel, lighter and stronger. However, despite these advances, there are still problems with leaks in pipelines generating great interest from the oil industry in view of the high costs incurred by financial services, potential risks and environmental costs.

Environmental disasters related to the oil spill in addition to degrading the environment, are responsible for spending millions of dollars in remediation.

In oil activities, particularly in transportation by pipelines, accidents have happened, causing financial and environmental losses.

There are currently a variety of techniques available for the detection of leaks, ranging from simple physical inspection by acoustic methods. Zhang [1] classified the detection methods into three categories: observation (perhaps the simplest and most ancient, is conducted through a visual inspection noting any ponding on the soil surface or anomalous growth of vegetation), direct detection (devices are used for the detection and location of leakage) and indirect detection (software is used based on mathematical models which allow to perform detection by means of data flow like pressure, temperature, mass flow rate, etc.).

The faster the identification of a leak in a pipeline, the faster valves are closed and the pumps will stop and, consequently, the greater the chances of avoiding a catastrophe are. However, in order to do the detection and precise identification of the position of a leak, it is necessary to know the behavior(s) of fluid(s) within the duct which allows determination of pressure drop between two points being evaluated.

According to Buiatti [2] the kind of leaks appearing in pipe networks can be divided into two classes:

1) Leak by "breaking" the tube—Occurs less frequently, but it is dangerous due to the amount of product spilled in the vicinity of the leak. However, these disrup-

tions are easy to detect because they are accompanied by pressure losses and volumetric differences.

2) Leakage of small proportions—Small leaks around 5 liters per hour are difficult to detect due to their sizes and can cause large losses of products to get noticed. Maybe provided by corrosion, fatigue of the material that makes up the pipe or by failure is in welds. There are few methods capable of detecting leaks of this order.

The occurrence of a leak surely results in a disturbance in the behavior of pressure in the pipe and this disturbance can describe the location, the size and the altitude of the leak. A dynamic associated with the leak is propagated along the pipe from the point of the leak and with a variable speed, being perceived in different moments by sensors on the pipe. The location of the leak in the pipe is calculated from the time interval between the perceptions of the dynamic data from sensors coupled to the distance between the sensors. Azevedo [3] proposed the division of the temporal evolution of a leak in three steps:

1) Pre-Leak: it corresponds to the behavior of the flow in the pipe before the leak phenomenon and reflects the normal steady-flow conditions;

2) Transitional: it corresponds to the behavior of accommodation of the flow in the pipe from the initial leak until the moment of reaching a new steady state;

3) Post-Leak: it corresponds to the behavior of the flow in the pipe after the occurrence of a leak and the stabilization of the flow conditions, reflecting the flow stationary conditions on pipes with the presence of a leak

Different works related to leakage in the pipe has been reported in the literature [4-8].

Agbakwuru [4] discusses about the challenges and technologies available for viewing and examination of leaks in pipelines located in the sea and under the conditions of poor visibility. It is proposed the use of Remotely Operated Vehicles with an optical eye as a tool to perform visual inspection of the pipeline and assist in the challenge of repairing the oil spill. The techniques used for pipelines submerged in water of poor visibility are based on tools used: hydrophone/Acoustic sensor, direct hydrocarbon leak detection, fluorescence. This author did a research about the oil/gas industry in the Nigeria Niger Delta Region. The causes of pipeline leaks were classified in operational, structural, unitended and intended damages. The Autor has shown the big values of leakes caused by sabotage in the Delta Region from the year 2005 to 2011. To reduce cost and improve effectiveness of systems of ducts in Delta Niger, it was proposed a combination of two methods of leak detection: the impressed alternating cycle current (IACC) for pipes with a small diameter (4 to 8 inches) and short length and the acoustic method for pipes with large diameter and long length.

Braga [5] evaluated air and water flow in a pipe with 1250 meters length. Coupled pipe flow is a data acquisition system, where this detection system has four pressure transducers connected to a computer equipped with an ADA converter board (Analog-digital-Analog). Three different types of flow regimes were studied: isolated bubbles of air without leak of the fluid, flow of a single air bubble in the presence of leak, and continuous flow air-water with leak. The experiments showed that the presence of air in the pipe generates reflections of waves (spring effect) that interferes with the detection of the leak. The author concluded that depending on the amount of air in the pipe, this can act as a damper of the shock wave, reducing the impact produced by leakage and thus, decreasing the sensitivity of the system.

The pressure transducers along the pipe convert pressure variations into variation of an electrical magnitude (voltage or current) to the PLC (programmable logic controller) which can work with this information. To drain compressed air Bezerra [6] built a pipe in PVC (polyvinyl chloride) with a diameter of 2.5 cm and a length of 5.56 meters. Two transducers were placed along the pipe and the transient pressure (pressure variation in the pipe with time) was measured simultaneously. The method detected the leak in the pipe, but we observed a difficulty in obtaining the location of the leak and this was explained by the fact that the two transients occur with a very small time difference, which was justified by the fact that the pipe had a short length. Bezerra [6] concluded that the program was successful in locating the leak which was necessary for the transducers installed at a distance farther than 12.24 meters from each other.

Garcia *et al*. [7] represented with the aid of commercial software COMSOL Multiphysics 4.0a®, a section of a pipe with 2 feet in length, 10 centimeters in diameter and a orifice leak of 4 mm in diameter. A cylindrical volume was linked perpendicularly to orifice surface to observe the leak jet. At the beginning, the pipe was completely filled with water at an absolute pressure of 101,325 Pa and the cylindrical volume had air at an absolute pressure of 0 Pa. Initially a leak test with a bench trial was held, where there was water inside the pipe and air was injected to study the flow behavior in the region of leakage. The numerical simulations were performed with the help of a mesh with 21.376 tetrahedral elements, with injections of air with a speed of 2 m/s and 10 m/s during a time interval of 1 second, considering laminar flow. This author analyzed the magnitude of the velocity profiles in a central plane of the section of duct and observed that in the region of the leak there was an increase in speed, and, in the simulation with the higher speed, the values in the leakage region reached 425.5 m/s featuring a supersonic behavior in this region.

Sousa *et al*. [8] simulated a rising flow of an isothermic water/oil two-phase mixture. The domain of flow is a vertical duct 8 meters long and a diameter of 15 cm, with an orifice leak of 6 mm located at 4 m of inlet section. The mesh was done on the domain and has 327,327 hexahedral elements. The simulation was performed in ANSYS CFX 11.0. The authors varied the volumetric fraction of oil in the inlet flow duct in amounts of 0.75 to 1 and it was noted that the smaller was the oil fraction, the major was the water fraction in the mixture and, consequently, the major was the mass flow through the orifice leak. The inlet velocities of mixture were varied and it was possible to realize that from 0.75 to 1.5 m/s the rate between the mass flow of leakage and total mass flow in the inlet section decreased with the increase of the velocity, but between 1.5 and 2 m/s the effect was the opposite due to inertial forces of flow being smaller than the forces caused by pressure drop from inlet to the leak. Complementing the authors' conclusions, it is observed that the emergence of the leak causes a pressure drop which is directly proportional to the volume fraction of water in the mixture and the flow rate of the same.

In this context, the proposed research is to contribute to the study of leakage and help to understand the phenomenon involved via techniques of computational fluid dynamics. This work aims to study numerically the two-phase flow (water-oil) in a horizontal tube in the presence of leakage.

2. Methodology

2.1. Study Domain

To study the influence of leak in the hydrodynamics of the oil-water two-phase flow in a horizontal pipe, was adopted a study domain. The representation of the computational domain was done with the aid of ICEM-CFD 12.1 with points, curves and surfaces (**Figure 1**). The study area consists of a horizontal pipe with 10 meters long and 20 inches in diameter. The tube contains two holes of 1.6 cm diameter at the points (x = 5 m, y = 0.1 m, z = 0 m) and (x = 7.5 m, y = 0.1 m, z = 0 m) and a third hole of 3 cm diameter at the point (x = 5 m, y = −0.1 m, z = 0 m).

The representative mesh of the pipe was generated using the block strategy, which a single block is initially created involving the area of study and then the block is subdivided into several others. This strategy allows greater control of mesh refinement in regions desirable. The mesh was refined on the geometry and contains approximately 327,000 control volumes as shown in **Figure 2**.

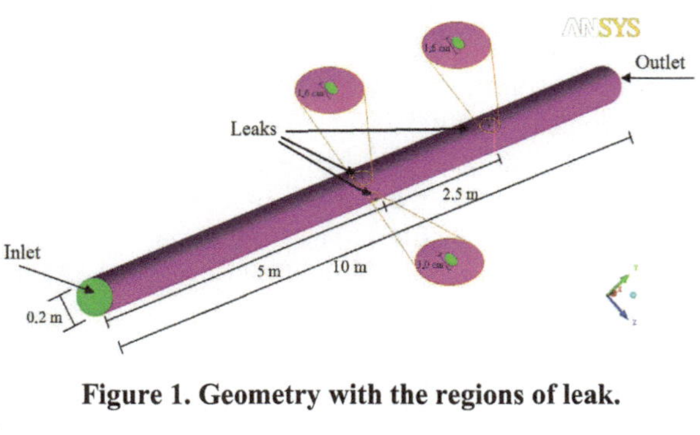

Figure 1. Geometry with the regions of leak.

2.2. Mathematical Modeling

In the studied flows, both monophase (oil) and the two-phase (water-oil) flows, they are based on the following considerations:

1) Transient and laminar flow;

2) The fluid is newtonian and incompressible and the physicochemical properties are constants;

3) Isothermal process;

4) There is no occurrence of chemical reactions;

5) It was considered the gravitational effect;

6) No transfer of mass and momentum at the interface water and oil phases;

7) The interfacial forces of no drag (forces of lift, wall lubrication, virtual mass, pressure and turbulent dispersion of solid) were neglected.

The equations that describe the flow was withdraw from CFX 12.1®.

The equation of mass conservation is:

$$\frac{\partial(f_\alpha \rho_\alpha)}{\partial t} + \nabla \cdot (f_\alpha \rho_\alpha \vec{U}_\alpha) = 0 \qquad (1)$$

where the the greek subscript α represents the involved phase in two-phase mixture of water/oil; f, ρ, and $\vec{U}$ are respectively the volumetric fraction, density and velocity vector. For the phase α, the velocity vector is given by $\vec{U}\alpha = (u, v, w)$.

The equation of momentum is:

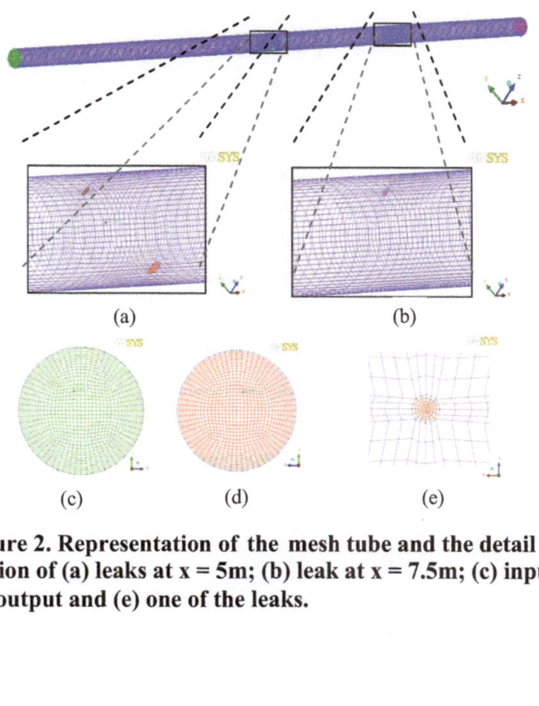

Figure 2. Representation of the mesh tube and the detail of section of (a) leaks at x = 5m; (b) leak at x = 7.5m; (c) input; (d) output and (e) one of the leaks.

$$\frac{\partial\left(f_\alpha \rho_\alpha \vec{U}_\alpha\right)}{\partial t}+\nabla \cdot\left[f_\alpha\left(\rho_\alpha \vec{U}_\alpha \otimes \vec{U}_\alpha\right)\right]= -f_\alpha \nabla p_\alpha+\nabla \cdot\left\{f_\alpha \mu_\alpha\left[\nabla \vec{U}_\alpha+\left(\nabla \vec{U}_\alpha\right)^T\right]\right\}+\vec{S}_{M\alpha}+\vec{M}_\alpha \tag{2}$$

where p is the pressure, $\vec{S}_{M\alpha}$ represents the external forces acting on the system per unit volume, $\vec{M}_\alpha$ describes the overall force per unit volume (interfacial drag forces).

The interfacial drag force is given by:

$$\vec{M}_\alpha=\frac{3}{4}\frac{C_d}{d_p} f_\beta \rho_\alpha\left|\vec{U}_\beta-\vec{U}_\alpha\right|\left(\vec{U}_\beta-\vec{U}_\alpha\right) \tag{3}$$

where d_p is the particle diameter and C_d is the drag coefficient calculated using the Schiller-Neumann correlation, as follows:

$$C_d=\max\left[\frac{24}{\mathrm{Re}}\left(1+0.15\,\mathrm{Re}^{0.687}\right),0.44\right] \tag{4}$$

It was adopted the model of particle (Eulerian-Eulerian approaches) where oil behaves as a continuous fluid and water as spherical particles dispersed.

The simulations occurred in transient regimes and time steps were applied with 15 iterations each. In a period 0 to 0.01 second the flow time step used was 0.001 second; from 0.01 to 0.1 second the time step value increased to 0.01 second and after elapsed time 0.01 second the time step became 0.1 second.

The initial and boundary conditions are:

1) At the inlet:

a) The monophase flow has the characteristics:

i) Speed of oil-u_x = 0.2 m/s;

ii) Entrance length: Le = 1.43 m.

b) For the two-phase flow (water-oil) we have:

i) The speed of the mix flow of water and oil-u_x = 0.2 m/s;

ii) Volume fraction-f_{oil} (**Table 1**);

2) At the output: prescribed static pressure equal to 101,325 Pa;

3) At the wall:

a) No slip: $u_x = u_y = u_z = 0$;

b) The smooth tube;

c) The static walls.

d) In the leak: average static pressure with value 101,325 Pa.

4) Pressure of reference: 0.0 Pa.

2.3. Physico-Chemical Properties

The properties of the fluids used in the present work, are shown in **Table 2**.

2.4. Case Studies

The simulations were performed to analyze the influence of the oil volume fraction in the amount of oil leaked, as the disorder caused by the presence of two leaks in hydrodynamic flow. The influence of orifice size of leak in the monophase flow of oil and two-phase flow of oil and water were also studied. The constructed geometry can be maintained in its original position or undergo a 180° degrees rotation around the x axis, depending on each case, so that in all cases studied the opening of the leakage always occurs at the bottom of the pipe. The case studies are listed in **Table 1**. The simulations were investigated using computers with Quad-Core Intel Xeon Processor E5430 Dual 2.66GHz with 8GB of RAM.

The leakage in 5 m from the fluid inlet section (hole diameter = 1.6 cm) is called Leak 1, the leakage in 7.5 m

Table 1. Data used in the simulations.

Case	Volumetric fraction		Pressure at the leakage Sections [Pa]		
	Oil f_{oil}	Water f_{water}	Leak 1	Leak 2	Leak 3
1	1.000	0.000	101,325	-	-
2	1.000	0.000	101,325	101,325	-
3	0.990	0.005	101,325	-	-
4	0.990	0.005	101,325	101,325	-
5	0.975	0.025	101,325	-	-
6	0.975	0.025	101,325	101,325	-
7	0.950	0.050	101,325	-	-
8	0.950	0.050	101,325	101,325	-
9	0.925	0.075	101,325	-	-
10	0.925	0.075	101,325	101,325	-
11	0.900	0.100	101,325	-	-
12	0.900	0.100	101,325	101,325	-
13	1.000	0.000	-	-	101,325
14	0.950	0.050	-	-	101,325
15	0.900	0.100	-	-	101,325

Table 2. Fluid properties.

Properties	Continuous phase (oil)*	Disperse phase (water)**
Density (kg/m³)	868.7	997
Molar mass (kg/kmol)	105.47	18.02
Viscosity (Pa·s)	0.17	0.0008899
Particle diameter (mm)		0.001
Surface tension (N/m)	0.03	

*Source: Cunha [9]; **Source: Welty *et al.* [10].

from the fluid inlet section (hole diameter = 1.6 cm) is called Leak 2 and the leakage in 5 m from the fluid inlet section (hole diameter = 3.0 cm) is called Leak 3.

3. Results and Discussion

3.1. Fluid Volumetric Flow Rate Analysis

Figure 3 shows the evolution of the oil mass flow rate on the leak located at 5 m from inlet, with diameter of 1.6 cm, located in bottom side of pipe (**Figure 1** rotated 180° around x axis). **Figure 3(a)** shows the evolution for cases where only the leak located at 5 m from inlet is activated (Cases 1, 3, 5, 7, 9 and 11) and the **Figure 3(b)** shows the same trend for the cases where the leaks at 5 m and at 7.5 m from inlet, with the same dimensions, are enabled (Cases 2, 4, 6, 8, 10 and 12). It is observed a similar behavior of the curves. With the decrease in volume fraction of oil entering the duct, it is observed a reduction of oil flow through the orifice leak. At the initial instants of the process, oil mass flow rate increase quickly until 0.4 s, when this time reaches the equilibrium condition (steady-state flow).

Figure 4 was done for continue the analysis of the oil mass flow rate at the hole cited before. It's notable that the presence of the second leak disturbs the behavior of the flow, changing the oil mass flow rate on the first leak. The solid line corresponds to the oil mass flow rate for the cases when only the leak at 5 meters from the input section is activated. The dashed line corresponds to the oil mass flow rate in the same region, but when the first

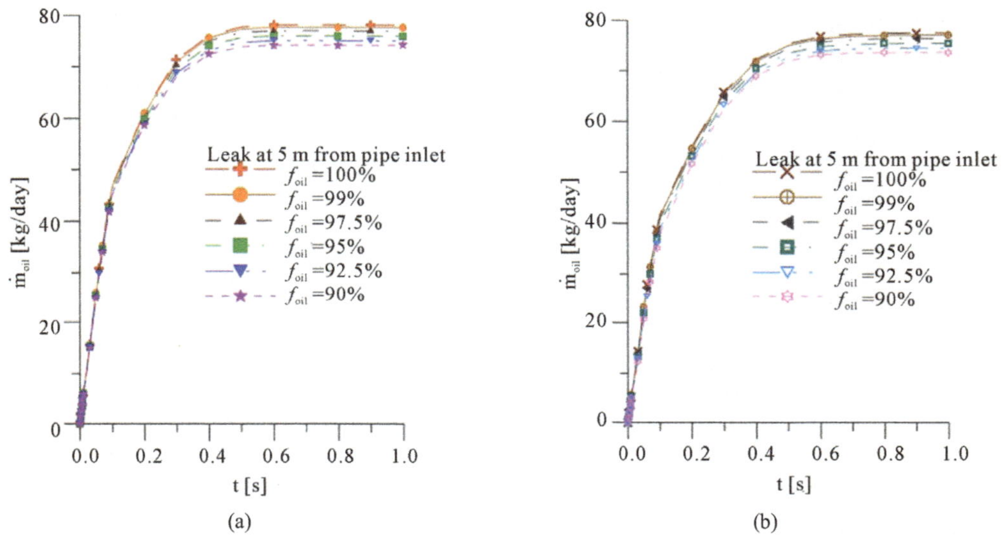

Figure 3. Oil mass flow in the orifice leak with diameter of 1.6 cm located at 5 m from the inlet versus time for different oil volumetric fractions injected into the pipe (a) Cases 1, 3, 5, 7, 9 and 11 and (b) Cases 2, 4, 6, 8, 10 and 12.

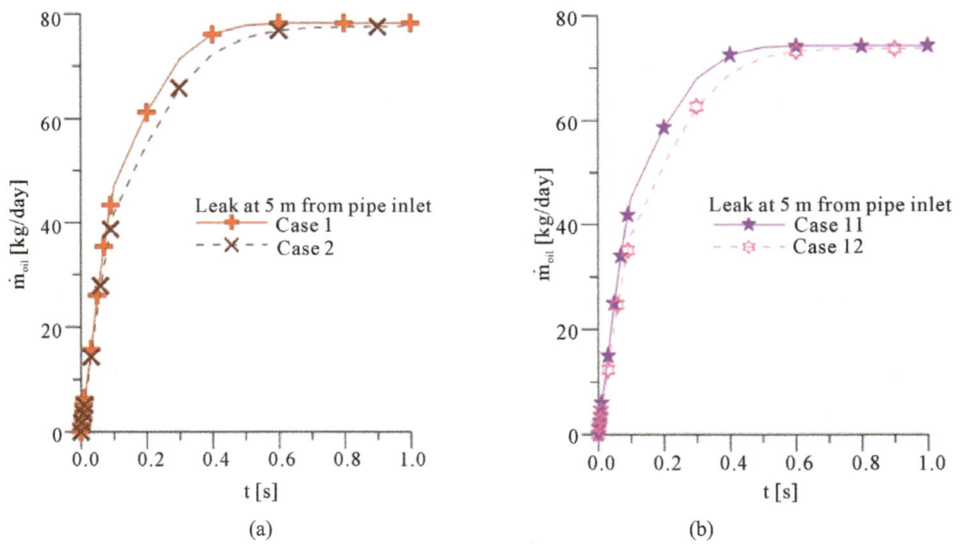

Figure 4. Oil mass flow rate in the orifice leak with diameter of 1.6 cm located at 5 m from the inlet versus time for different oil volumetric fractions injected into the pipe (a) Cases 1 and 2 and (b) Cases 11 and 12.

and second leaks are activated simultaneously. Comparing the two curves mentioned above, for the Cases 1 and 2 (**Figure 4(a)**), we see that the flow described by the continuous curve is stabilized higher and faster than the represented by the dashed curve flow. The same behavior is observed for the Cases 11 and 12 (**Figure 4(b)**).

For a better view of the graph (**Figures 5** and **10**), we adopted the dimensionless time as follows:

$$t^* = t / t_{\text{total}} \quad (5)$$

were t is the time and t_{total} is the total time of simulation. For the cases 01, 07 e 11 were adopted the total time of 1 second. For the cases 13, 14 and 15 the value is 1.8 s.

To analyse the effect of the orifice leak diameter on the oil mass flow rate, the cases 1, 7, 9, 13-15 are studied (**Figure 5**). To study the cases 13-15, where the leak is located in the bottom side of the pipe, the original position is maintained like the **Figure 1** shows, *i.e.*, the pipe is not rotated.

By increasing the diameter of the leak, it is observed that there is an increase in the oil mass flow rate in the leak, the continuous curve of **Figure 5(a)** is equivalent to the oil mass flow in Case 1, where the leak with a diameter of 1.6 cm and located at 5 meters from the input section is activated. The dashed curve refers to the oil mass flow rate in Case 13 where the leak with a diameter of 3.0 cm and located at 5 meters from the input section is activated. It is easily understood that the volumetric flow in the orifice with a diameter of 3.0 cm (dashed curve) is higher than for an orifice with a diameter of 1.6 cm (continuous curve). It is also interesting to note that by increasing the diameter of the orifice, more time is necessary to the oil mass flow rate to achieve the equilibrium condition. The same behavior is observed for Cases 7 and 14 (**Figure 5(b)**) and the Cases 11 and 15 (**Figure 5(c)**).

Figure 6 illustrates the result of the total mass flow rate in leakage as a function of oil volumetric fraction at the entrance of pipe, after 1 second of flow for leak with

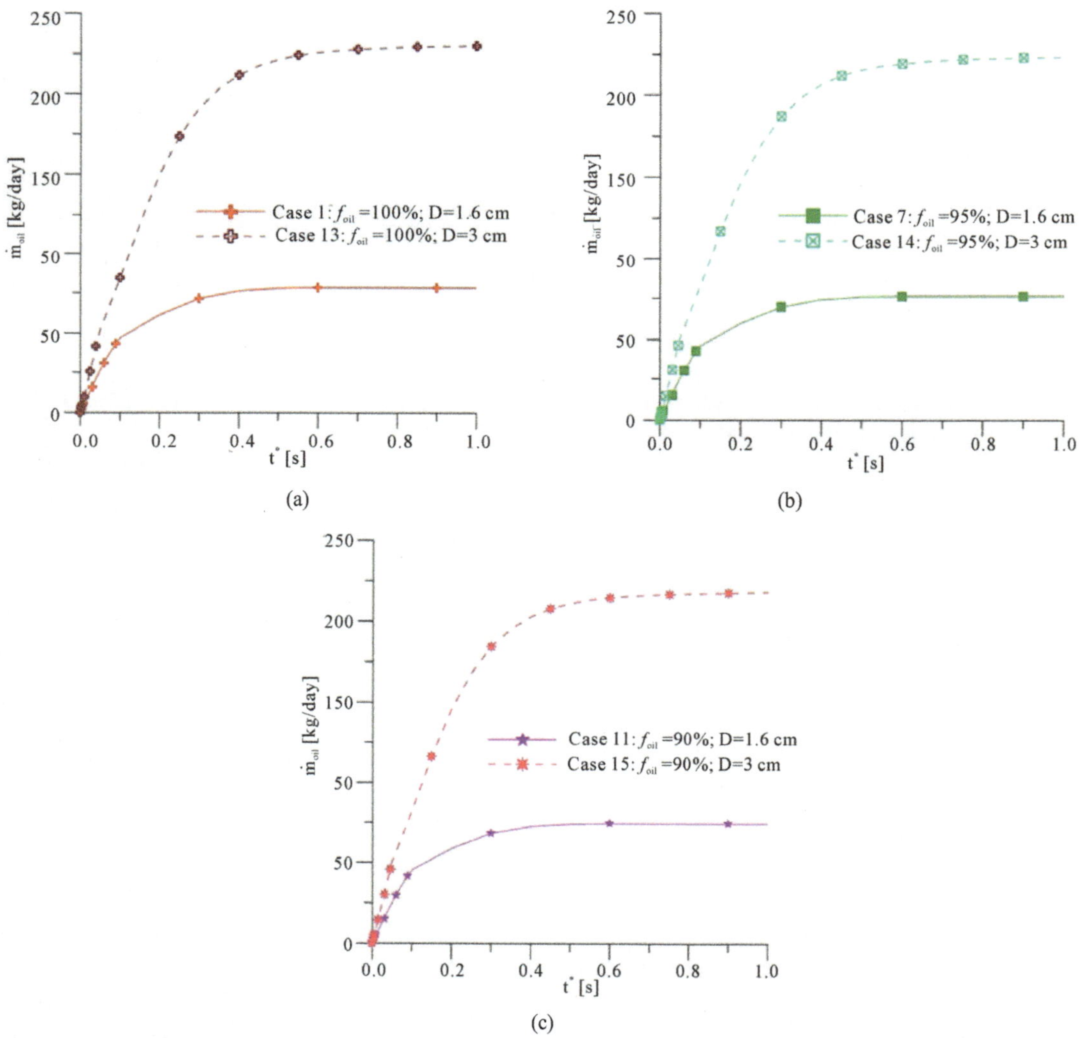

Figure 5. Oil mass flow rate in the leak versus dimensionless time for the leaks with different diameters.

a diameter of 1.6 cm and after 2 seconds of flow for leak with a diameter of 3.0 cm. It's possible to see that the smaller oil volumetric fraction in the mixture, the higher is the total mass flow rate leaving the orifice, which can be attributed to a reduction in viscosity of the mixture. As expected, the bigger the orifice leak the greater the loss of the fluids.

3.2. Pressure Drop Analysis

Figure 7(a) shows the evolution of the pressure drop between the entrance and the exit of the pipe as a function of time for different feeds of oil volume flow fractions when one leak is active (Cases 1, 3, 5, 7, 9 and 11). The similar behavior is observed, but with different magnitudes, for each value of the oil volumetric fraction and after 0.6 s of leaking there is a stabilization of the behavior of pressure. One fast drop in the pressure is verified at the initial instants of the flow due to presence of the leak.

The same behavior pressure drop between the inlet and outlet section is observed for cases where there are two active leaks, **Figure 7(b)** (Cases 2, 4, 6, 8, 10 and 12). However it is observed that the stabilization of the behavior of pressure occurs only after 0.7 s leak started. This behavior is explained by the fact two leaks in a same pipe cause a greater change as compared with the same pipe containing only one leak.

In order to observe the behavior of pressure along the duct, taken up average pressure values in different planes yz distributed along the duct (0, 1, 2, 3, 4, 4.5, 5, 5.5, 6, 7, 7.5, 8, 9 and 10 m). With these average pressures were traced curves represented in **Figure 8** for Cases 1, 3, 5, 7, 9 and 11 and **Figure 9** for the cases 2, 4, 6, 8, 10 and 12 which shows the behavior of pressure along the duct. These curves were taken at the beginning of the leak (t = 0.001 s), t = 0.05 s, t = 0.1 s, t = 0.3 s and at the time that the behavior of the flow has been stabilized (t = 1 s). In all cases, the behavior change in pressure before and after the leak can be observed clearly.

By activating the leaks of 1.6 cm and 3 cm diameter, it is expected that they disturb the behavior of the pressure

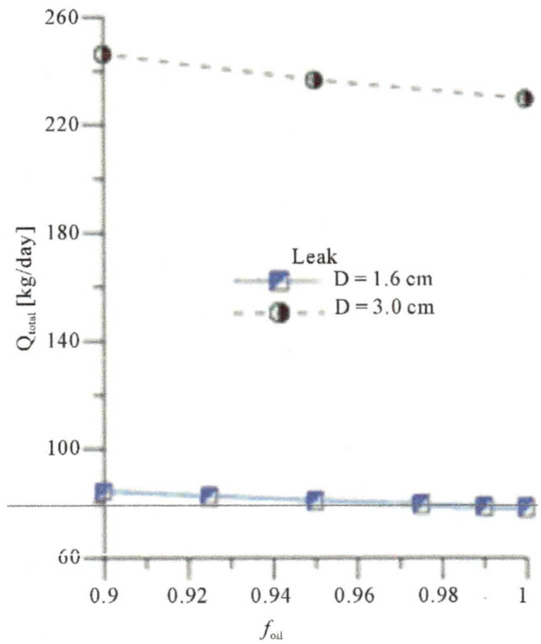

Figure 6. Oil-water mixture mass flow rate in the leak as a function of the oil volumetric fraction at the pipe inlet for leak with different diameters.

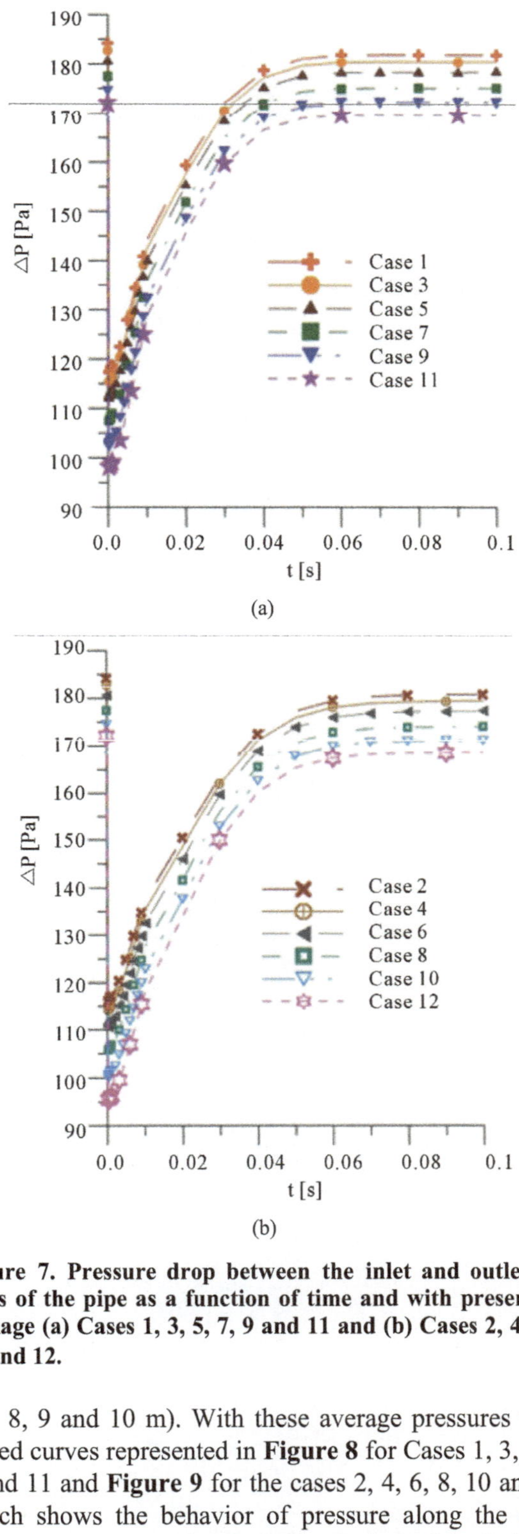

Figure 7. Pressure drop between the inlet and outlet sections of the pipe as a function of time and with presence of leakage (a) Cases 1, 3, 5, 7, 9 and 11 and (b) Cases 2, 4, 6, 8, 10 and 12.

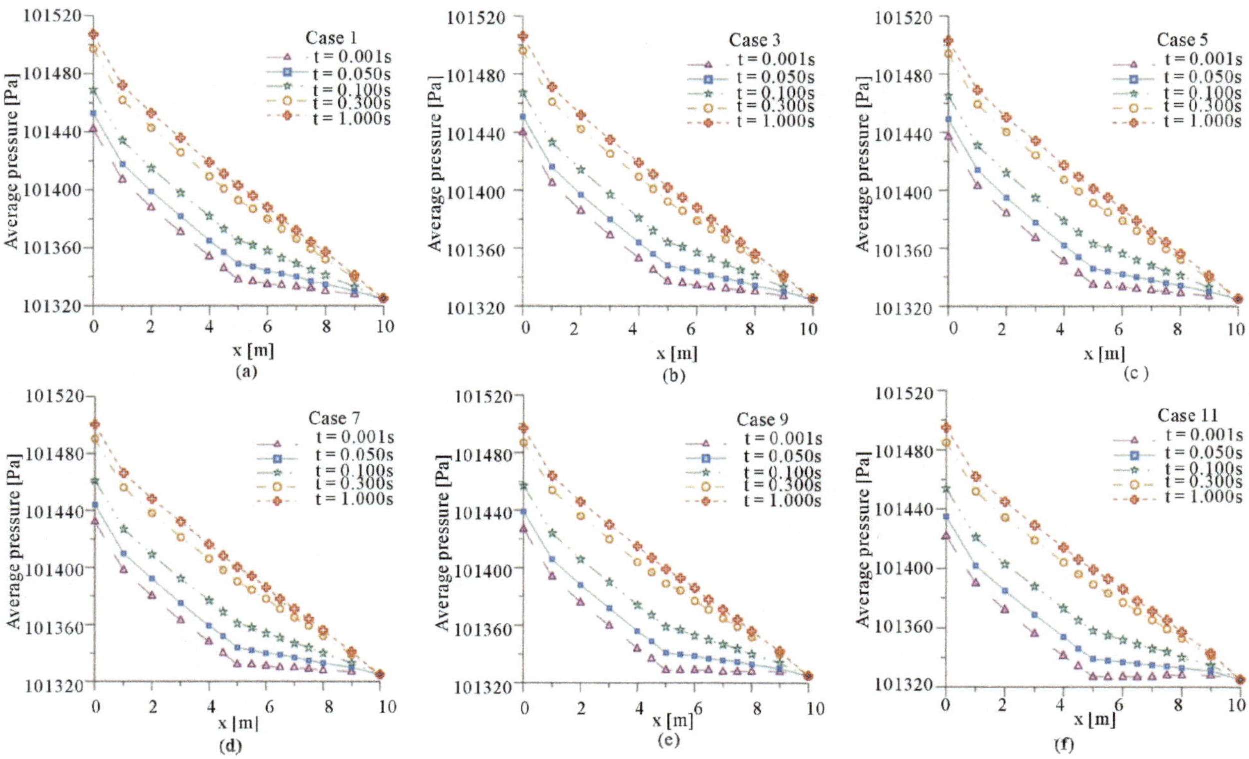

Figure 8. Average pressure along the duct at different flow time for the cases with only one leak.

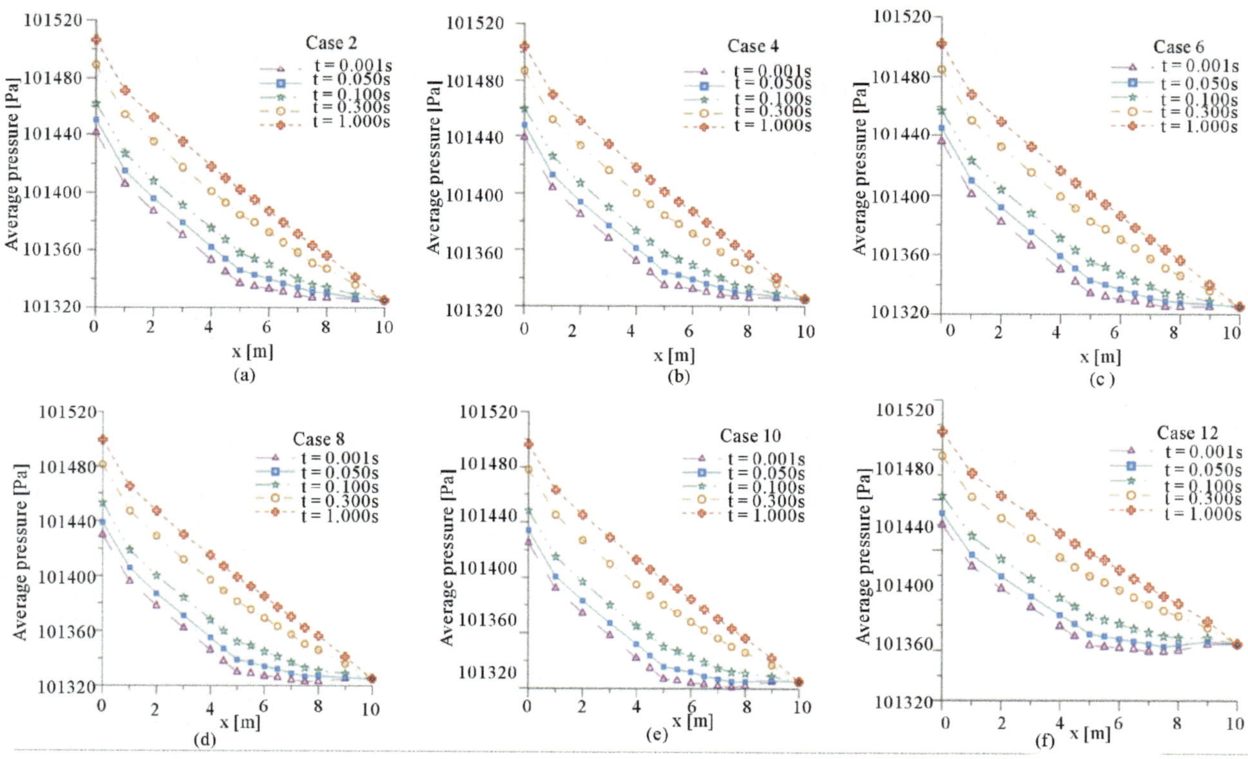

Figure 9. Average pressure along the duct at different flow time for the cases with two leaks.

in the duct. The results have proven that after a certain period of time the pressure in the duct turn back to stabilize. **Figure 10** (oil monophase flow) shows that by varying the diameter of the leak, the pressure stabilization period also varies. The greater the leakage, the greater will be the disturbance, and therefore the time

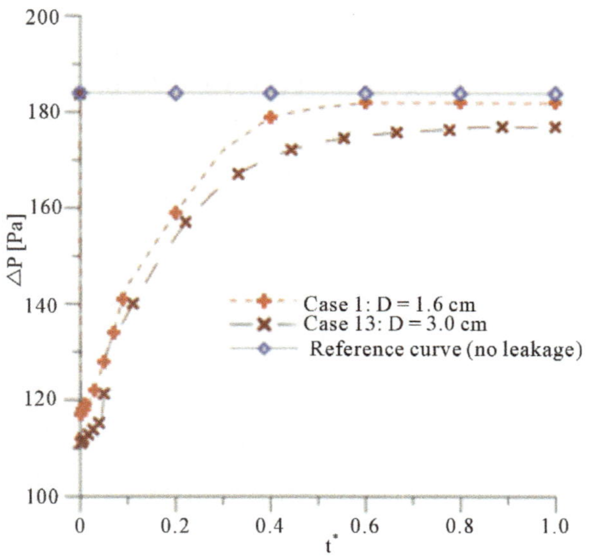

Figure 10. Average pressure drop between the inlet and outlet sections as a function of dimensionless time in the presence of leakage (Cases 1 and 13) and in the absence of leakage (reference curve).

taken to stabilize the pressure in a new value.

4. Conclusions

With the results of numerical simulation of single-phase flow (oil) and two-phase flow (water-oil) in pipes, it can be concluded that:

1) The proposed mathematical model was able to predict the leak in a horizontal pipe, evaluating the effect of the position, sizes and numbers of the leak on the fluid dynamic behavior of the mono and two-phase flows;

2) The volume fraction of oil in two-phase mixture injected into the duct affected the amount of oil leaked;

3) The presence of a second leak on downstream of the initial leakage on duct affected the oil flow of the first one, which became established with a lower flow rate compared to the situation where the first leak occured alone;

4) By increasing the diameter of the leak orifice, the greater the volume of fluid leaving the orifice was;

5) There was a slight variation of the pressure values before and after the leak and this varied depending on the fluid composition in the flow;

The time required for the pressure reaches stability after the leak depending on the size and locations of the leak in the pipe.

5. Acknowledgements

The authors would like to express their thanks to Capes, CNPq, FINEP, PETROBRAS, ANP/UFCG/PRH-25 and RPCMOD for supporting this work, and also grateful to the authors of the references cited in this paper that helped in the improvement of quality.

REFERENCES

[1] J. Zhang, "Designing a Cost Effective and Reliable Pipeline Leak Detection System," *Pipeline Reliability Conference*, Houston, 19-22 November 1996, 11 Pages.

[2] C. M. Buiatti, "Monitoring of Tubes by Computational Thecniques On-Line," Dissertation, Department of Chemical Engineering, State University of Campinas, Campinas, 1995.

[3] F. M. de Azevedo, "Proposal of Algorithm for Leak Detection in Pipelines Using Frequencial Analysis of Signals Pressure," Dissertation, Federal University of Rio Grande do Norte, Natal, 2009.

[4] J. Agbakwuru, "Pipeline Potential Leak Detection Technologies: Assessment and Perspective in the Nigeria Niger Delta Region," *Journal of Environmental Protection*, Vol. 2, No. 8, 2011, pp. 1055-1061.

[5] C. F. Braga, "Leak Detection by Computer 'On-Line' in Tubes Transporting Mixture Gás-Liquid," Master Dissertation, Department of Chemical Engineering, State University of Campinas, Campinas, 2001.

[6] B. A. F. Bezerra, "Leak Detection in Gas Pipelines by the Method of Transient Pressure Using CLP and Sensors," Monograph (Conclusion of Specialization Course), Federal University of Pernambuco, Recife, 2008.

[7] F. M. Garcia, J. Giaretton, M. B. Quadri and A. Bolzan, "Fluidodynamic Simulation of Leaking Water in a Section of Duct for Applications in Oil & Gas Industry," *Proceedings of the 6th Brazilian Congress of Research and Development in Oil and Gas*, Florianópolis, 9-13 October 2011, pp. 1-8.

[8] J. V. N. de Sousa, A. G. B. de Lima and S. R. de Farias Neto, "Numerical Analysis of Heavy Oil-Water Flow and Leak Detection in Vertical Pipeline," *Advances in Chemical Engineering and S cience*, Vol. 3, No. 1, 2013, pp. 9-15.

[9] A. L. Cunha, "Advanced Non-Isothermal Recovery of Heavy Oil in Petroleum Reservoirs by Numerical Simulation," Master Dissertation, Department of Chemical Engineering, Federal University of Campina Grande, Campina Grande, 2010.

[10] J. R. Welty, C. E. Wicks and R. Wilson, "Fundamentals of Momentum, Heat, and Mass Transfer," 3th Edition, John Wiley & Sons, Hoboken, 1984.

11

Interaction of Bubbles with Vortex Ring Launched into Bubble Plume

Tomomi Uchiyama[1], Sou Kusamichi[2]
[1]EcoTopia Science Institute, Nagoya University, Nagoya, Japan
[2]Bridgestone Cycle, Ageo, Japan

ABSTRACT

This study is concerned with an experimental exploration for the interactions of bubbles with a vortex ring launched vertically upward into a bubble plume. A vortex ring launcher, composed of a cylinder and a piston, is mounted at the bottom of a water tank. Small hydrogen bubbles are released into still water from a cathode, which is wound around the cylinder outlet, by the electrolysis of water. The bubbles rise by the buoyant force and induce a bubble plume. The water in the cylinder is discharged into the bubble plume by the piston, resulting in a laminar vortex ring convecting along the central axis of the plume. Just after the launch of the vortex ring, the bubbles are spirally entrained into the vortex ring with the roll up of the shear layer. The void fraction within the vortex ring increases with the convection of the vortex ring until a certain displacement of the vortex ring, where the reduction occurs. The vortex ring convects with a constant velocity higher than that in still water. The entrained bubbles reduce the strength of the vortex ring.

Keywords: Vortex Ring; Bubble Plume; Bubble Entrainment; Vorticity; Visualization; PIV

1. Introduction

Gas bubbles released into a liquid induce the liquid flow as they rise due to the buoyant force. Such bubble-driven flow or bubble plume can attain the active contact and mixing between the gas and liquid phases. Therefore, it is utilized in engineering devices handling chemical reaction, bio-process, and coal liquefaction. It is also employed in environmental applications for ice prevention in lakes and purification of water. A number of researches have thus far been performed on bubble plumes. Various methods to predict the entrained liquid flow rate [1,2] and the plume characteristics [3,4] have been proposed. The relation between the meandering motion of rising bubble and the bubble flow rate has also been investigated [5]. The authors [6] carried out a numerical simulation for a plane bubble plume, and they analyzed the large-scale vortical structures as well as the bubble meandering motion. To heighten the performance and efficiency of the devices utilizing bubble plumes, it is effective to increase the interfacial area concentration of the plume. But there are few attempts at such increment.

In single-phase mixing layer, it is well known that the mixing and momentum transport of the fluid is dominated by the large-scale organized eddies. Rightley and Lasheras [7] performed an experimental study on a water mixing layer laden with small air bubbles, and elucidated that the bubble distribution is chiefly governed by the large-scale eddies. Druzhinin and Elghobashi [8] and Yang *et al.* [9] conducted numerical investigations on bubble-laden mixing layers, and reported that the bubbles concentrate preferentially around the center of large-scale eddy. In single-phase jet, there are also organized large-scale eddies near the nozzle outlet. Milenkovic *et al.* [10,11] carried out experiments on a bubble-laden jet to study the bubble entrainment into the large-scale eddies generated by disturbances near the nozzle.

Vortex ring can transport mass and momentum through its convection with the self-induced velocity. The transport ability is so high that it has received much attention, and some attempts at applying a vortex ring to transport particles are reported. Domon *et al.* [12] performed an experimental study on the transport of solid particles in water. A vortex ring was loaded with small resin particles at the launch into still water, and the behavior of the particles convected with the vortex ring was observed. Yanagida *et al.* [13] proposed an olfactory display method for a virtual reality of the next generation. Scented particles, charged into a vortex ring, were efficiently delivered to a specific user's nose by the convection of the vortex ring through the air. Yagami and Uchiyama [14] numerically simulated the convection of glass particles by a vortex ring in the air, and demonstrated that the particles are successfully transported when their Stokes number is small. Uchiyama [15] conducted a numerical simulation of a water jet laden with small air bubbles, and made clear that a vortex ring induced near the nozzle outlet by an axisymmetric disturbance involves the bubbles and convects with the bubbles. The abovementioned results suggest that a vortex ring is usefully employed for the control of bubble motion. Therefore, a vortex ring launched into a bubble plume may increase the interfacial area concentration. As a vortex ring convects with inducing the liquid flow, it may also enhance the mixing between the two phases. But there are few researches on a vortex ring laden with bubbles. Sridhar and Katz [16] investigated experimentally the motion of a bubble entrained into a vortex ring, and showed that the motion calculated by the measured lift and drag forces agrees well with the experimentally visualized result. Sridhar and Katz [17] also studied the effect of entrained bubbles on the strength and vortical structure of a vortex ring, and made clear that the deformation of the vortex ring occurs even when a few bubbles are entrained. These experimental studies did not examine the changes of dimension and convection velocity of vortex ring caused by the bubble. The number of bubbles was so small, and the bubble volume fraction within the vortex ring was not measured. Consequently, the interactions between a vortex ring and the entrained bubbles are not fully clarified.

The objective of this study is to explore experimentally the interactions of bubbles with a vortex ring launched vertically upward into an annular bubble plume. A vortex ring launcher, composed of a cylinder and a piston, is mounted at the bottom of a water tank. Small hydrogen bubbles are released into still water from a cathode, which is wound around the cylinder outlet, by the electrolysis of water. The bubbles rise by the buoyant force and induce a bubble plume. The water in the cylinder is discharged into the bubble plume by the piston, resulting in a vortex ring convecting vertically upward in the plume. In this study, a laminar vortex ring is launched into a bubble plume. The mean bubble diameter is 0.2 mm, and the bubble flow rate is 4.1 mm^3/s. The behavior of the vortex ring and the bubble motion are investigated in a region, of which vertical height from the cylinder outlet is six times the inner diameter of the cylinder.

2. Experimental Set-Up and Method

2.1. Experimental Set-Up

Figure 1 outlines the experimental set-up. The experiment is conducted in a water tank made of transparent acrylic resin. The width and depth of the tank are 300 mm, while the height is 1000 mm. The top of the tank is open to the atmosphere. A vortex ring launcher, composed of a cylinder and a piston, is mounted at the bottom of the tank. The cylinder centerline is parallel to the vertical direction. A cathode is wound around the cylinder outlet, and an anode is placed on the tank wall. Applying a DC voltage between the electrodes, small hydrogen bubbles are released from the cathode. The bubbles rise by the buoyant force and induce the water flow around them, resulting in an annular bubble plume.

A vortex ring is launched vertically upward into the bubble plume by discharging the water in the cylinder with the piston. The push of the piston is performed with a slider connected to an AC servomotor. The velocity and stroke are controlled with a personal computer.

The vortex ring and the bubbles are visualized by a laser light sheet, of which power, wavelength and thickness are 100 mW, 532 nm and 1 mm, respectively. Their images in the vertical plane passing through the cylinder centerline are captured by a video camera. The spatial resolution, the frame rate and the shutter speed are 640 × 480 pixels, 200 fps and 1/200 s, respectively.

The water and bubble velocities are measured by using a PIV system. Nylon particles (mean diameter: 80 μm, specific weight: 1.02) are laden as the tracer for the water flow. Therefore, the bubbles should be removed from the visualized images when measuring the water velocity. In this experiment, the remove was properly performed by

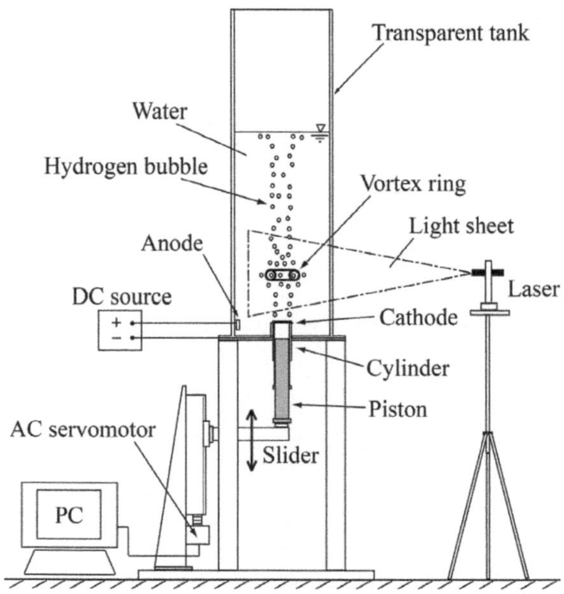

Figure 1. Experimental set-up.

setting the camera aperture at an appropriate level, because the brightness of the bubble differs greatly from that of the tracer particle.

2.2. Launch of Vortex Ring

Figure 2 shows the close up for the vortex ring launcher mounted at the bottom of the water tank. The water depth is 300 mm. The piston stroke L_0 is 100 mm, and the top dead center is 46 mm below the cylinder outlet. The inner diameter of cylinder D_0 is 42.5 mm, while the outer one is 57.8 mm. The height of cylinder outlet from the tank bottom is 45 mm. The origin of the vertical (z) and radial (r) axes is set at the center of cylinder outlet.

Figure 3 presents the time variation for the piston motion, where the displacement z_p and the velocity U_0 (= dz_p/dt) are plotted. The motion time T_0 is 0.748s. The velocity profile is trapezoid, and the maximum velocity U_m (=154 mm/s) is maintained for $0.64T_0$.

When a fluid is discharged from a cylinder into the still fluid by a piston, a laminar vortex ring or a turbulent one is launched depending on the condition for the discharge. Glezer [18] identified the conditions that lead to the generation of the laminar and turbulent vortex rings as shown in **Figure 4**. The identification is based on the cylinder inner diameter D_0, the piston stroke L_0 and the circulation of vortex ring Γ_0. Γ_0 is computed by the following equation:

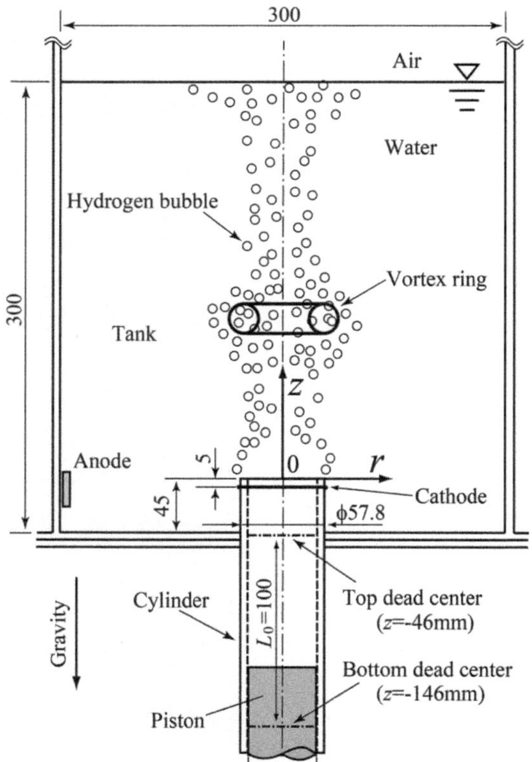

Figure 2. Vortex ring launcher and water tank.

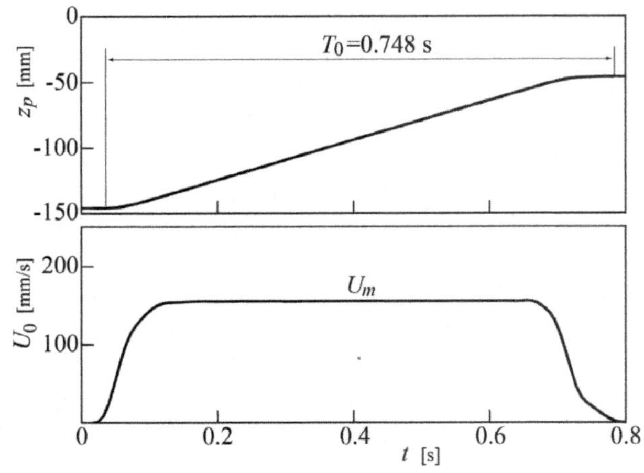

Figure 3. Displacement and velocity profiles of piston.

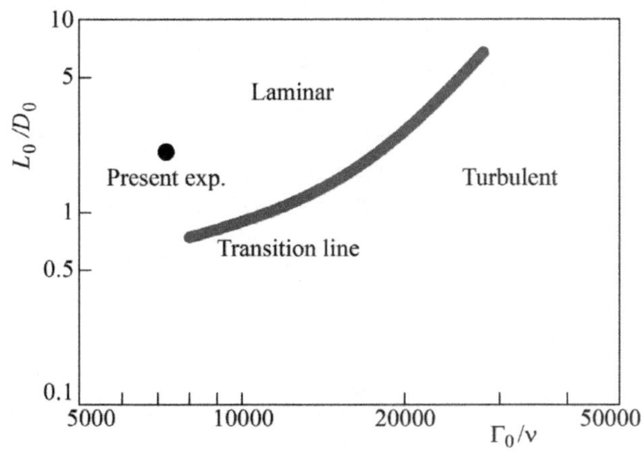

Figure 4. Condition for launch of vortex ring.

$$\Gamma_0 = \int_0^{T_0} \frac{U_0^2}{2} dt \tag{1}$$

where T_0 and U_0 are the motion time and velocity of the piston respectively.

The circulation Γ_0 calculated from Equation (1) results in Γ_0/ν = 7252. According to the identification of Glezer in **Figure 4**, it is found that a laminar vortex ring is launched into still water in this experiment.

2.3. Launch of Vortex Ring

Small hydrogen bubbles are released by the electrolysis of water. A copper wire of diameter 0.1 mm is employed for the cathode. It is wound around the cylinder of the vortex ring launcher, as illustrated in **Figure 2**. The height is 5 mm below the cylinder outlet.

The hydrogen bubbles released from the circular cathode are spherical, and their diameter is less than 0.5 mm. The voltage between the electrodes varies the amount of the bubbles released per unit time. This study sets the voltage at 60 V. The bubble volume release rate, measured by downward displacement of water, is 4.1

mm^3/s, while the bubble mean diameter is 0.2 mm.

When the terminal velocity v_t for a single bubble with diameter 0.2 mm rising in still water is estimated from the numerical calculation, it is 16.3 mm/s.

3. Experimental Results and Discussions

3.1. Behavior of Vortex Ring Launched into Still Water

First, the behavior of the vortex ring in still water is investigated to examine the characteristics of the vortex ring in this experimental set-up. A visualized image for the central vertical cross-section of the vortex ring is presented in **Figure 5**. It is acquired by adding water paint to the launched water in the cylinder. The mean diameter and volume fraction of the paint are about 50 μm and 0.006 respectively. A vortex pattern is clearly observed. A shear layer originates at the boundary between the water discharged from the cylinder and the still water in the tank, and the paint is entrained into the vortex core with the roll up of the shear layer. A laminar distribution of the paint is found behind the vortex ring. There are no active mixings. The visualized image indicates the characteristics of a laminar vortex ring. This is parallel to the Glezer's identification [18] demonstrated in **Figure 4**.

On the visualized image of the central vertical cross-section of the vortex ring shown in **Figure 5**, the z coordinate of the center for vortex core is regarded as the displacement of vortex ring z_v, while the distance between the centers of the vortex cores is defined as the diameter of vortex ring D.

The displacement of vortex ring z_v changes with the lapse of time as shown in **Figure 6**, where the piston maximum velocity U_m and the cylinder inner diameter D_0 are used for the nondimensional expressions. The time when the piston reaches the top dead center is set at $t = 0$. The change of z_v is almost linear, indicating that the vortex ring rises with a constant velocity in still water. The rising velocity, dz_v/dt, is 0.41 U_m. The rise almost ceases at $z_v/D_0 \fallingdotseq 5.8$. Because the vortex ring is affected by the water surface at $z/D_0 = 6$.

The change in the diameter D is plotted against the displacement z_v in **Figure 7**. The diameter remains almost unaltered at $z_v/D_0 \leqq 5.8$. The marked increment at $z_v/D_0 > 5.8$ is attributable to the effect of the water surface.

This study measures the water velocity distribution on the vertical central cross-section of the vortex ring at four displacements of $z_v/D_0 = 0.78$, 2.77, 4.1, 5.01, as mentioned later. The measurement section includes the vortex core. The radial and vertical widths are $2.23D_0$ and $1.59D_0$ respectively. When the vorticity is calculated at each displacement, the circulation Γ can be estimated from the distribution. The nondimensional circulation Γ/v is plotted by the open symbols in **Figure 8**. The circulation remains almost unaltered in the region investigated by this study. When the piston reaches the top dead

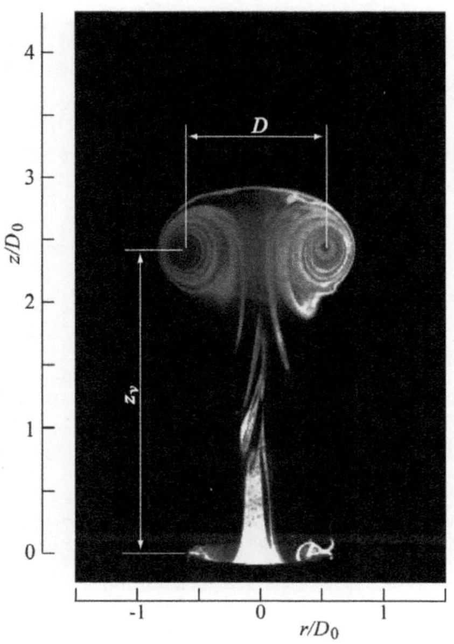

Figure 5. Visualized image of vortex ring launched into still water.

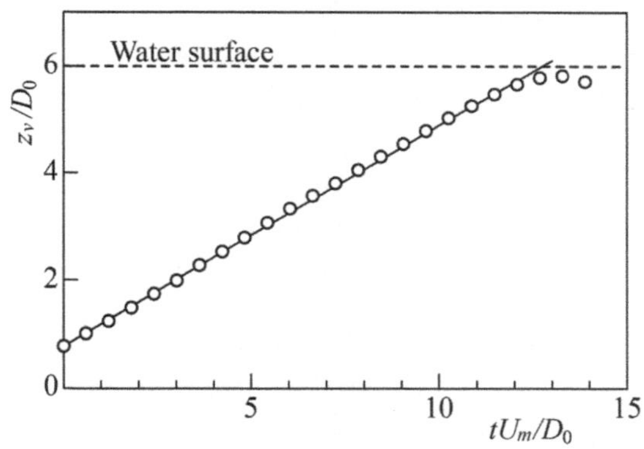

Figure 6. Time variation for displacement of vortex ring in still water.

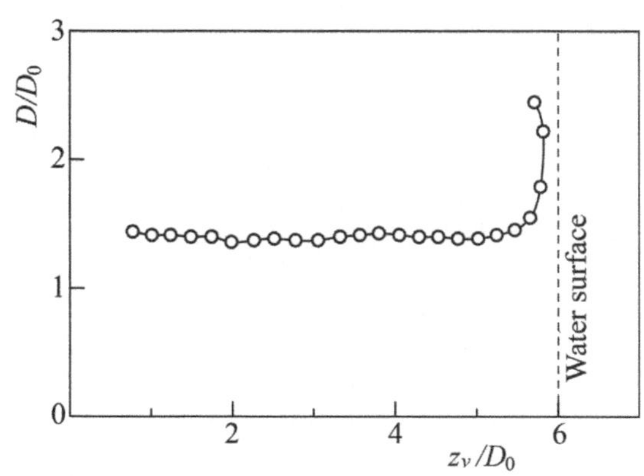

Figure 7. Axial evolution for diameter of vortex ring in still water.

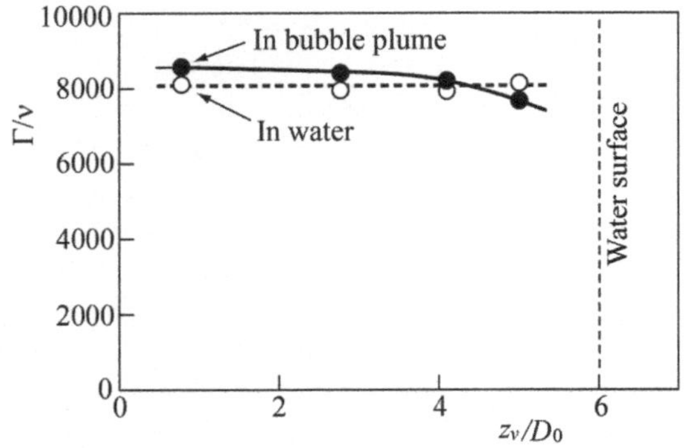

Figure 8. Axial change in circulation around vortex core.

center ($t = 0$), the displacement z_v/D_0 is 0.78. The circulation Γ/ν at this displacement is 8063. It is slightly larger than the value, $\Gamma/\nu = 7252$, estimated from Equation (1). Similar result was also reported by Glezer [18] and Gharib *et al.* [19]. Equation (1) is based on a slug model assuming that the water velocity is uniform within the cylinder cross-section. The difference from the measurement is attributable to this assumption.

3.2. Characteristics of Bubble Plume

Figure 9 illustrates the bubble distribution for the temporally developed bubble plume, where the distribution on the vertical plane passing through the plume centerline is depicted. The bubbles, generated from the cathode wound around the cylinder outlet, rise due to the buoyant force. They hardly move straight in the vertical direction. They shift toward the plume centerline near the cathode ($z/D_0 \leqq 1$), and meander slightly with their rise at $z/D_0 > 1$. The shift is caused by the fact that the water outside of the plume flows toward the centerline to satisfy the mass conservation low of the water. Because the rising bubbles induce an upward water flow around the centerline.

When the time-averaged water velocity $\overline{u}$ is measured, the axial (vertical) component $\overline{u_z}$ changes in the radial direction as plotted in **Figure 10**, where the distributions at four elevations from the cylinder outlet are presented. At $z/D_0 = 0.78$, $\overline{u_z}$ takes its maximum value at $r/D_0 = 0.26$. This is because the bubbles released from the cylinder tip shift toward the centerline and they concentrate at this radial position as seen in **Figure 9**, and accordingly an upward water flow with higher vertical velocity is induced there. At $z/D_0 = 2.77$ and 4.1, $\overline{u_z}$ reaches the maximum on the centerline of the plume. The velocity $\overline{u_z}$ increases with the elevation z. One can confirm the entrainment of the water into the plume. A spatially developing buoyant jet occurs. The velocity lowers at $z/D_0 = 5.01$ owing to the effect of the water surface.

The time-averaged bubble velocity $\overline{v}$ is measured at the radial positions of $r/D_0 = \pm 0.25$. It changes in the vertical (z) direction as shown in **Figure 11**, where the

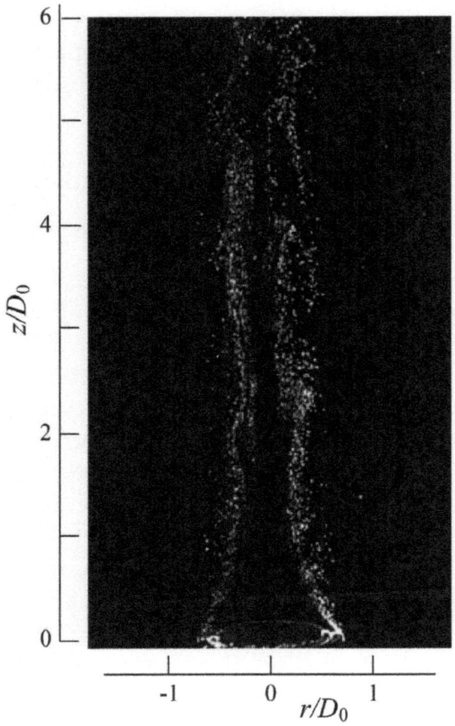

Figure 9. Visualized image of bubbles in bubble plume without vortex ring.

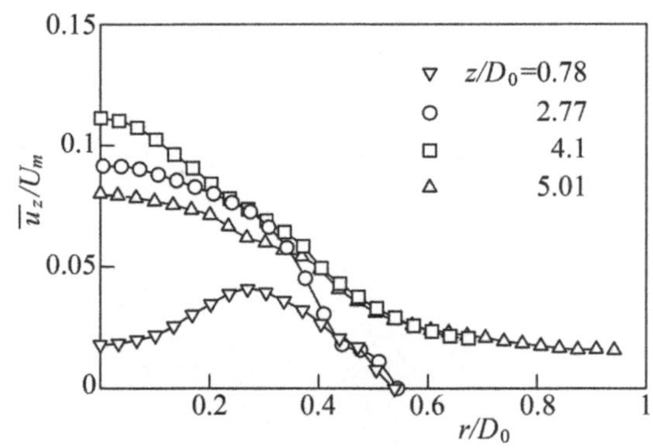

Figure 10. Time-averaged water velocity in bubble plume without vortex ring.

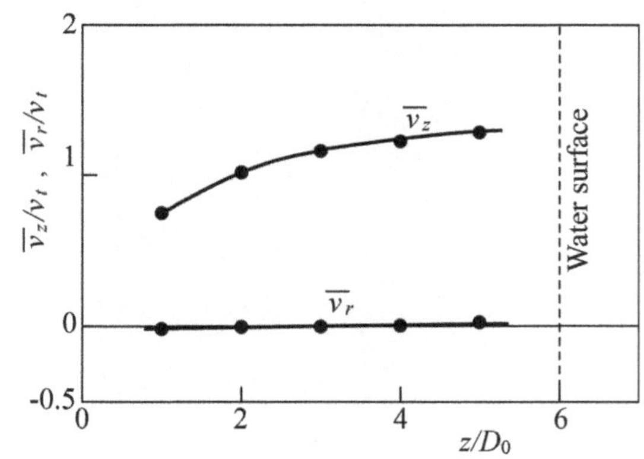

Figure 11. Time-averaged bubble velocity in bubble plume at $r/D_0 = \pm 0.25$.

bubble terminal velocity v_t is used for the nondimensional expression. The radial positions correspond to the regions where the bubbles concentrate as seen in **Figure 9**. The axial (vertical) velocity $\overline{v_z}$ heightens with increasing the elevation, demonstrating that the plume is spatially developing in the measured region. But it shows a tendency to be saturated. The velocity $\overline{v_z}$ is higher than the bubble terminal velocity at $z/D_0 \geqq 2$. This is owing to the fact that the bubbles rise relative to the vertically upward water flow induced by them. The radial (horizontal) velocity $\overline{v_r}$ is much lower than $\overline{v_z}$, and the change in the vertical direction is small.

3.3. Behavior of Vortex Ring Launched into Bubble Plume and Bubble Distribution

When the vortex ring is vertically launched into the bubble plume along the centerline, the bubbles distribute as shown in **Figure 12**. The distributions on the central vertical cross-section at four displacements of vortex ring z_v are presented. At $z_v/D_0 = 0.78$, the bubbles above the vortex ring tend to move to the rear of the vortex ring along the vortex core, and they are being spirally entrained into the vortex core. The bubble entrainment is caused by the roll up of the shear layer originating from the cylinder tip. This displacement is at the time $tU_m/D_0 = 0$ when the piston reaches the top dead center. At $z_v/D_0 = 2.77$, the spiral entrainment further proceeds. The bubble volume fraction within the cross-section reaches its maximum value at this displacement as mentioned later. At $z_v/D_0 = 4.1$, the number of bubbles in the vortex ring around the center axis ($r/D_0 = 0$) is smaller than that at $z_v/D_0 = 2.77$, and a lot of bubbles distribute behind the vortex ring. Because the bubbles moving to the rear of vortex ring are hardly entrained into the vortex ring due to the cease of the roll up for the shear layer, and accordingly they are left behind the vortex ring. When the vortex ring rises more ($z_v/D_0 = 5.01$), the amount of bubbles inside the vortex ring increases again due to the collision between the vortex ring and the bubbles accumulated near the water surface. A marked increment occurs around the center axis.

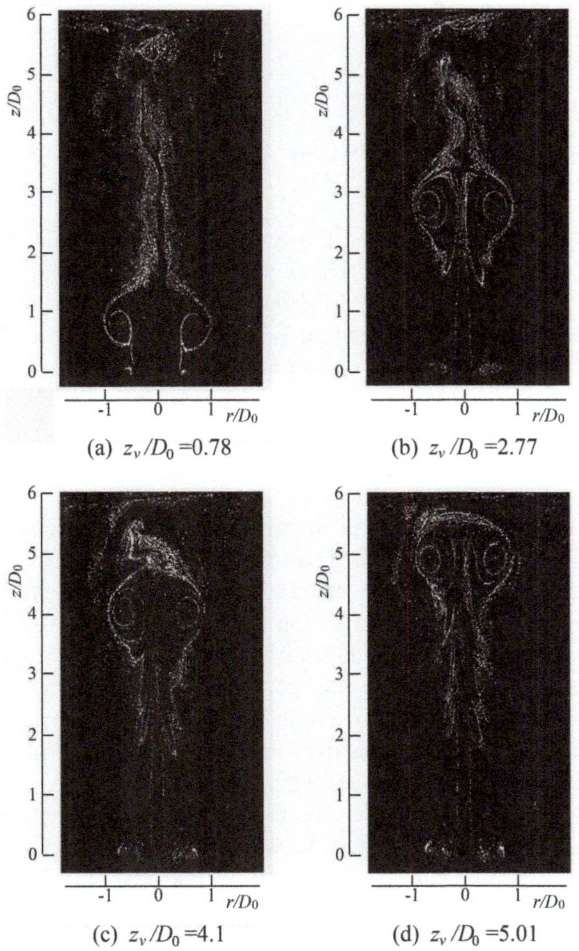

Figure 12. Visualized image of bubbles around vortex ring launched into bubble plume.

On the visualized image of the bubble distribution within the central vertical cross-section of the vortex ring, the z coordinate of the center for vortex core is regarded as the displacement z_v, while the distance between the centers of the vortex cores is defined as the diameter D. **Figure 13** indicates the image at $z_v/D_0 = 1.22$. It is confirmed that z_v and D are clearly identified.

Figure 14 shows the time variation for z_v. The displacement in the bubble plume is almost the same as that in still water at $tU_m/D_0 \leqq 2.42$. But it is larger at $tU_m/D_0 > 2.42$, indicating that the rising velocity of the vortex ring in the bubble plume is higher. Because the vortex ring is affected with the vertically upward water flow induced by the rising bubbles as explained later. When $tU_m/D_0 > 2.42$, the time variation of z_v is linear. The vortex ring rises at a constant velocity.

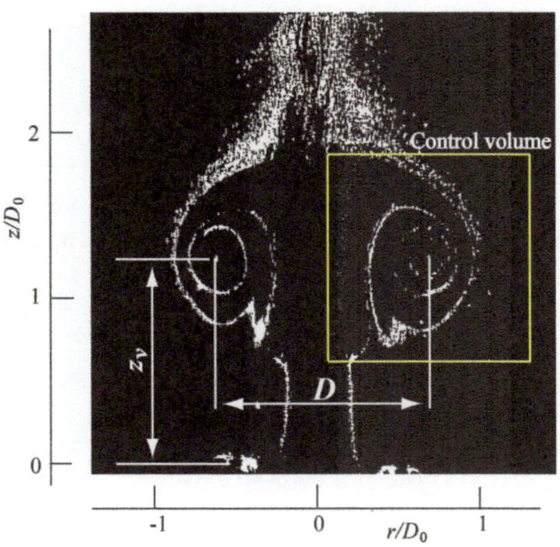

Figure 13. Displacement and diameter of vortex ring launched into bubble plume.

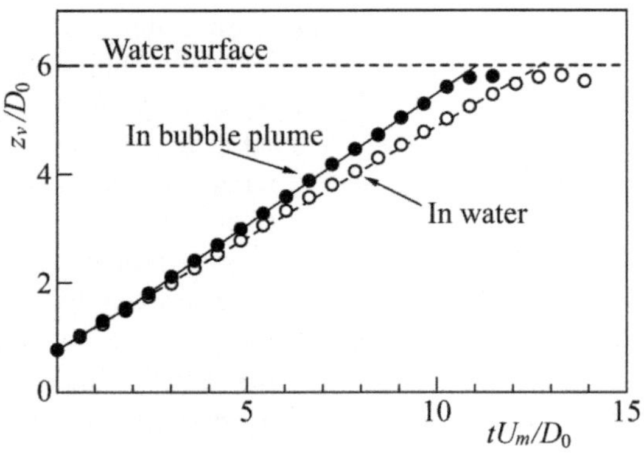

Figure 14. Time variation for displacement of vortex ring launched into bubble plume.

The diameter of vortex ring D changes as a function of z_v as presented in **Figure 15**. The diameter D in the bubble plume decreases with the rise of vortex ring. Thus, the diameter is smaller than that in still water. A buoyant jet is generated as shown in **Figure 9**, and there exists an upward water flow around the centerline ($r/D_0 = 0$). Therefore, the water is entrained into the bubble plume, and the water flowing toward the centerline reduces the D value.

3.4. Void Fraction in Central Vertical Cross-Section of Vortex Ring

To grasp the change in the amount of bubbles inside the vortex ring with the convection of the vortex ring, the void fraction α in a rectangular region surrounding the vortex core is calculated. Binarized images for the bubble distribution are used for the calculation. The region is depicted in **Figure 13**. The width and the height are $1.25D_0$, and the centroid is fixed at the center of the vortex core.

Figure 16 shows the relation between the void fraction and the displacement of vortex ring z_v. At $0.78 \leqq z_v/D_0 \leqq 2.77$, α increases monotonically. This is because the bubbles are being spirally entrained into the vortex ring with the roll up of the velocity shear layer, as seen in **Figures 12(a)** and **(b)**. But α decreases at $2.77 \leqq z_v/D_0 \leqq 4.1$ where the strength of the shear layer lessens. Because the bubbles are left behind the vortex ring, as seen in **Figure 12(c)**. At $z_v/D_0 > 4.1$, α increases again (**Figure 12(d)**). This is attributable to the fact that the vortex ring collides with the bubbles accumulated near the water surface. The maximum value of α at $z_v/D_0 = 2.77$ is 2.43 times as large as that near the cylinder outlet ($z_v/D_0 = 0.78$).

3.5. Distributions for Velocity and Vorticity of Water

The distribution of water velocity on the central vertical cross-section of the vortex ring is shown in **Figure 17**,

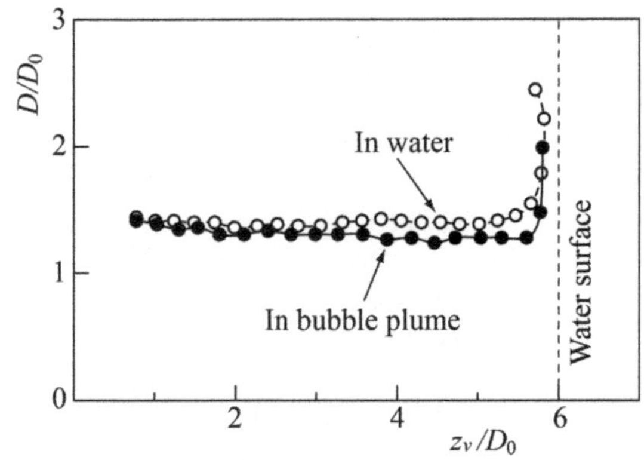

Figure 15. Axial evolution for diameter of vortex ring launched into bubble plume.

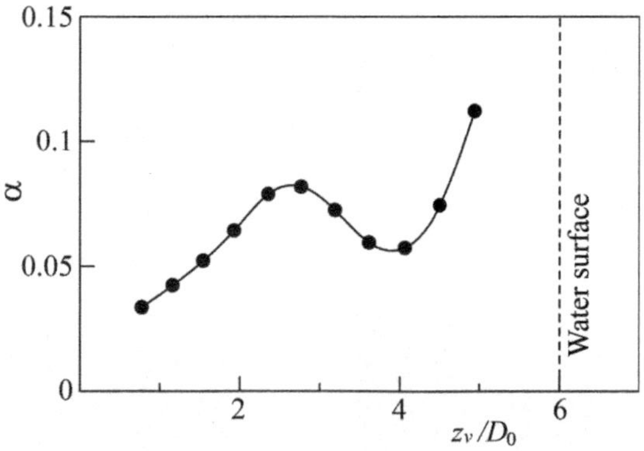

Figure 16. Axial evolution for void fraction in cross-section of vortex ring.

where the ensemble-averaged velocity $\hat{\mathbf{u}}$ at four displacements of vortex ring is presented. The distribution in still water and that in the bubble plume are compared, and the bubble distribution is also presented. When the vortex ring is launched into still water, the vortex core remains nearly circular. The water velocity scarcely depends on the displacement of the vortex ring. Even when the vortex ring is launched into the bubble plume, the vortex core exists. The vortex core at $z_v/D_0 = 0.78$ and 2.77 is not circular, but deforms with the convection of vortex ring. The deformation is reconfirmed from the contour lines for the magnitude of velocity shown in **Figure 18**. The deformation is mainly caused by the shear flow of water induced by the rising bubbles. It may also be attributable to the fact that the time variation of the bubble distribution around the vortex core changes markedly due to the active bubble entrainment. Sridhar and Katz [17] made clear that a vortex ring deforms even when a few bubbles are entrained. The present authors elucidate the variation of the water velocity in the vortex core with the convection of vortex ring. They also grasp the change of the shape for vortex core.

Figure 19 presents the axial (vertical) component of the ensemble-averaged water velocity $\hat{u}_z$, where the distribution on the horizontal line passing through the

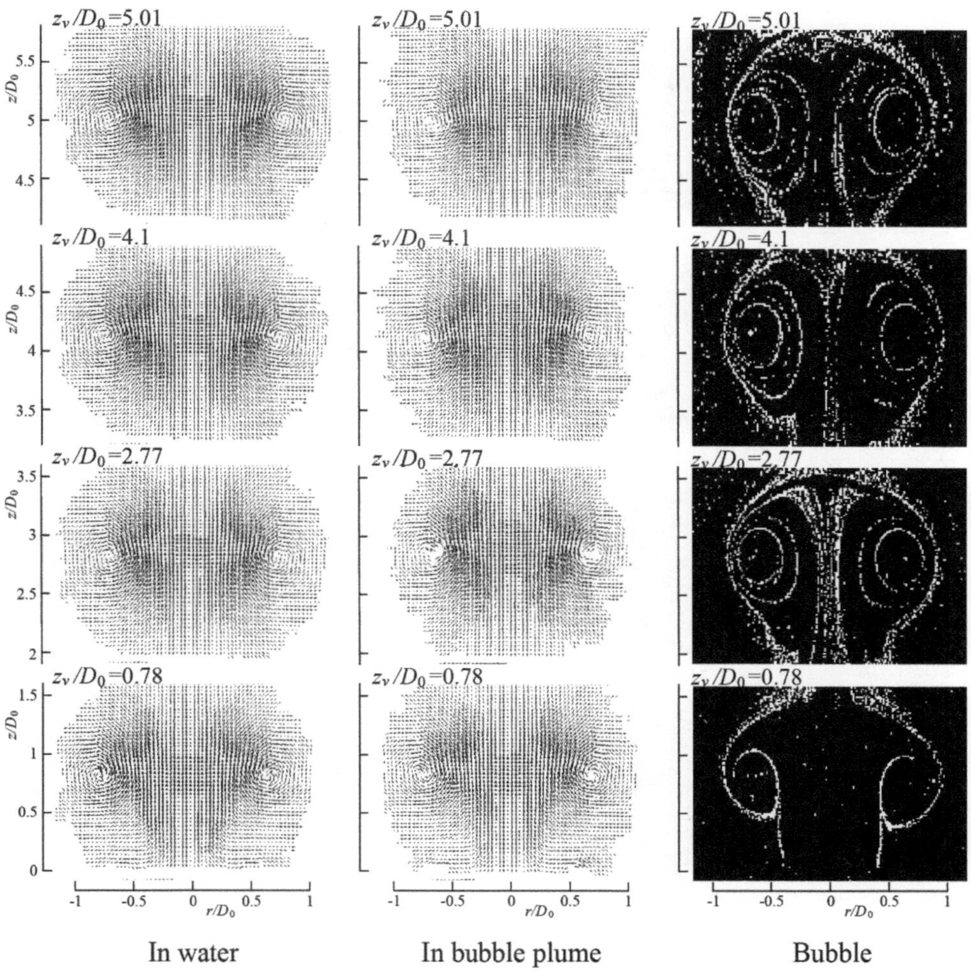

Figure 17. Water velocity around vortex ring.

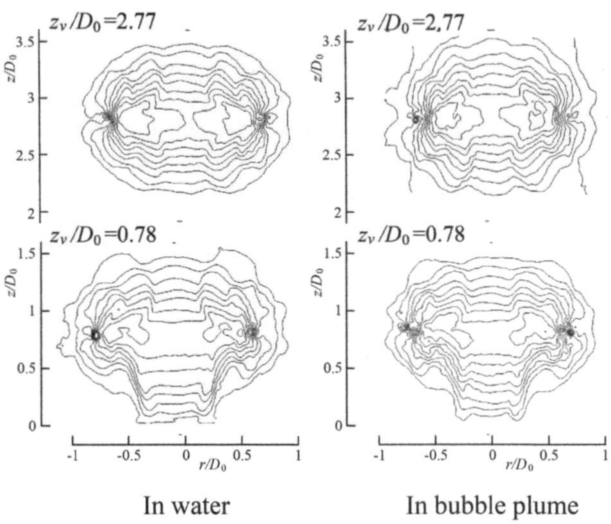

Figure 18. Contour lines for magnitude of water velocity around vortex core.

centers of the vortex cores shown in **Figure 17** is plotted. The velocity remains unaltered between the vortex cores. The velocity $\hat{u}_z$ in the bubble plume is higher than that in still water at z_v/D_0 = 2.77 and 4.1. This is due to the upward water flow of the plume. The velocity gradient around the center of the vortex core in the bubble plume is smaller than that in still water.

The distribution for the ensemble-averaged vorticity $\hat{\omega}$ on the central vertical cross-section of vortex ring is shown in **Figure 20**, where the bubbles are superimposed on the result in the bubble plume. The absolute value of the vorticity around the vortex core in the bubble plume is lower than that in still water at every displacement of the vortex ring. Because the friction acts on the interface between the bubbles and the water. Ruetsch and Meiburg [20] conducted a numerical simulation of a plane shear layer laden with bubbles. They highlighted that the bubbles concentrate around the vortex center and that the concentration reduces the strength of the vortex strength.

The present experiment grasps the variation in the vorticity distribution with the convection of vortex ring, and therefore it elucidates the dynamic interaction between the vortex ring and the entrained bubbles.

The circulation Γ around the vortex core in the bubble plume is calculated from the vorticity field. The nondi-

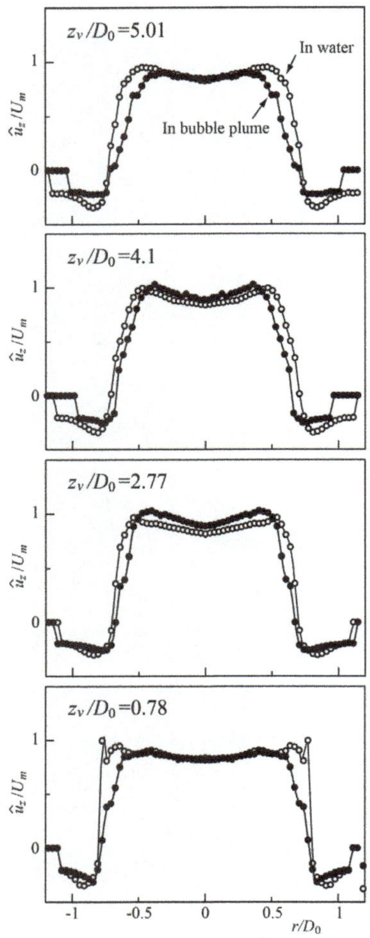

Figure 19. Water axial velocity on horizontal line passing through vortex cores.

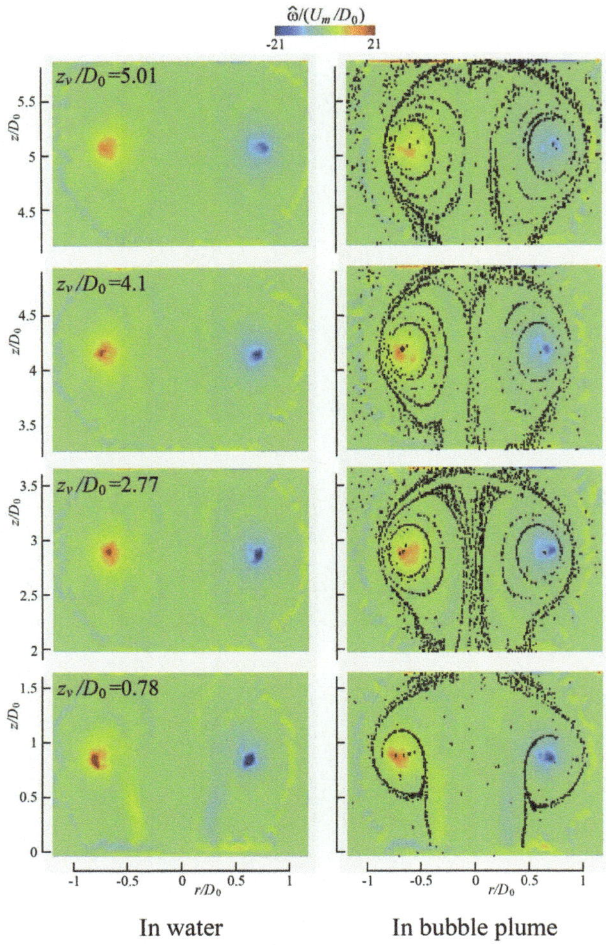

Figure 20. Vorticity distribution around vortex ring.

mensional circulation Γ/ν is plotted by the painted symbols in **Figure 8**. At $z_v/D_0 \leqq 4$, the circulation in the bubble plume is higher than that in still water. Because the effect of the upward water flow is superimposed on Γ. The circulation in the bubble plume lessens with the rise of vortex ring. Because the absolute value of the vorticity in the vortex ring decreases owing to the friction between the water and the entrained bubbles.

The vorticity $\hat{\omega}$ on the horizontal line passing through the centers of the vortex cores is plotted in **Figure 21**. At $z_v/D_0 = 0.78$, the maximum value for the vorticity in the water is markedly high. One can find the occurrence of a velocity shear layer from the cylinder tip. But the maximum value in the bubble plume at the displacement is not so high. This is attributable to the fact that the velocity gradient of the shear layer is reduced by the bubbles. The absolute value of $\hat{\omega}$ lessens greatly around the vortex core, indicating that the vorticity distribution is flattened by the entrained bubbles.

4. Conclusions

The interactions of bubbles with a vortex ring launched into a bubble plume were investigated. Laminar vortex ring was vertically launched into bubble plume. The mean bubble diameter is 0.2 mm. The results are summarized as follows:

1) The bubbles are spirally entrained into the vortex ring just after the launch of vortex ring. When the vortex ring reaches a certain height, the amount of the bubbles around the center axis of vortex ring lessens, and the bubbles moving to the rear of the vortex ring are left behind the vortex ring.

2) The rising velocity of vortex ring in the bubble plume is higher than that in still water. This is because the vortex ring is affected by the vertically upward water flow induced by the bubbles. The velocity of vortex ring is almost constant.

3) The diameter of vortex ring reduces with the rise of the vortex ring. The bubble plume is generated by the rising bubbles, and there exists the vertically upward water flow around the centerline. Therefore the water is entrained into the plume, and the water flowing toward the centerline reduces the diameter.

4) The void fraction in the central vertical cross-section of vortex ring increases until a certain displacement

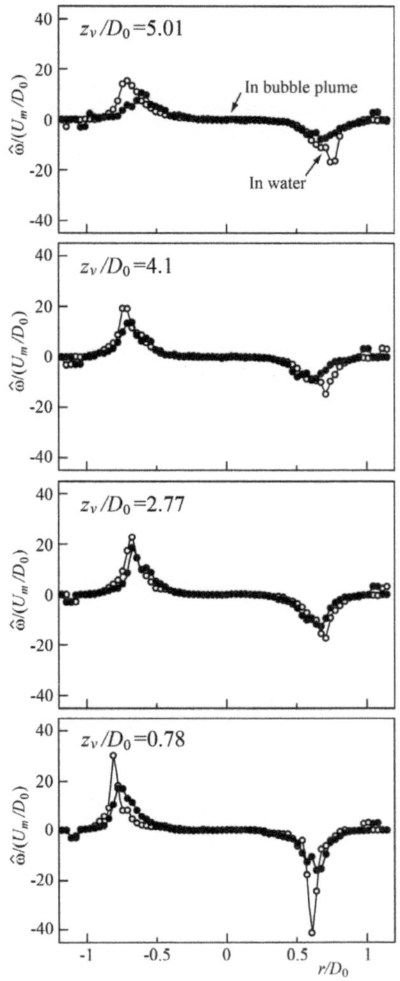

Figure 21. Vorticity on horizontal line passing through vortex cores.

of vortex ring, where the reduction occurs. It heightens again when the vortex ring approaches the water surface. Because the vortex ring collides with the bubbles accumulated near the water surface.

5) The vortex core deforms with the convection of vortex ring. The entrained bubbles cause the decrement of the absolute value for the vorticity around the vortex core, reducing the strength of vortex ring.

REFERENCES

[1] N. A. Hussain and R. Siegel, "Liquid Jet Pumped by Rising Gas Bubbles," *Journal of Fluids Engineering*, Vol. 98, No. 1, 1976, pp. 49-57.

[2] A. M. Leitch and W. D. Baines, "Liquid Volume Flux in a Weak Bubble Plume," *Journal of Fluid Mechanics*, Vol. 205, No. 1, 1989, pp. 77-98.

[3] J. H. Milgram, "Mean Flow in Round Bubble Plumes," *Journal of Fluid Mechanics*, Vol. 133, No. 1, 1983, pp. 345-376.

[4] S. A. Socolofsky and E. E. Adams, "Role of Slip Velocity in the Behavior of Stratified Multiphase Plumes," *Journal of Hydraulic Engineering*, Vol. 131, No. 4, 2005, pp. 273-282.

[5] M. Alam and V. H. Arakeri, "Observations on Transition in Plane Bubble Plumes," *Journal of Fluid Mechanics*, Vol. 254, No. 1, 1993, pp. 363-374.

[6] T. Uchiyama and S. Matsumura, "Three-Dimensional Vortex Method for the Simulation of Bubbly Flow," *Journal of Fluids Engineering*, Vol. 132, No. 10, 2010, pp. 1-8.

[7] P. M. Rightley and J. C. Lasheras, "Bubble Dispersion and Interphase Coupling in a Free-Shear Flow," *Journal of Fluid Mechanics*, Vol. 412, No. 1, 2000, pp. 21-59.

[8] O. A. Druzhinin and S. E. Elghobashi, "Direct Numerical Simulation of a Three-Dimensional Spatially Developing Bubble-Laden Mixing Layer with Two-Way Coupling," *Journal of Fluid Mechanics*, Vol. 429, 2001, pp. 23-61.

[9] X. Yang, N. H. Thomas, L. J. Guo and Y. Hou, "Two-Way Coupled Bubble Laden Mixing Layer," *Chemical Engineering Science*, Vol. 57, No. 4, 2002, pp. 555-564.

[10] R. Z. Milenkovic, B. Sigg and G. Yadigaroglu, "Study of Periodically Excited Bubbly Jets by PIV and Double Optical Sensors," *International Journal of Heat and Fluid Flow*, Vol. 26, No. 6, 2005, pp. 922-930.

[11] R. Z. Milenkovic, B. Sigg and G. Yadigaroglu, "Bubble Clustering and Trapping in Large Vortices, Part 1: Triggered Bubbly Jets Investigated by Phase Averaging," *International Journal of Multiphase Flow*, Vol. 33, No. 10, 2007, pp. 1088-1110.

[12] K. Domon, O. Ishihara and S. Watanabe, "Mass Transport by a Vortex Ring," *Journal of the Physical Society of Japan*, Vol. 69, No. 1, 2000, pp. 120-123.

[13] Y. Yanagida, S. Kawato, H. Noma, A. Tomono and N. Tetsutani, "Projection-Based Olfactory Display with Nose Tracking," *Proceedings of IEEE Virtual Reality*, 2004, Chicago, pp. 43-50.

[14] H. Yagami and T. Uchiyama, "Numerical Simulation for the Transport of Solid Particles with a Vortex Ring," *Advanced Powder Technology*, Vol. 22, No. 1, 2011, pp. 115-123.

[15] T. Uchiyama, "Three-Dimensional Vortex Simulation of Bubble Dispersion in Excited Round Jet," *Chemical Engineering Science*, Vol. 59, No. 7, 2004, pp. 1403-1413.

[16] G. Sridhar and J. Katz, "Drag and Lift Forces on Microscopic Bubbles Entrained by a Vortex," *Physics of Fluids*, Vol. 7, No. 2, 1995, pp. 389-399.

[17] G. Sridhar and J. Katz, "Effect of Entrained Bubbles on

the Structure of Vortex Rings," *Journal of Fluid Mechanics*, Vol. 397, 1999, pp. 171-202.

[18] A. Glezer, "The Formation of Vortex Rings," *Physics of Fluids*, Vol. 31, No. 12, 1988, pp. 3532-3542.

[19] M. Gharib, E. Rambod and K. Shariff, "A Universal Time Scale for Vortex Ring Formation," *Journal of Fluid Mechanics*, Vol. 360, 1998, pp. 121-140.

[20] G. R. Ruetsch and E. Meiburg, "Two-Way Coupling in Shear Layers with Dilute Bubble Concentrations," *Physics of Fluids*, Vol. 6, No. 8, 1994, pp. 2656-2670.

12

Evaluation of Methane Yield on Mesophilic-Dry Anaerobic Digestion of Piggery Manure Mixed with Chaff for Agricultural Area

Dong-Heui Kwak[1], Mi-Sug Kim[1], Jae-Seung Kim[2], Young-Youl Oh[3], Soon-Ok Noh[4,5], Byung-Ok So[4,5], Su-Young Jung[4,5], Su-Jin Jung[4,5], Soo-Wan Chae[4,5*]

[1]Department of Environmental and Chemical Engineering, Seonam University, Namwon, Republic of Korea
[2]Department of Environmental Engineering, Chonbuk National University, Jeonju, Republic of Korea
[3]Yoyo Korea Agricultural Association, Jeongehb, Republic of Korea
[4]Clinical Trial Center for Functional Foods, Chonbuk National University Hospital, Jeonju, Republic of Korea
[5]Department of Medical Nutrition Therapy, Chonbuk National University, Medical School, Jeonju, Republic of Korea

ABSTRACT

A mesophilic-dry anaerobic digestion process is valid in treating high-concentration substrates containing low moisture content. It has merits of lower wastewater discharge and lower heat capacity required in maintaining reactor temperature as compared with a thermophilic-wet anaerobic digestion process. In fact, chaff can be easily obtained in farming areas and used as a mixture substrate as one of bulking agents for controlling moisture and supplying carbon. For this reason, this study applies the chaff to improve livestock manure, which contains high moisture content and is discharged from domestic pig farms. This study aims at verifying its feasibility for improving methane production efficiency on a basis of BMP (Biochemical Methane Potential) assay obtained through a series of experiments. Finding results were methane gas production and gas production per volatile solid (VS) added, and methane gas production among biogas production were increased as the chaff added in the piggery manure was increased. According to experimental results for improving the methane production efficiency, mixture of the chaff and the piggery manure played an important role in controlling the moisture content and improving the methane gas production rate, and also verified its feasibility in the mesophilic-dry anaerobic digestion process indicating relatively less difficulty for operation and management.

Keywords: Anaerobic Digestion; Methane; Chaff; Biogas; Piggery Manure

1. Introduction

A methane fermentation process has an advantage in treating organic contaminants for preventing environmental pollution when comparing a conventional aerobic treatment process. Naturally, the methane fermentation process has a combination limit of processes but it is relatively useful in aspects of energy production and resource collection. The methane fermentation process such as an anaerobic digestion process is a skill studied and used for a long time and is recently being magnified in a situation as an international concern focusing on climate change control and renewable energy demand. Especially, a biogas plant, one of the methane fermentation skills, has been used in many countries and known as one of effective strategy techniques for bio-fuel production [1].

Domestic livestock manure emission classified by livestock types has a component ratio as follows; 57.6% piggery manure (740,000 m^3/d) and 42.4% cow manure (540,000 m^3/d) [2], and dairy cow manure of the cow manure is emitted in the overcrowded area such as farms but few Korean native cattle are raised in small farmers and there are many bad cases in collecting and in treating Korean native cattle manure. The organic content is a raw matter to produce the methane and exists in the piggery manure as low as 2% to 5%. Thus, utilization of the piggery manure as a substrate is low and also it is known well that a stable operation of an anaerobic digestion tank is difficult because a fluctuation range of the organic

*Corresponding author.

content is large periodically and the moisture content is high enough [3].

The anaerobic digestion process is divided into a wet process and a dry process according to solid content or moisture content of the substrate used. Until the mid-1980s, the wet process has been mainly applied in the field using waste matters within 10% solid content as the substrate. With EU as the center rapidly from the 1990s, however, the dry process has been developed to digest organic waste matters containing the solid content over 20% [4]. In treating the high-concentration substrate having low moisture content, the dry process requires low heat capacity to maintain the reactor temperature and discharges low wastewater after treatment [5]. However, it is not valid to put the livestock manure into the dry process directly because the livestock manure emitted from domestic piggery farms contains a great deal of moisture and the solid content of the livestock manure is very low. Meanwhile, the bulking agents such as rice straws, chaff, dead leaves fragments, sawdust, etc. are easy to obtain in the farm area and such agricultural by-products have been used as the bulking agent for composing manure from old times and also as the carbon supplement for maintaining the proper C/N ratio. Practically, most of domestic livestock farms are located in the farming settlement that is producing a great deal of the bulking agent. In adopting the biogas plant in the domestic farming areas, the dry anaerobic digestion process is in a more advantageous situation than the wet process when considering realistic conditions, In Europe recently, studies on the dry digestion operation for the municipal organic solid waste are actively proceeding to reduce waste amounts for landfill and to produce the bio-energy [6,7]. However, previous studies mainly present that the operation results for high temperature (50°C - 60°C) conditions and continuous operation cases are also pretty rare [8].

To analyze the ultimate methane production rate (mL/g-VS_{added}) caused by organic matters as the substrate for the anaerobic digestion process, it measures the methane amounts produced during the anaerobic batch incubation period and cumulative methane formula can be used to determine the methane production yield based on the observed data. Representative models such as Modified Gompertz model or Exponential model are used to analyze experiment data obtained through the methane production potential test [9,10]. Using those models described in Equations (1) and (2) as below, comparative studies are variously proceeding to determine the ultimate methane production yield of the substrates related to diverse components [11,12].

Modified Gompertz Model Equation [13]:

$$M = M_o \times \exp\left\{-\exp\left[\frac{R_m \times e}{M_o}(\lambda - t) + 1\right]\right\} \qquad (1)$$

where, M: cumulative methane production yield (mL-CH_4/g-VS)

t: incubation time of an anaerobic digestion tank (days)

M_o: ultimate methane production yield (mL-CH_4/g-VS)

R_m: maximum methane production rate (mL-CH_4/g-VS·day)

e: exp (1) = 2.71828182

λ: lag phase, days

Exponential Model Equation:

$$B = B_o\left(1 - e^{-kt}\right) \qquad (2)$$

where, B: cumulative methane production yield (mL-CH_4/g-VS)

t: incubation time of an anaerobic digestion tank (days)

B_o: ultimate methane production yield (mL-CH_4/g-VS)

k: 1st order reaction rate constant (day^{-1}).

With the purpose of energy resource recovery through the methane production due to the piggery manure in the farming area, this study conducts a series of experiments with the dry anaerobic digestion using the chaff. The chaff is easily obtained in the farming area as one of the byproducts and as the substrate to mix with the manure as well as to control the moisture. The anaerobic digestion has been conducted in the single-phase mesophilic condition, BMP (biochemical methane potential) assay has been applied to estimate the methane production potential due to the several mixture ratios between the piggery manure and the chaff based on the experimental results, and this study has been accomplished to find a way to improve the methane production efficiency.

2. Materials and Methods

2.1. Experimental Equipments and Operation Conditions

In experiments, a batchwise reactor of methane yield is prepared for single-phase anaerobic digestion as shown in **Figure 1**. For a dry methane production process of livestock manure as a main substrate, typical experimental conditions were adopted to examine gas production yield and responses characteristics. Experiments were set up to control pH if necessary and to incubate for 40 days as controlling to keep typical temperature for mesophilic digestion in a range of 35°C ± 1°C [14], and an additional agitator was excluded in the experiments.

Figure 1 describes a schematic diagram of a batchwise single phase digester for methane yield. 0.5 L serum bottles were set up to an incubator at a constant temperature. Operation conditions were monitored for every serum bottle in different mix proportions between piggery manure as a main substrate and chaff as a mixture substrate to control the moisture. Major items such as pH change and the gas production yield from the serum bottle were monitored due to each condition. The operator also

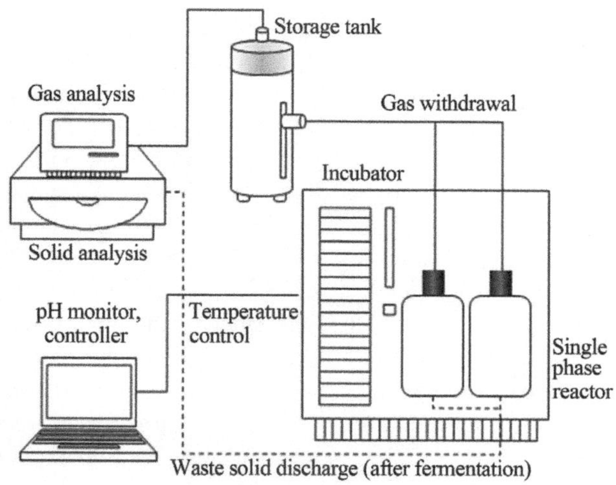

Figure 1. Schematic diagram of batchwise single phase digester for methane yield.

measured generating-capacity and methane content in gas collected in a teflon bag through an exhaust pipe in the middle of a gas-tight rubber stopper of the serum bottle at a constant time interval, every 3 days. Additionally, taking small amounts from all of samples and putting them into each 50 mL-tube, the operator measured its weight per VS (volatile solid) changed with operating under the same conditions.

Table 1 presents operation conditions and a substrate composition in the single phase digester. For two operation parameters such as the substrate concentration and the solid content, the methane production was estimated in different mixing ratios between the piggery manure and the chaff. All samples except a control group (marked as Run 1) were added trace elements in order to minimize unstable effects of microbial growth due to lack of essential elements in the anaerobic digester.

2.2. Sample and Analysis Method

Livestock manure has been concerned in domestic water management as a non-point source. Among the livestock manure, piggery manure is newly applied as a main substrate for methane production. To improve the methane production, chaff in powder form (below than 100 mesh) is also mixed with the piggery manure to control the moisture as well as to improve the methane production. Major properties of the piggery manure are shown in **Table 2** and chemical properties of all samples (Run 1 - Run 5) are described in **Table 3**. Also, **Table 4** presents trace elements (mineral salts and trace metals) added for the safe operating condition. Water quality and solid matters were analyzed by the standard method (2005) in this study.

The piggery manure samples were collected in the retaining tank for gathering and mixing the manure of the pigsty before inputting washing water of a pig farm.

Table 1. Operation condition and substrate composition of single phase digester.

Samples	Mixing ratio between manure and chaff	Substrate and solid content			
		Piggery manure (g)	Powder of chaff* (g/L)	Solid content* (%)	Trace elements
Run 1	2.50:1 (control)	300	120	35.4	non-spiked
Run 2	4.17:1	300	72	27.2	spiked
Run 3	3.13:1	300	95	30.6	spiked
Run 4	2.78:1	300	108	33.1	spiked
Run 5	2.50:1	300	120	35.7	spiked

Note: *Converted value into concentration.

Table 2. Chemical composition of piggery wastewater used as main substrate in this study.

Description	Unit	Measured values
pH		8.7
BOD	mg/L	2205
COD_{cr}	mg/L	2221
$SCOD_{cr}$	mg/L	1126
T-N	mg/L	3439
T-P	mg/L	121.8
NH_3-N	mg/L	1304
P_4^{3-}-P	mg/L	4.3
Alkalinity	mg/L as $CaCO_3$	5800
Fixed solid	mg/L (%)	5900 (0.59)
Volatile solid	mg/L (%)	3300 (0.33)

Table 3. Chemical composition of experimental substrates (mixture of piggery manure and chaff).

Description	Unit	Run 1	Run 2	Run 3	Run 4	Run 5
pH		8.1	7.9	7.8	8.0	8.0
COD_{cr}	mg/kg	16,587	10,092	14,309	15,447	16,760
T-N	mg/kg	1583	3286	2988	1848	1643
T-P	mg/kg	151.4	189.2	169.2	159.2	153.2
NH_3-N	mg/kg	1122.6	1553.1	1416.0	1401.2	1204.8
Fixed solid	%	4.071	6.129	5.707	5.147	4.571
Volatile solid	%	3.048	2.984	2.879	2.966	3.071

Characteristics of target wastewater compiled from the piggery farm were indicated high pH around 8.7 and high alkalinity about 5800 mg/L as $CaCO_3$, which were typical in the livestock wastewater. The wastewater was measured in BOD (2205 mg/L) and $TCOD_{Cr}$ (2221 mg/L) and contained soluble COD_{Cr} ratio around 50.7%. Total solid concentration was 0.92% with 99.1% water content and the solid consisted of VS (35.9%) and inorganic solid (64.1%). Fraction of nutrients compared with the organic was as follows; COD:NH_3-N:PO_4^{3-}-P = 51.7:30.3:1.0.

Table 3 describes chemical compositions of experimental substrates in five different sample groups from Run 1 to Run 5 classified by different mixing ratios between the piggery manure and the chaff at identical conditions. The solid content was ranged from 27.2% to 35.4% and COD_{cr} was distributed from 10.092 mg/L to 16,760 mg/L. Ratios of the organic matter and nutrients (COD:T-N:T-P) for the sample groups (Run 1 - Run 5) were 109.6:10.5:1, 53.3:17.4:1, 84.6: 17.7:1, 97.0:11.6:1, and 109.4:10.7:1, respectively.

2.3. BMP Assay

A BMP (biochemical methane potential) assay was developed by Owen *et al.* [10] to evaluate potential efficiency for biodegradability of target livestock manure in an anaerobic process and it was analyzing organic concentration converted into CH_4. In this study, a serum bottle was filled with a target sample, covered with a butyl rubber septum, sealed with a reinforced plastic lid, and kept in an incubator at a constant temperature 35°C to induce anaerobic degradation. Gas producing capacity and its composition were analyzed every time interval and then a methane production rate was recorded. The control group (Run 1) was examined under the same conditions to modify other gas capacity and other effects occurred from the target experimental samples. Unlike the target experimental groups, the control group did not add trace elements consisting of mineral salts and trace metals. The gas producing capacity was measured using a glass syringe in the constant time interval and the methane production rate was analyzed using Gas Chro- matography-Mass Spectrometer, GC-MS (6890 N Network GC system).

Table 4. Composition of mineral salts and trace metals.

Concentration of mineral salts (mg/L)		Concentration of trace metals (mg/L)			
NH_4Cl	0.53	$MnCl_2 \cdot 4H_2O$	0.0005	$NaMoO_4 \cdot 2H_2O$	0.00001
$CaCl_2 \cdot 2H_2O$	0.075	H_3BO_3	0.00005	$CoCl_2 \cdot 6H_2O$	0.0005
$MgCl_2 \cdot 6H_2O$	0.1	$ZnCl_2$	0.00005	$NiCl_2 \cdot 6H_2O$	0.00005
$FeCl_2 \cdot 4H_2O$	0.02	$CuCl_2$	0.00003	Na_2SeO_3	0.00005

3. Result and Discussion

3.1. Experimental Result of Dry Mix Digestion

Piggery manure has had a difficult time in proceeding wet anaerobic digestion because of its high moisture content. For improving the anaerobic digestion of the piggery manure, it may be mixed with chaff obtained easily in farming areas. **Table 5** presents main experimental results of gas production and methane yields through 40 days digestion operating period for five sample groups.

In anaerobic degradation reactions of organic matters, substrate concentrations and physical and chemical compositions can affect on a reaction rate of hydrolysis

Table 5. Experiment results of methane yields for five types of mixed substrates.

Descriptions	Samples				
	Run 1	Run 2	Run 3	Run 4	Run 5
Cumulative volume of biogas yield (L)	3.21	3.48	3.69	4.19	4.76
Average daily biogas yield (L)	0.708	0.780	0.825	0.933	1.040
Maximum daily biogas yield (L)	1.070	1.160	1.230	1.397	1.587
Average methane fraction in biogas (%)	0.639	0.680	0.692	0.689	0.720
Maximum methane fraction in biogas (%)	0.735	0.782	0.791	0.842	0.849
Volatile solid (added) (g)	12.8	11.1	11.4	12.1	12.9
VS removal fraction (-)	0.667	0.843	0.808	0.833	0.797
Total biogas production rate (L/g-$VS_{removed}$)	0.376	0.372	0.396	0.416	0.463
Average daily biogas production rate (L/g-$VS_{removed}$·day)	0.102	0.109	0.121	0.116	0.139
Maximum daily biogas production rate (L/g-$VS_{removed}$·day)	0.198	0.201	0.208	0.216	0.235
Terminate duration of methane yield (days)	37 - 40	34 - 37	37 - 40	37 - 40	37 - 40

and acid formation. Hence, this experiment set an equal amount of the livestock manure for all of samples to prevent the methane production rate from reducing.

For considering the biogas production as shown in **Table 5**, total gas production and the gas production yield per VS added were also increased when the more chaff was mixed with the piggery manure. As mentioned above, the methane gas fraction in the biogas swung upward as the more chaff was added in the mixture. The addition of the chaff obviously improved the methane production rate and reduced the moisture content. In **Figure 5**, Run 1 and Run 2 used the constant mix proportion of the chaff but Run 1 without the trace elements produced less gas amounts compared to Run 2.

VS removal efficiency was estimated depending on measurement results of VS change amounts and was distinct from the change of the gas production yield. Although the VS was measured for the same sample under the identical conditions, in fact, there was a limitation in conforming tiny change of weight when the 0.5 L-serum bottle was measured directly. Thus, it may be possible to have some error because this study has measured a 0.05 L-additional container at the constant time interval.

Figure 2 indicates cumulative gas production rates of five sample groups for the digestion period. A large amount of gas was produced when the large amount of the chaff was added. Also, the daily gas production rate was slightly rapid at the sample group with the large amount of the chaff as shown in **Figure 3** and a maximum gas production peak was appeared at 10 - 15 days after the digestion was started. Therefore, the sample group with the larger chaff content indicated the more gas production yield and the faster gas production rate, and the earlier maximum gas production peak.

3.2. Ultimate Methane Productivity Evaluation

Theoretical total methane productivities of piggery manure (516 L/g-VS by the hog and 530 L/g-VS by the sow) were higher than those of the cow manure (469 L/g-VS). The ultimate methane yield was in order as follows; 356 L/g-VS by the hog, 275 L/g-VS by the sow, and 148 L/g-VS by the milk cow. Also, the straw and the chaff were known to have much more methane yields as compared with the animal manure [15].

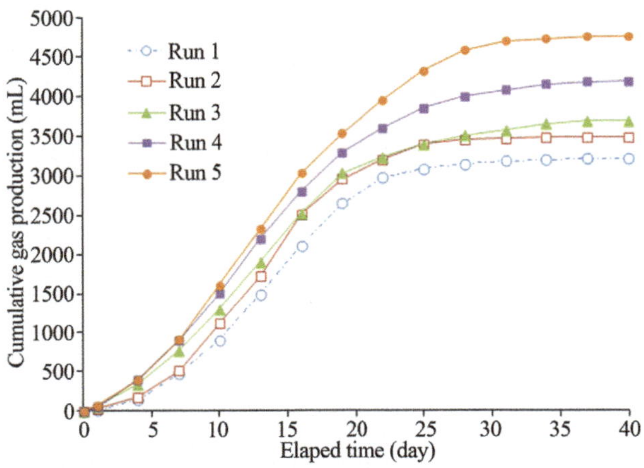

Figure 2. Comparison of cumulative gas yield from five types of samples.

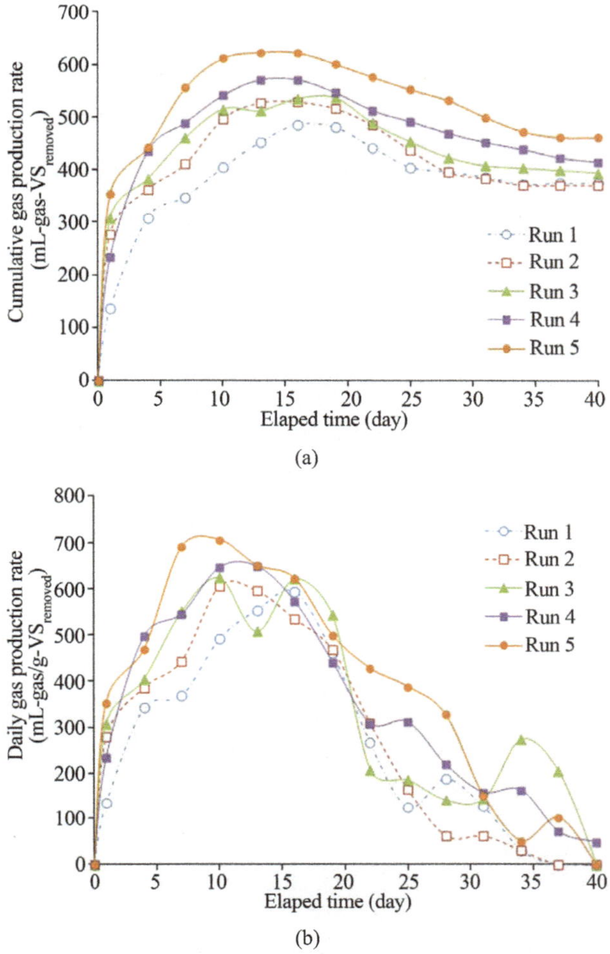

Figure 3. Distribution of gas production rate from five types of samples (3 days period). (a) Cumulative gas production rate; (b) Daily gas production rate.

Modified Gompertz Model (MGM) and Exponential Model (EM) were used to evaluate the cumulative methane yield on a basis of the BMP assay operated for 40 days under the experimental conditions for five sample groups. By comparison of the ultimate methane yields (cumulative methane yields at the final time, 40 days) evaluated by two models, MGM (168.39 - 314.08 mL-CH_4/g-VS_{added}) determined relatively lower ranges than EM (167.62 - 312.47 mL-CH_4/g-VS_{added}).

In **Table 6**, for the mesophilic-dry digestion experiment, the ultimate methane yield was observed in the range of 163.02 mL - 313.27 mL-CH_4/g-VS_{added} which was applied in two models as M_o and B_o. The methane yield was also increased when the chaff mixing ratio was

Table 6. Summary of kinetic parameters predicted by two models.

Model	Parameters	Samples				
		Run 1	Run 2	Run 3	Run 4	Run 5
Modified Gompertz Model	Cumulative methane yield, M (mL-CH_4/g-VS_{added})	168.49	245.33	256.65	291.22	314.08
	Ultimate methane yield, M_o (mL-CH_4/g-VS_{added})	168.02	245.17	256.03	291.57	313.27
	Maximum methane production rate, R_m (mL-CH_4/g-VS_{added}·day)	15.98	16.97	16.90	17.58	19.77
	Lag phase, λ (day)	3.0	3.0	3.0	3.0	3.0
	Determination coefficient, R^2 (-)	0.94	0.98	0.99	0.99	0.98
Exponential Model	Cumulative methane yield, B (mL-CH_4/g-VS_{added})	167.62	245.04	254.79	291.14	312.47
	Ultimate methane yield, B_o (mL-CH_4/g-VS_{added})	168.02	245.17	256.03	291.57	313.27
	Reaction rate constant, k (/day)	0.15	0.19	0.13	0.16	0.15
	Determination coefficient, R^2 (-)	0.89	0.84	0.92	0.85	0.87

Note) *M_o or B_o is the ultimate methane yield observed during the experiment.

high as shown in the case of total gas production rate. Also, Run 1 without the trace elements indicated the lower cumulative methane yield (77.15 mL-CH_4/g-VS_{added}) than that for Run 2 with the trace elements.

The results of BMP assay simulated by two types of models showed a difference and the modified Gompertz Model was better fit compared with the Exponential Model as shown in **Figure 4**.

In the experiment, the ultimate methane yield per the inserted VS was distributed in the range of 163.02 mL - 313.27 mL-CH_4/g-VS_{added}. In previous studies, mixture digestion of food waste and livestock manure was tested in the thermophilic-wet digestion process and the ultimate methane yield per the inserted VS was ranged from 313.35 to 377.43 mL-CH4/g-VS_{added} [16], which was relatively higher than that obtained in this study, and similar to that of the single livestock manure tested at the thermophilic digestion and its result presented as 241 mL/g-VS_{added} [17]. In addition, typical experimental values obtained from foreign countries were 250 L/g-VS for the cow manure including the straw and 279 L/g-VS [18] for the horse-manure with the straw estimated by the thermophilic-wet anaerobic digestion process, and 318 mL/g-VS_{added} [19] for sewage sludge in similar. However, comparative evaluation represented little because BMP assay were varied greatly due to the operation conditions such as nutriment component and addition amount, heat, agitation, and so on [20].

According to the evaluation result of the ultimate methane yield, the mixture of the livestock manure and the chaff in the mesophilic anaerobic digestion process has an advantage when comparing with the only livestock manure containing high moisture content at the

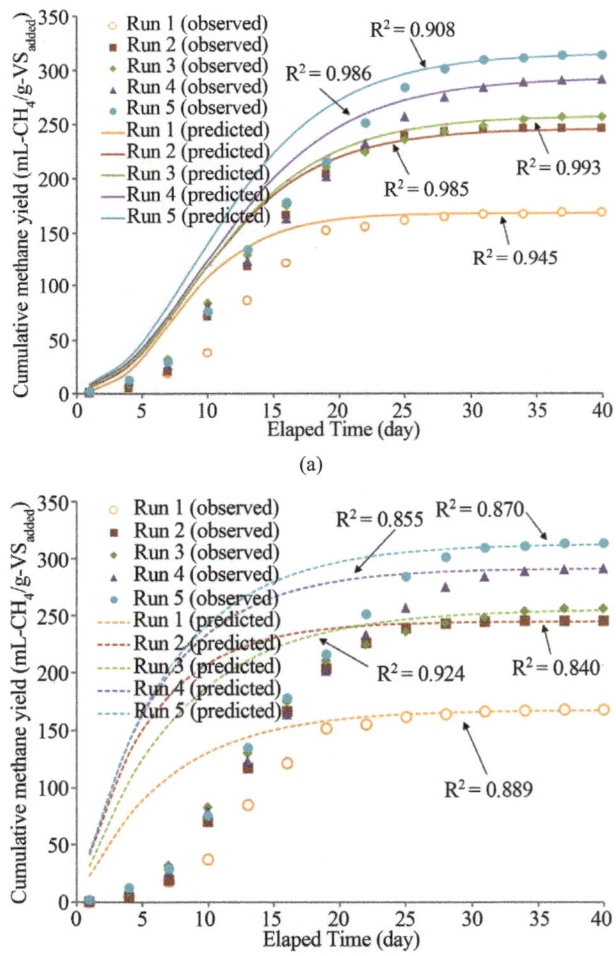

(a)

(b)

Figure 4. Biochemical methane potential assay simulated by two models. a) Modified Gompertz Model; b) Exponential Model.

thermophilic digestion process in common. Also, it is verified that the trace elements are required to add when the chaff to mix with the livestock manure is not sufficient. Therefore, further study is required more detailed research.

3.3. Mixture Ratio of Manure and Chaff

On the domestic side, food waste as a mixture substrate of livestock manure has been mainly used to solve a pending issue and to improve methane production efficiency. Domestic studies related to optimal mixing ratio of the livestock manure and the food waste have been conducted but the study using herbal plants such as chaff and straw is rare. Because rural areas mainly produce the livestock manure and urban areas discharge the food waste mostly, facilities and energy for transportation are required if the food waste is considered as the substrate mixing with the animal manure. The herbal plants like the chaff are easily obtained in the rural area and mostly used to mix with the livestock manure as shown in developed countries having many cases for biogas plants [21].

With that background, this study verified that the methane production rate was varied due to the amount of the chaff mixed with the livestock manure. As the amount of the chaff was enlarged to expand biogas production and to increase methane production efficiency, it was naturally converted to the dry anaerobic digestion process caused by the lower moisture content and it had little regard for wastewater emission. Coverse *et al.* studied on the methane production using the high rate of organic loading as inserting carbohydrate mixture compounds [22]. According to their study, the maximum methane production rate was high at the thermophilic digestion but overall methane production yield was larger at the mesophilic digestion. A previous study applied the straw as the mixture substance had reported the most methane production was yielded from the mixture sample of 3% straw and the livestock manure containing 5% solid [23].

To verify the optimal mixing ratio of the piggery manure (SM) as the main compound and the chaff (SB) as the additional mixing compound in this study, the ultimate methane production yield and the maximum production rate due to SM/SB ratio are presented in **Figure 5**. As shown in **Figure 5(a)**, the ultimate methane production yield and the maximum methane production rate were increased less than 3.5 of SM/SB ratio. The SM/SB ratio is remarkably varied due to the livestock manure composition changed by the stock farmers and the discharging season (or time) and its role may be slightly changed due to the type of the chaff and the condition of the chaff. Therefore, it is necessary to develop a standard model using the more accurate SM/SB ratio examined by several parameters such as solid content, livestock manure types, chaff types, dry conditions of the chaff including moisture content and woody presence, and so on.

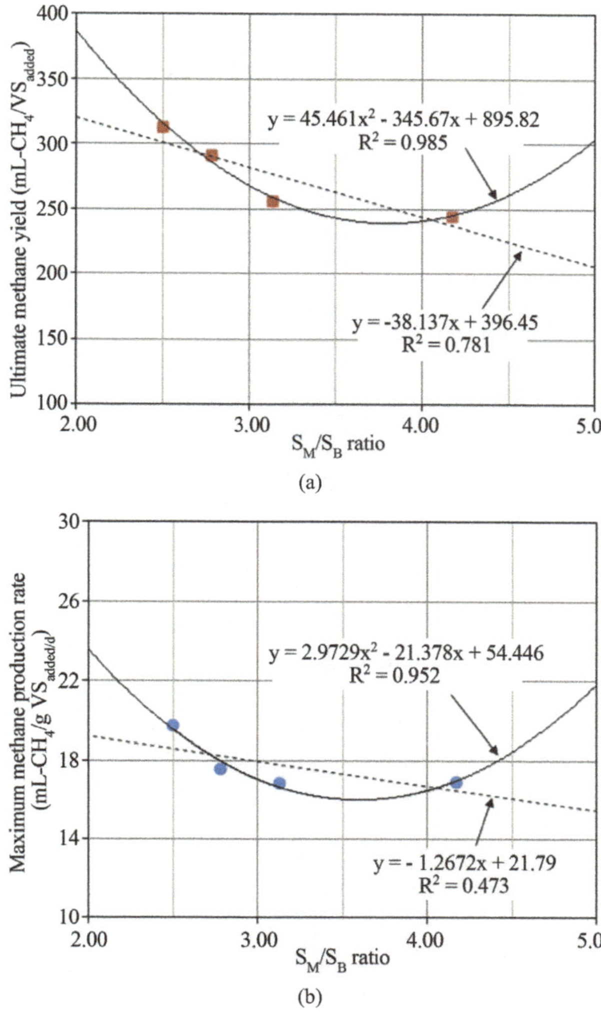

Figure 5. Change of methane yield coefficients in terms of M/H ratio. a) Ultimate methane yield; b) Maximum methane production rate.

4. Conclusions

Considering domestic waste and natural water management, piggery manure is mostly required to be treated among non-point sources of the stock raising system and can also be a main compound of methane production. With that background, this study performed a series of experiments on mesophilic-dry anaerobic digestion of the main compound, piggery manure, mixed with chaff, a mixture compound or substrate, for supplying carbon and controlling moisture. Also, this study conducted comparative evaluation of methane production potential produced from the anaerobic digestion process due to a mixing ratio of the piggery manure and the chaff, and obtained following conclusions after the careful consideration and emphasis on the reasonable mixing ratio:

1) The larger amounts of the chaff resulted in the larger methane production yields as well as the larger gas production yields per volatile solid (VS) inserted. The methane ratio in the produced biogas was also increased as the amount of the chaff was increased.

2) When using the same mixture ratio of the chaff, the gas production from the sample without trace elements was less than that from the sample with trace elements. Therefore, one can say that the trace elements are important when producing the methane from the mixture compounds of the piggery manure and the chaff.

3) For the mesophilic-dry digestion, the cumulative methane production yield per the inserted VS was ranged in 163.02 mL - 313.27 mL-CH4/g-VS_{added}. Similar to the measurement result of total gas production yield, the methane production yield was increased with the higher mixture ratio of the chaff.

4) For the comparative evaluation of the BMP assay, the ultimate methane yield using Modified Gompertz Model was calculated as 168.39 - 314.08 mL-CH_4/g-VS_{added}. It was slightly lower than that in the thermophilic-wet digestion process for the mixture of the domestic food waste and livestock manure and similar to that in the thermophilic-wet digestion process of the sewage activated sludge.

5) When the optimal mixture ratio (SM/SB) between the piggery manure (SM) and the chaff (SB) was less than 3.5, the ultimate methane production yield (M_o) and the maximum production rate (R_m) were decreased.

In conclusion, the mixture of the chaff to the piggery manure controls the moisture content and improves the methane production rate. Also, this study verifies the feasibility of the mesophilic-dry anaerobic digestion process, which is less difficult in operating and maintaining.

5. Acknowledgements

This study was supported by a grant (CUHBRI-2012-0 2-007) of the CNUH-BRI as the project, Development of Future Oriented Healthcare Model and Fundamental Technology for Agro-Medical System, and partly supported by a grant (2013 research project) of Jeonbuk Green Environment Center. Also, we appreciate to Dr. Kyung-Yub Hwang (Research Fellow Emeritus) in Korea Institute of Science and Technology who gives us professional advice to complete this paper and experiments.

REFERENCES

[1] P. Börjesson and B. Mattiasson, "Biogas as a Resource-Efficient Vehicle Fuel," *Trends Biotechnology*, Vol. 26, No. 1, 2008, pp. 7-13.

[2] H. J. Park, M. K. Song and C. K. Na, "Pretreatment Efficiency of Piggery Wastewater Using Coagulation-MAP Sedimentation," *Journal of Korea Society of Waste Management*, Vol. 27, 2010, pp. 457-466.

[3] Y. M. Yoon, Y. J. Kim and C. H. Kim, "The Evaluation of Economical Efficiency to Composting and Liquefying Process of Biomass Discharged in Pig Breeding," *Agriculture Economics*, Vol. 31, 2009, pp. 39-62.

[4] D. Bolzonella, L. Innocenti, P. Pavan, P. Traverso and F. Cecchi, "Semi-Dry Thermophilic Anaerobic Digestion of the Organic Fraction of Municipal Solid Waste: Focusing on the Start-Up Phase," *Bioresource Technology*, Vol. 86, No. 2, 2003, pp. 123-129.

[5] P. Pavan, P. Battistoni and J. Mata-Alvarez, "Performance of Thermophilic Semi-Dry Anaerobic Digestion Process Changing the Feed Biodegradability," *Water Science and Technology*, Vol. 41, 2000, pp. 75-81.

[6] N. Forster-Carneiro, M. Perez and L. I. Romero, "Anaerobic Digestion of Municipal Solid Wastes: Dry Thermophilic Performance," *Bioresource Technology*, Vol. 99, No. 17, 2008, pp. 8180-8184.

[7] B. Montero, J. L. Garcia-Morales, D. Sales and R. Solera, "Analysis of Methanogenic Activity in a Thermophilicdry Anaerobic Reactor: Use of Fluorescent *in Situ* Hybridization," *Waste Management*, Vol. 29, No. 3, 2009, pp. 1144-1151.

[8] S. E. Oh, M. K. Lee and D. H. Kim, "Continuous Mesophilic-Dry Anaerobic Digestion of organic Solid Waste," *Journal of Korean Society of Environmental Engineers*, Vol. 31, 2009, pp. 341-345.

[9] J. J. Lay, Y. Y. Li and T. Noike, "Development of Bacterial Population and Methanogenic Activity in a Laboratory-Scale Landfill Bioreactor," *Water Research*, Vol. 32, No. 12, 1998, pp. 3673-3679.

[10] W. F. Owen, D. C. Stuckey, J. B. Healy, L. Y. Young and P. L. McCarty, "Bioassay for Monitoring Biochemical Methane Potential and Anaerobic Toxicity," *Water Research*, Vol. 13, No. 6, 1979, pp. 485-492.

[11] R. S. Daniel and J. M. Tiedje, "General Method for Determining Anaerobic Biodegradation Potential," *Applied and Environmental Microbiology*, Vol. 47, 1984, pp. 850-857.

[12] I. Angelidaki, M. Alves, D. Bolzonella, L. Borzacconi, J. L. Campos, A. J. Guwy, S. Kaalyuzhnyi, P. Jenicek and J. B. van Lier, "Defining the Biomethane Potential (BMP) of Solid Organic Wastes and Energy Crops: A Proposed Protocol for Batch Assays," *Water Science and Technology*, Vol. 59, No. 5, 2009, pp. 927-934.

[13] M. H. Zwietering, I. Jongenburger, F. M. Rombouts and K. van't Riet, "Modeling of the Bacterial Growth Curve," *Applied and Environmental Microbiology*, Vol. 56, No. 6, 1990, pp. 1875-1881.

[14] T. L. Hansen, J. E. Schmidt, I. Angelidaki, E. Marca, J.

Cour Jansen, H. Mosboek and T. H. Christensen, "Method for Determination of Methane Potentials of Solid Organic Waste," *Waste Management*, Vol. 24, No. 4, 2004, pp. 393-400.

[15] H. B. Møllera, S. G. Sommera and B. K. Ahringb, "Methane Productivity of Manure, Straw and Solid Fractions of Manure," *Biomass and Bioenergy*, Vol. 26, No. 5, 2004, pp. 485-495.

[16] J. K. Park, S. R. Jeong, J. H. Kang, Y. M. Ahn, H. E. Jin and N. H. Lee, "A Study on Optimization Condition for Anaerobic Co-Digestion of Food Waste with Livestock Wastes," *Journal of Korea Society of Waste Management*, Vol. 29, 2012, pp. 356-364.

[17] S. H. Kim, H. C. Kim, C. H. Kim and Y. M. Yoon, "The Measurement of Biochemical Methane Potential in the Several Organic Waste Resources," *Korean Journal of Soil Science and Fertilizer*, Vol. 43, 2010, pp. 356-362.

[18] S. Aslanzadeh, M. J. Taherzadeh and I. S. Horvath, "Pretreatment of Straw Fraction of Manure for Improved Biogas Production," *Bio-Resources*, Vol. 6, 2011, pp. 5193-5205.

[19] J. G. Lin, Y. S. Ma, A. C. Chao and C. L. Huang, "BMP Tests on Chemically Pretreated Sludge," *Bioresources Technology*, Vol. 68, No. 2, 1999, pp. 187-192.

[20] P. Shanmugam and N. J. Horan, "Simple and Rapid Methods to Evaluate Methane Potential and Biomass Yield for a Range of Mixed Solid Wastes," *Bioresource Technology*, Vol. 100, No. 1, 2008, pp. 471-474.

[21] D. Jackowiak, D. Bassard, A. Pauss and T. Ribeiro, "Optimization of a Microwave Pretreatment of Wheat Straw for Methane Production," *Bioresource Technology*, Vol. 102, No. 12, 2011, pp. 6750-6756.

[22] J. C Converse, R. E. Graves and G. W. Evans, "Anaerobic Degradation of Dairy Manure under Mesophilic and Thernophilic Temperatures," *Transactions of the ASAE*, Vol. 20, 1977, pp. 336-340.

[23] J. E. Robbins, M. T. Armold and S. L. Lacher, "Methane Production from Cattle Waste and Delignified Strawt," *Infection and Immunity*, Vol. 38, 1979, pp. 175-177.

Coating of Medical-Grade PVC Material with ZnO for Antibacterial Application

Huaxiang Lin[1], Luyao Ding[1], Weihua Deng[1], Xuxu Wang[1*], Jinlin Long[1], Qun Lin[2]
[1]Research Institute of Photocatalysis, Fujian Provincial Key Laboratory of Photocatalysis, State Key Laboratory Breeding Base, Fuzhou University, Fuzhou, China
[2]First Affiliated Hospital, Fujian Medical University, Fuzhou, China

ABSTRACT

The ZnO sol well-crystallized was prepared by the sol-gel method. The ZnO films were coated on medical-grade PVC surface by the improved organic-inorganic interfacial adhesion method. The physical and photocatalytic properties of the samples were characterized by XRD, SEM, DRS spectra and measured by the photodegradation reaction of Rhodamine B (RhB) and anti-bacteria for *Escherichia coli* (*E. co li*), respectively. The results show that pretreatment of PVC by the mix solution of THF-PVC helps to improve the amount and adhesion strength of ZnO suspension to PVC surface. The photocatalytic and antibacterial properties of the THF-ZnO/PVC film are better than that of the ZnO/PVC and neat PVC. Under UV irradiation, the THF-ZnO/PVC film shows the best antibacterial properties with 99% kill rate of bacteria.

Keywords: ZnO; PVC; Photocatalysis; Antibacterial

1. Introduction

The increasing use of polymer materials such as polyethylene, polyurethanes, and poly (vinyl chloride) (PVC) in the hospital care has led to a concomitant increase in the incidence of biomaterial-related infections (BRI) [1]. Adhesion of bacteria to biomaterials led to the formation of biofilm on the surface, which plays a crucial role in the pathogenesis of the BRI [1-3]. The growth and production of biofilm protect the bacteria from the host defence mechanisms and external agents as the drug treatments [1,2,4-7], which makes the cure of the bacterial infections quite difficult and requires either higher doses or more potent antibiotics.

In order to efficiently prevent or reduce biofilm formation, many efforts have been done to enhance the antibacterial properties of biomaterials. Some efforts such as modifying the physicochemical properties of biomaterial surface, coating with silver, azidation treatment, antibiotic impregnation into the polymer matrix, have been examined in recent years [8-12].

Recently, coating semiconductor photocatalyst on materials' surface for antibacterial activity has attracted great interest due to the following advantages when compared with other methods: 1) excellent photocatalytic and antimicrobial activity, 2) nontoxicity and bio-compatibility, 3) strong physicochemical stability and durable antibacterial properties. The antibacterial mechanism of semiconductor photocatalysis is mainly based on its UV photoactive property. Taking ZnO as an example was shown in Equations (1)-(4) [13].

$$ZnO + h\nu \rightarrow e^- + h^+ \quad (1)$$

$$h^+ + OH^- \rightarrow \bullet OH \quad (2)$$

$$e^- + O_2 \rightarrow O_2 \bullet^- \quad (3)$$

$$O_2 \bullet^- + 2H^+ + e^- \rightarrow H_2O_2 \quad (4)$$

When ZnO was exposed to UV light with the wavelength less than 388 nm, the semiconductor surface generated valence band holes and conduction band electrons as in Equation (1). The holes could react with water or surface hydroxyls (H_2O/OH^-) to generate hydroxyl radicals (HO•), while conduction band electrons reacted with adsorbed molecular oxygen (O_2) to generate superoxide radical (O_2•). The radicals (O_2• and HO•) can destroy and shrink bacterial cells by reacting with the organism of bacterial cells in the photocatalytic process [14, 15]. Among various oxides semiconductor photocatalysis,

*Corresponding author.

TiO_2 and ZnO are generally considered to be the most suitable for antibacterial agents [15-18]. TiO_2 film coated on different materials' surface is effective at killing bacteria and viruses under UV irradiation [15,16]. ZnO as antibacterial agents has been reported exhibiting higher antibacterial activity than other metal oxide nanoparticles [19]. It has supposed that ZnO not only kills bacteria under UV irradiation bacteria but also inhibits the bacterial growth under normal visible lighting conditions [20]. To extend the usage of antibacterial agents, many studies have been done to coat ZnO film on many materials' surface that was widely used in life such as glass, ceramic, stainless steel, polymer, and so on. In general, to obtain highly adherent and antibacterial activity of ZnO film, heating the materials that have been modified by ZnO film at high temperature is required. However, one of the main barriers to coat ZnO on polymer is the low heat resistance of polymer. Low heat resistance of polymer disables the routine coating method to be used to modify inorganic ZnO film on polymer.

Recently, organic-inorganic composite membrane by mixing inorganic and organic materials has been supposed to use because of its simplicity, mild conditions and stability [21,22]. In this work, a similar organic-inorganic interfacial adhesion method was used to coat ZnO film on polymer materials. First, suspension of well-crystallized ZnO was prepared by sol-gel at low temperature. Then, to improve the adhesion strength of ZnO to PVC, the mix solution of PVC and THF was used to pretreat the surface of PVC sheet. After pretreated by the mix solution of PVC and THF, PVC sheets were coated with ZnO film by dip-coating with suspension ZnO. The photocatalytic and antibacterial activity of ZnO film was investigated. To evaluate the antibacterial activity after modified by ZnO, the performance of ZnO film was compared with neat PVC under similar operating conditions.

2. Materials and Methods

2.1. Material

Medical-grade PVC sheets were purchased from Tyco Healthcare International Trading (Shanghai) Co., Ltd., and were washed with absolute alcohol before use. *Escherichia coli* were cultured at 37°C in Luria-Bertani (LB) medium (pH 7.0) containing tryptone 1%, yeast extract 0.5%, sodium chloride 1%. When tested, the bacterial concentration was adjusted to the required concentration (10^4 - 10^5 cfu/ml). Other reagents were of analytical grade and used as received.

2.2. Preparation of ZnO Sol

$Zn(Ac)_2 \cdot 2H_2O$ was dissolved in methanol to obtained a 0.075 M solution of Zn (Ac). Ammonia was dropped into the solution to adjust the pH to 8 and then stirred the solution at 65°C for 2 h to obtain ZnO sol.

2.3. Coating ZnO on PVC Surface

The ZnO film was prepared by dip-coating method with ZnO suspension as precursor. The PVC sheets (1.5 × 4.5 cm) were pre-immersed in THF-PVC solution for 10s. Immediately, the PVC samples were dipped into ZnO colloidal sol and were slowly pulled out of the sol at a speed of 1200 mm/h in an ambient atmosphere. The ZnO gel film on PVC were dried in an oven at 60°C for 30 min and then were dried at 120°C for 2 h. After seven such coating steps, the transparent ZnO film on PVC were obtained. The PVC sheets without pretreatment were coated with ZnO under similar operating conditions for comparison. For ease of presentation, the ZnO film with pretreatment were labelled as THF-ZnO/PVC and that without pretreatment were labelled as ZnO/PVC.

2.4. Structure and Morphology Characterizations

The crystal structure of the sample was characterized by X-ray diffraction (XRD) (Bruker D8 Advance powder diffractometer). The scanning electron microscopy (SEM) was performed on a Philips XL30 ESEM system. The UV-Vis spectra were obtained in the diffuse reflection mode using a Carry 500 spectrometer at a resolution of 2 nm. The spectra of the PVC, ZnO/PVC and THF-ZnO/PVC sheets were recorded in the ultraviolet and visible in the wavelength range from 250 to 800 nm.

2.5. Photocatalytic and Antibacterial Property

2.5.1. Photocatalytic Activity

The photocatalytic activities of PVC, ZnO/PVC and THF-ZnO/PVC were determined by monitoring the degradation of organic dye Rhodamine B (RhB). A quartz tube reactor with condenser pipe was used and four 4W-UV lamps with wavelength at 365 nm were fixed around the reactor. The PVC sheets (1.5 × 4.5 cm) were immersed in 20 ml aqueous solution of RhB (1×10^{-5} mol/dm^3) and kept for 12 h in black prior to irradiation. The concentration of RhB at different irradiation times was monitored by Varian UV-260 spectrometer.

2.5.2. Bacterial Adhension

The neat PVC, ZnO/PVC and THF-ZnO/PVC sheets were immersed in the aqueous solution of *Escherichia coli* (*E. coli*), and kept at 37°C for 24 h. The sheets were taken out and rinsed gently with sterile phosphate buffered saline (PBS) to remove the non-adherent bacteria. Then, the bacteria adhered on the sheets were eluted into 5 ml of sterile PBS in an ultrasonic cleaner for 5 min. The number of the eluted bacteria was then determined by colony counts (CFU). The adherent number was expressed by the ratio of the total adherent bacterial to the area of the measured sample.

2.5.3. Antibacterial Property

ZnO/PVC sheets (2.5 × 2.5 cm) were placed on sterile plates, and then 0.5 ml broth inoculated with 10^4 cfu/ml of *E. coli* was added onto the surfaces. The samples were irradiated with an 8 W UV lamp (with wavelength at 365 nm) for 150 min. After irradiated, the sheets were rinsed with sterile phosphate buffered saline (PBS) and the number of viable bacteria was determined by colony counts (CFU). The blank PVC sample was also test for comparison. The antibacterial property of PVC, ZnO/PVC and THF-ZnO/PVC was valued by the kill percentage, a ratio of the dead number of the bacterial to the initial number of the bacterial.

3. Results and Discussion

3.1. Structure and Morphology

The ZnO nanoparticles were prepared by simple sol-gel method. The size analysis of ZnO sol indicated that most of ZnO colloid particles are located in the range of 50 - 70 nm (**Figure 1**). ZnO colloid particles were dried into powder at 120°C for X-ray diffraction（XRD）analysis. Several obvious peaks are observed at 2θ values at 31.6, 34.3, 36.1, 47.3, 56.4, 62.5 and 67.7, which correspond to wurtzite structure of hexagonal ZnO [23] (**Figure 2**). No other phases are observed in the patterns, indicating that the sample that prepared by this method are pure wurtzite ZnO.

Figure 3 shows the SEM images of neat PVC, ZnO/PVC and THF-ZnO/PVC. The neat PVC surface is smooth with little flaws. After coated only with ZnO colloid, the morphology of ZnO/PVC become rougher and a small ZnO nanostructure sheet spread on the ZnO/PVC surface. However, when PVC were pretreated by THF-PVC solution and subsequently coated with ZnO

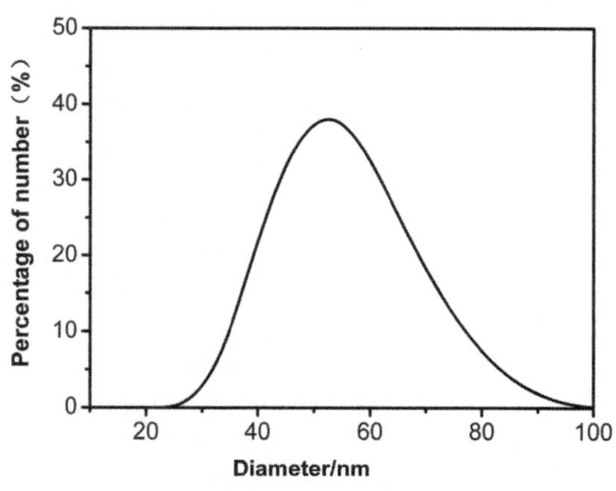

Figure 1. Size distribution of ZnO colloid particles.

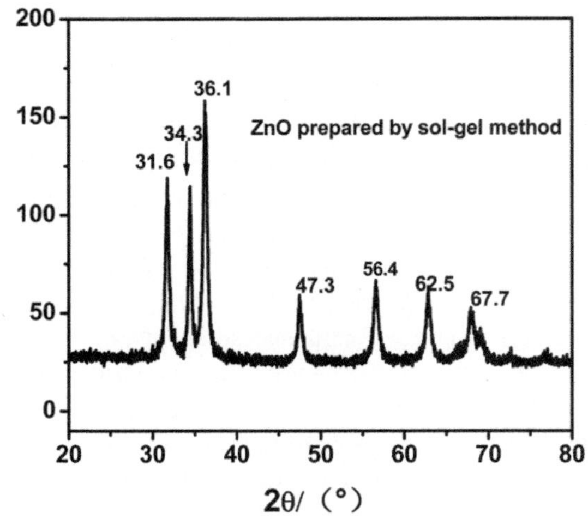

Figure 2. XRD spectra of ZnO.

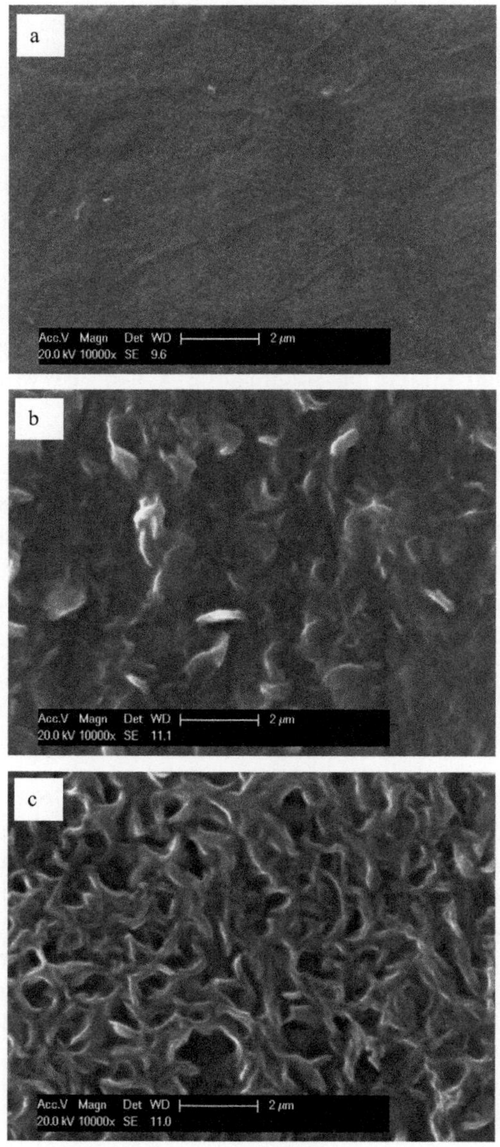

Figure 3. SEM images of different PVC surface; a) neat PVC; b) ZnO/PVC; THF-ZnO/PVC.

colloid, many ZnO nanostructure sheets stacked on THF-ZnO/PVC surface closely and the surface of THF-ZnO/PVC was composed by irregularity cross-linked ZnO sheets. This result indicated that pretreated PVC by THF-PVC solution improves the adherent amount of ZnO on PVC surface. The Optical absorption property of PVC, ZnO/PVC and THF-ZnO/PVC were examined with UV-Vis reflectance spectra (DRS), as shown in **Figure 4**. The DRS spectra show that ZnO/PVC and THF-ZnO/PVC exhibit significant enhancement of light absorption in the UV region than nude PVC, which is attributed to band-gap excitation of ZnO. Another difference in the DRS spectra is that the absorption baseline of the resulting samples in visible light region is decrease in the order of PVC > ZnO/PVC > THF-ZnO/PVC. This is originnated from the transparent nature of the samples. ZnO deposited on PVC reduces transparency of PVC, which causes more of incident light reflect and consequently enhances the baseline absorption in visible light region. The more amount of ZnO coated on PVC, the lower baseline absorption of the sample.

3.2. Photocatalytic Activity

Figure 5 shows the change in photocatalytic conversions

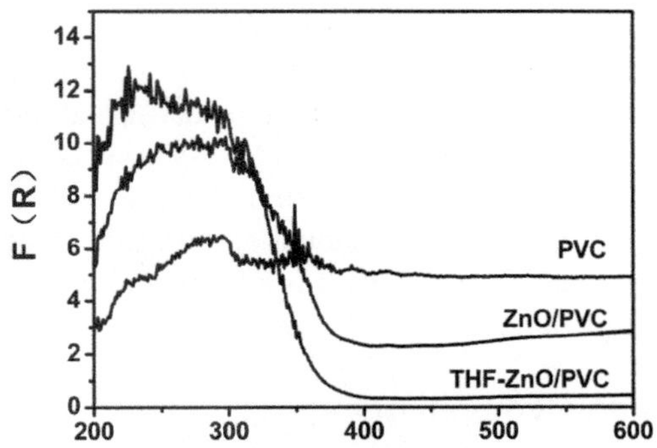

Figure 4. UV-vis diffuse reflectance spectra of PVC, ZnO /PVC and THF-ZnO/PVC.

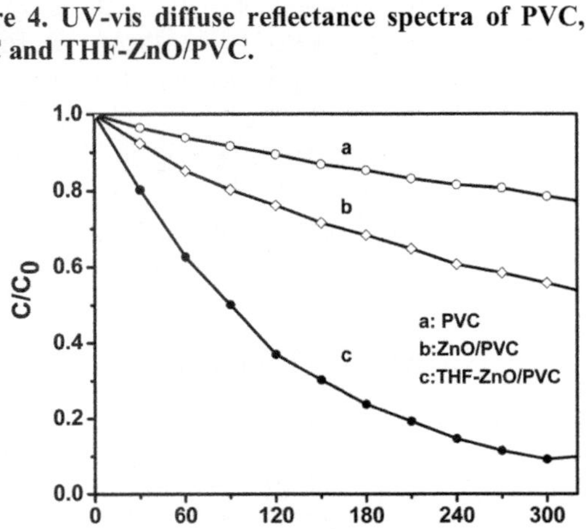

Figure 5. Photocatalytic degradation conversion of RhB on PVC, ZnO/PVC and THF-ZnO/PVC.

of RhB with reaction time on the nude PVC, ZnO/PVC and THF-ZnO/PVC. As can been seen from these spectra, the nude PVC sheet exhibits no photocatalytic activity. Only about 25% of RhB are photodecomposed following about 300 min of illumination. The photocatalytic activity of ZnO/PVC samples exhibit a little better than that of neat PVC, with the conversion rate of RhB increases to about 50% of RhB upon about 300 min of illumination. The pretreatment of THF-PVC solution for PVC has much great effect on the photocatalytic activity of the resulting sample. As can be seen form spectra c in **Figure 5**, the degraded rate of RhB on THF-ZnO/PVC is greatly increased and about 90% of RhB are degraded upon about 300 min of illumination. The result shows that pretreatment of THF-PVC for PVC is beneficial for the coating of ZnO on PVC surface.

3.3. Bacterial Adhesion and Sterilization Behavior

Figure 6 shows the adherent number of *E. c oli* to the surface of nude PVC, ZnO/PVC and THF-ZnO/PVC under dark condition. After culture bacteria on the sample surface in dark for 18h, the number of *E. coli* adhered on nude PVC is similar to initial number of *E. coli* (**Figure 6(d)**), showing that the nude PVC material is inertial to prevent adhesion of bacterial. The adherent number of *E. coli* on ZnO/PVC decreases a little when compared with that of nude PVC as shown in **Figures 6(a)** and **(b)**, suggesting that coating ZnO on PVC helps to enhance antibacterial activity of PVC. Especially, when the PVC was pretreated with mix solution of THF-PVC and subsequently coated with ZnO, the antibacterial activity of THF-ZnO/PVC increases greatly with the adherent number of *E. c oli* is almost zero (**Figure 6(c)**). The result suggests that THF-ZnO/PVC can effective prevent the adhesion and growth of bacterial on PVC surface.

Figure 7 shows the kill percentage of *E. coli* on PVC, ZnO/PVC and THF-ZnO/PVC under UV irradiation for 150 min. PVC exhibits no antibacterial activity under UV light. And ZnO/PVC shows a little antibacterial activity with the survival ratio of *E. coli* on ZnO/PVC is 46.8% after irradiation for 150 min. However, the THF-ZnO/PVC exhibits stronger antibacterial property. More than 99% of the bacteria were killed under UV irradiation for 150 min.

It has been reported that it is the photocatalytic property of metal oxide that is responsible for the excellent sterilization properties [15,24]. In other words, the higher the photocatalytic property, the higher the sterilization activity, and vice versa. This proposition was identified by the photocatalytic property and sterilization activity of TiO_2 [15]. In the present work, the photocatalytic property and sterilization activity of THF-ZnO/PVC are much better than that of ZnO/PVC. The main reason is may be

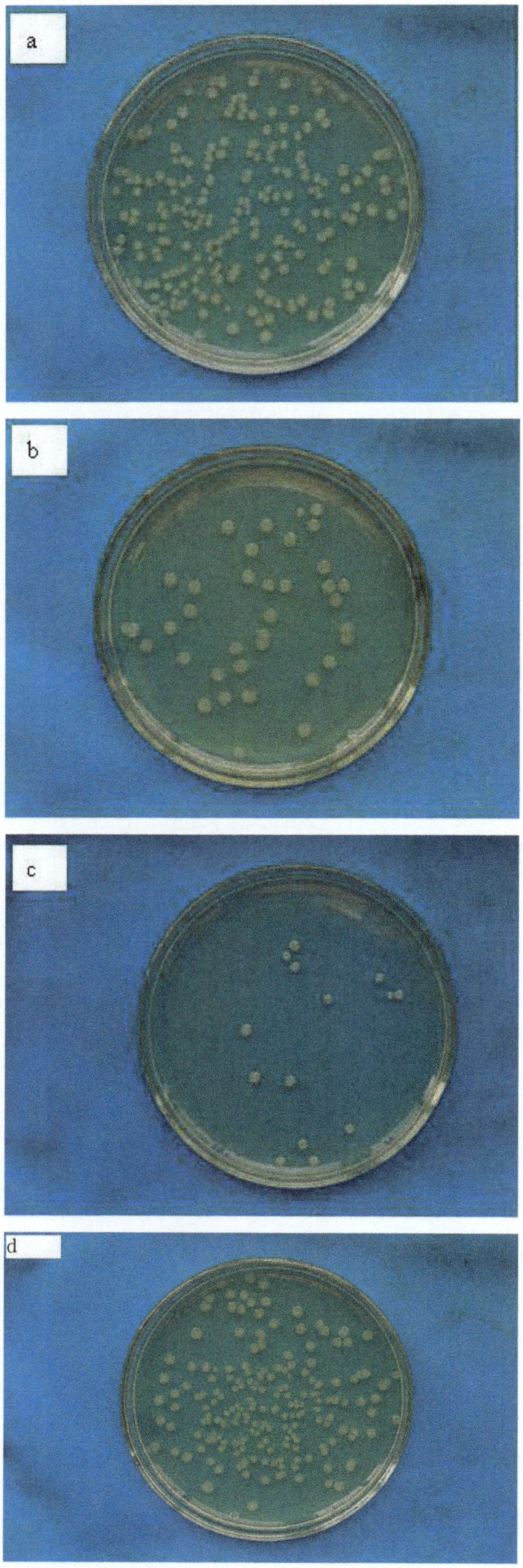

Figure 6. Bacteria adherent number on different PVC surface under dark condition. a) neat PVC; b) ZnO/PVC; THF-ZnO/PVC; d) initial number.

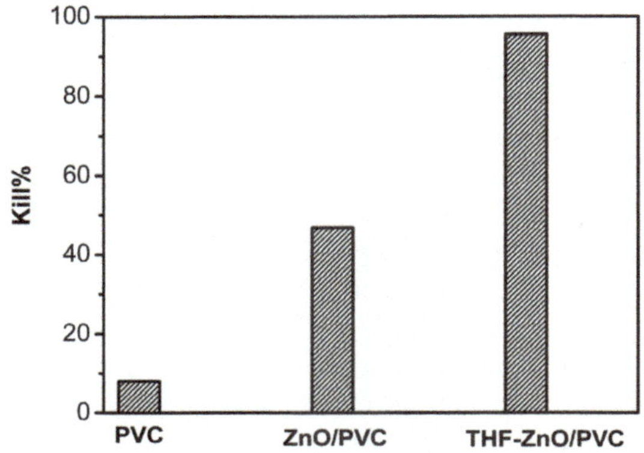

Figure 7. Kill percentage of bacteria on the samples under UV irradiation for 150 min.

that the amount of ZnO on THF-ZnO/PVC is larger than that on ZnO/PVC. Under UV irradiation, the amount of electron and hole produced on THF-ZnO/PVC were higher than that on ZnO/PVC that resulted in higher concentration of radicals (O_2• and HO•) on THF-ZnO/PVC.

4. Conclusion

This work infers that the antibacterial properties of medical-grade PVC material can be enhanced by coating with ZnO. After pretreatment by THF-PVC solution and subsequently coated with ZnO, the PVC surface can not only inhibit the growth of bacteria on dark conditions effectively but also kill bacteria under UV irradiation. The result provides a convenient method that can enhance the adhesive strength and the amount of inorganic oxide film coated on organic materials. It is also easy to operate and can be applied in various substrates.

5. Acknowledgements

This work was financially supported by the NSFC (Grants Nos. 21003021, 21173044 and 21203029), the great Science and Technology Project of Fujian Province of P. R. China (2012Y4003), the National Science Foundation for Fostering Talents in Basic Research of China (No. J1103303), and National Basic Research Program of China (973 Program, No. 2012CB722607) and the Science and Technology project of Fuzhou University (2011-XQ-09).

REFERENCES

[1] S. Lakshmi, S. S. Pradeep and J. A. Kumar, "Bacterial Adhesion onto Azidated Poly (Vinyl Chloride) Surfaces," *Journal of Biomedical Materials Research*, Vol. 61, No. 1, 2002, pp. 26-32.

[2] P. Vergidis and R. Patel, "Novel Approaches to the Diagnosis, Prevention, and Treatment of Medical Device-Associated Infection," *Infectious Disease Clin ics of North*

America, Vol. 26, No. 1, 2012, pp. 173-186.

[3] D. S. Jones, J. G. McGovern and D. A. Woolfson, "Role of Physiological Conditions in the Oropharynx on the Adherence of Respiratory Bacterial Isolates to Endotracheal Tube Poly (Vinyl Chloride)," *Biomaterials*, Vol. 18, No. 6, 1997, pp. 503-510.

[4] Y. H. An and J. R. Friedman, "Prevention of Sepsis in Total Joint Arthroplasty," *Journal of Hospita l Infection*, Vol. 33, No. 2, 1996, pp. 93-108.

[5] Y. H. An and J. R. Friedman, "Concise Review of Mechanisms of Bacterial Adhesion to Biomaterial Surfaces," *Journal of Biomedical Materials Research*, Vol. 43, No. 3, 1998, pp. 338-348.

[6] A. G. Gristina, "Biomaterial-Centered Infection: Microbial Adhesion versus Tissue Integration," *Science*, Vol. 237, No. 4822, 1987, pp. 1588-1595.

[7] A. G. Gristina, C. D. Hobgood and L. X. Webb, "Adhesive Colonization of Biomaterials and Antibiotic Resistance," *Biomaterials*, Vol. 8, No. 6, 1987, pp. 423-426.

[8] N. P. Desai, S. F. A. Hossainy and J. A. Hubbei, "Surface-Immobilized Polyethylene Oxide for Bacterial Repellence," *Biomaterials*, Vol. 13, No. 7, 1992, pp. 417-420.

[9] K. D. Park, Y. S. Kim and D. K. Hun, "Bacterial Adhesion on PEG Modified Polyurethane Surfaces," *Biomaterials*, Vol. 19, No. 7-9, 1998, pp. 851-859.

[10] Z. Zdanowski, B. Koul and E. Hallberg, "Inflence of Heparin Coating on *in Vi tro* Bacterial Adherence to Poly (Vinyl Chloride) Segments," *Journal of Bio materials Science, Polymer Edition*, Vol. 8, No. 11, 1997, pp. 825-832.

[11] D. J. Balazs, K. Triandafillu and Y. Chevolot, "Surface Modification of PVC Endotracheal Tubes by Oxygen Glow Discharge to Reduce Bacterial Adhesion," *Surface and Interface Analysis*, Vol. 35, No. 3, 2003, pp. 301-309.

[12] K. Triandafillu, D. J. Balazs and B. D. Aronsson, "Adhesion of Pseudo-Monas Aeruginosa Strains to Untreated and Oxygen-Plasma Treated Poly (Vinyl Chloride) (PVC) from Endotracheal Intuba-tion Devices," *Biomaterials*, Vol. 24, No. 8, 2003, pp. 1507-1518.

[13] T. J. Whang, M. T. Hsieh and H. H. Chen, "Visible-Light Photocatalytic Degradation of Methylene Blue with Laser-Induced Ag/ZnO Nanoparticles," *Applied Surface Science*, Vol. 258, No. 7, 2012, pp. 2796-2801.

[14] W. K. Ho, J. C. Yu and S. C. Lee, "Photocatalytic Activity and Photo-Induced Hydrophilicity of Mesoporous TiO_2 Thin films Coated on Aluminum Substrate," *Applied Catalysis B: Environmental*, Vol. 73, No. 1-2, 2007, pp. 135-143.

[15] H. X. Lin, Z. T. Xu and X. X. Wang, "Photocatalytic and Antibacterial Properties of Medical-Grade PVC Material Coated With TiO_2 Film," *Journal of Biom edical Materials Research*, Vol. 87, No. 2, 2008, pp. 425-431.

[16] W. Y. Su, S. H. Wang and X. X. Wang, "Plasma Pre-Treatment and TiO_2 Coating of PMMA for the Improvement of Antibacterial Properties," *Surface and Coatings Technology*, Vol. 205, No. 2, 2010, pp. 465-469.

[17] Y. Li, W. Zhang and J. F. Niu, "Mechanism of Photogenerated Reactive Oxygen Species and Correlation with the Antibacterial Properties of Engineered Metal-Oxide Nanoparticles," *ACS Nano*, Vol. 6, No. 6, 2012, pp. 5164-5173.

[18] A. Thill, O. Zeyons, O. Spalla and F. Chauvat, "Cytotoxicity of CeO_2 Nanoparticles for *Escherichia coli.* Physico-Chemical Insight of the Cytotoxicity Mechanism," *Environmental Science & Technology*, Vol. 40, No. 19, 2006, pp. 6151-6156.

[19] K. H. Tam, A. B. Djurisic, C. M. N. Chan and Y. H. Leung, "Antibacterial Activity of ZnO Nanorods Prepared by a Hydrothermal Method," *Thin Solid Film* , Vol. 516, No. 18, 2008, pp. 6167-6174.

[20] N. Jones, B. Ray, K. T. Ranjit and A. C. Manna, "Antibacterial Activity of ZnO Nanoparticle Suspensions on a Broad Spectrum of Microorganisms," *FEMS Microbiology Letters*, Vol. 279, No. 1, 2008, pp. 71-76.

[21] Y. Yang, H. Zhang, P. Wang and Q. Zheng, "The Influence of Nano-Sized TiO_2 Fillers on the Morphologies and Properties of PSF UF Membrane," *Journal of Membrane Science*, Vol. 288, No. 1-2, 2007, pp. 231-238.

[22] R. A. Damodara, S. J. Youa and H. H. Chou, "Study the Self Cleaning, Antibacterial and Photocatalytic Properties of TiO_2 Entrapped PVDF Membranes," *Journal of Hazardous Materials*, Vol. 172, No. 2-3, 2009, pp. 1321-1328.

[23] S. W. Liu, C. Li and J. G. Yu, "Improved Visible-Light Photocatalytic Activity of Porous Carbon Self-Doped ZnO Nanosheet-Assembled Flowers," *Cryst Eng Comm*, Vol. 13, No. 7, 2011, pp. 2533-2541.

[24] M. Bekbolet, "Photocatalytic Bactericidal Activity of TiO_2 in Aqueous Suspensions of *E. coli*," *Water Science and Technology*, Vol. 35, No. 11-12, 1997, pp. 95-100.

A Comparison between Evolutionary Metaheuristics and Mathematical Optimization to Solve the Wells Placement Problem

Ghazi AlQahtani, Ahmed Alzahabi, Ernee Kozyreff, Ismael R. de Farias Jr., Mohamed Soliman
Texas Tech University, Lubbock, USA

ABSTRACT

The Wells Placement Problem (WPP) consists in choosing well locations within an oil reservoir grid to maximize the reservoir total oil production, subject to distance threshold between wells and number of wells cap constraints. A popular approach to WPP is Genetic Algorithms (GA). Alternatively, WPP has been approached in the literature through Mathematical Optimization. Here, we conduct a computational study of both methods and compare their solutions and performance. Our results indicate that, while GA can provide near-optimal solutions to instances of WPP, typically Mathematical Optimization provides better solutions within less computational time.

Keywords: Wells Placement; Genetic Algorithm; Integer Programming

1. Introduction

Consider an oil reservoir for which oil production totals over the entire reservoir area are given (or estimated). We are interested in the Wells Placement Problem (WPP), *i.e.*, the problem of maximizing the total oil production of the reservoir by selecting locations to build at most a certain number of vertical wells at a given minimum distance apart from one another.

WPP (and variations of it) has been the subject of a number of studies in the last decades [1-9]. The main approaches used to find solutions have been Genetic Algorithms (GA), which are a special type of Evolutionary Metaheuristics, and Integer Programming (IP), a subclass of Mathematical Optimization.

To the best of our knowledge, the earliest study of wells placement optimization using IP was the work conducted in 1974 by Rosenwald and Green [8], who developed a numerical optimization framework to select optimal positions of wells. Their objective was to minimize the difference between actual and desired flow curves by choosing optimum well locations from predefined location sets. In 1997, Vasantharajan and Cullick [9] introduced the term *reservoir quality* to represent reservoir connectivity adjusted by tortuosity, which was the surface area surrounding the flow by the volume contained. The well site selection process was performed through IP, with the goal of maximizing reservoir quality. Later, in 1999, Cullick *et al.* [4] improved the formulation of the problem by reducing its number of constraints.

On the other side, GA is the most widely used method to solve WPP optimization [1]. Guyaguler *et al.* [6] developed a hybrid method of GA linked with polytope search and coupled with a proxy generated by kriging. The objective function in this work was to maximize net present value by finding the optimum injection well locations. The output of this work indicates that using kriging as a proxy with the hybrid optimization framework reduced the computational expense considerably. This work was extended by Badru and Kabir [2] by including horizontal wells and linking the hybrid GA with experimental design to find the impact of uncertainty on recovery and the number of wells required. In 2006, Bangerth *et al.* [3] compared GA with four algorithms designed to solve WPP. GA was also compared with Covariance Matrix Adaptation-Evolution Strategy by Ding [5] to optimize well locations for different types of wells with calculated productivity index as the objective function. Onwunalu *et al.* [7] used GA combined with statistical proxy to establish a relationship between the variables and the objective function using unsupervised learning techniques.

Some of the articles cited above report computational results using either GA or IP, but we are not aware of a study that compares both. In this paper, we use GA and IP to solve instances of WPP, and report the computational results of those tests.

Specifically, our test reservoir sits on an $m \times n$ grid, where each of the mn sites in the grid represents a possible location for a well. For each $(i,j) \in I \times J$, where $I = \{1,\cdots,m\}$ and $J = \{1,\cdots,n\}$, a production total $p_{(i,j)}$ is associated with the site, and each pair of sites chosen needs to be at least D grid units apart from each other. Finally, the maximum number of well locations selected, N, is predetermined.

2. Evolutionary Metaheuristics

Evolutionary Metaheuristics use ideas of biology and Darwin's theory of evolution, such as reproduction, mutation, and "survival of the fittest", to provide solutions to problems that arise in other fields of study. Because one of the most used types of evolutionary metaheuristics in petroleum engineering is genetic algorithms, we developed a GA to tackle our problem of interest (for details on the structure and use of GA in general, see [10]).

2.1. Definition of Terms and Variables

Our GA starts with an *initial population* that is improved through *crossover*, *local search*, *mutation*, and the addition of new members to the population, which we call *intruders*. After a generation, the best fit individuals of the population are selected, the others are discarded, and a new generation begins. The algorithm runs for a predetermined number of generations, n_g, and returns the fittest individuals of the final population. The fitness of an individual, here, is the total oil production associated with it.

An individual y^z is represented by an N-array of ordered pairs (which we call *coordinates*), and denoted as

$$y^z = \left((i_1,j_1)^z,(i_2,j_2)^z,\cdots,(i_N,j_N)^z\right)$$

where $(i_N,j_N)^z$, $k \in \{1,\cdots,N\}$, represent the locations on the grid of the wells defined by y^z. Associated with y^z, the value p_z is the total oil production that results from choosing this individual, given by Equation (1).

$$p_z = \sum_{k=1}^{N} p_{(i_k,j_k)^z} \tag{1}$$

We denote the $m \times n$ production totals matrix by P, *i.e.*, an element of row i and column j of p is $p_{(i,j)}$, and we define $p_{(0,0)} = 0$ We let q be the sorted vector of the elements of P and its corresponding coordinates:

$$q = \left((q_1,(r_1,s_1)),(q_2,(r_2,s_2)),\cdots,(q_{mn},(r_{mn},s_{mn}))\right)$$

where

$$q_1 \ge q_2 \ge \cdots \ge q_{mn},$$

$$q_k = p_{(r_k,s_k)},$$

and

$$\{(r_1,s_1),\cdots,(r_{mn},s_{mn})\} = I \times J.$$

2.2. The Genetic Algorithm

The high level structure of our GA is as follows:

```
WPP-GA(P, m, n, D, N, n_g, n_p, n_c, n_i, Δ)
1  for z := 1 to n_p + 2n_c + n_i do
2  |  for k := 1 to N do
3  |  |  (i_k, j_k)^z := (0,0);
4  |  end
5  end
6  INITIAL-POPULATION;
7  for g := 1 to n_g do
8  |  CROSSOVER;
9  |  LOCAL-SEARCH;
10 |  MUTATION;
11 |  SELECTION;
12 end
13 return y^1;
```

At each generation g, n_p is the number of individuals in the initial population, n_c is the number of crossover operations performed, and n_i is the number of intruders created. The value Δ is used in the sub-routine **LOCAL-SEARCH**, and will be explained later.

Lines 1 - 5 initialize the variables of the algorithm, which are the individuals of the population, at zero. Line 6 creates the initial population, and lines 7 - 12 perform evolutionary sub-routines for n_g generations. Finally, line 13 returns the best fit individual found.

Throughout the algorithm, and for all individuals considered, it is verified whether the distance of all locations in the grid associated with the individual satisfy the minimum distance constraint. When it does, we say that the individual is *feasible*. This verification is done using the following sub-routine.

```
CHECK-FEASIBILITY(y^t)
f := 1;
for k := 2 to N do
|  if (i_k, j_k)^t = (0,0) then
|  |  continue;
|  end
|  (i_1, j_1) := (i_k, j_k)^t;
|  for ℓ := 1 to k − 1 do
|  |  if (i_ℓ, j_ℓ)^t = (0,0) then
|  |  |  continue;
|  |  end
|  |  (i_2, j_2) := (i_ℓ, j_ℓ)^t;
|  |  if sqrt((i_1 − i_2)^2 + (j_1 − j_2)^2) < D then
|  |  |  f := 0;
|  |  |  break;
|  |  end
|  end
|  if f = 0 then
|  |  break;
|  end
end
return f;
```

This routine takes as input an individual, y^t, and returns 1 if the individual is feasible, or 0 otherwise. The procedure simply computes the Euclidean distance between every pair of coordinates that define the individual, and checks whether the distance is at least D.

INITIAL-POPULATION(P, m, n, D, N, n_p)

```
1  (i_1, j_1)^1 := (r_1, s_1);
2  ℓ := 2;
3  for k := 2 to mn do
4      y^0 := ((i_1, j_1)^1, ..., (i_{ℓ-1}, j_{ℓ-1})^1, (r_k, s_k),
             (0,0), ..., (0,0));
5      if CHECK-FEASIBILITY(y^0) = 1 then
6          (i_ℓ, j_ℓ)^1 := (r_k, s_k);
7          ℓ := ℓ + 1;
8      end
9      if ℓ > N then
10         break;
11     end
12 end
13 for z := 2 to n_p do
14     (i_1, j_1)^z := (rand({1, ..., m}), rand({1, ..., n}));
15     ℓ := 2;
16     for k := 1 to 10N do
17         i* := rand({1, ..., m});
18         j* := rand({1, ..., n});
19         y^0 := ((i_1, j_1)^z, ..., (i_{ℓ-1}, j_{ℓ-1})^z, (i*, j*),
                 (0,0), ..., (0,0));
20         if CHECK-FEASIBILITY(y^0) = 1 then
21             (i_ℓ, j_ℓ)^z := (i*, j*);
22             ℓ := ℓ + 1;
23         end
24         if ℓ > N then
25             break;
26         end
27     end
28 end
29 return y^1, ..., y^{n_p};
```

The routine **INITIAL-POPULATION** can be divided into two main blocks: lines 1 - 12 and 13 - 29. The first block creates a "greedy" solution to the problem by choosing the coordinates of y^1, the first individual of the population, based on the vector q of sorted values of P. The second block creates the rest of the initial population by selecting random locations on the grid (using the function $\text{rand}(A)$, which returns a random element of a finite set A), with the condition that each individual created is feasible.

We note that, for large values of N, it may be hard, and perhaps even impossible, to find N locations on the grid such that the minimum distance between each pair of locations is D. Our algorithm makes $10N$ attempts to find such locations, and, in the case that it is not successful, some of the coordinates of the individual being created remain as $(0,0)$.

The sub-routine **CROSSOVER** performs the reproduction of individuals. In this routine, two individuals of the population $y^1,\cdots,y^{n_p}$ are selected at random (lines 2 - 3), and the crossover point, represented by d, is also selected at random (line 4). In lines 5 - 12, the crossover occurs, and two new individuals are created. These new individuals are then checked for feasibility, and, in the case that they are not feasible, they are "erased" from the population by setting all of their coordinates to $(0,0)$ (lines 13 - 21). The procedure is repeated n_c times, and the routine creates $2n_c$ new individuals.

CROSSOVER$(P, D, N, n_p, n_c, y^1, \ldots, y^{n_p})$

```
1  for k := 1 to n_c do
2      a_1 := rand({1, ..., n_p});
3      a_2 := rand({1, ..., n_p} \ {a_1});
4      d := rand({1, ..., N});
5      for ℓ := 1 to d do
6          (i_ℓ, j_ℓ)^{2k-1} := (i_ℓ, j_ℓ)^{a_1};
7          (i_ℓ, j_ℓ)^{2k} := (i_ℓ, j_ℓ)^{a_2};
8      end
9      for ℓ := d + 1 to N do
10         (i_ℓ, j_ℓ)^{2k-1} := (i_ℓ, j_ℓ)^{a_2};
11         (i_ℓ, j_ℓ)^{2k} := (i_ℓ, j_ℓ)^{a_1};
12     end
13     if CHECK-FEASIBILITY(y^{2k-1}) = 0 then
14         for ℓ := 1 to N do
15             (i_ℓ, j_ℓ)^{2k-1} := (0,0);
16         end
17     end
18     if CHECK-FEASIBILITY(y^{2k}) = 0 then
19         for ℓ := 1 to N do
20             (i_ℓ, j_ℓ)^{2k} := (0,0);
21         end
22     end
23 end
24 return y^{n_p+1}, ..., y^{n_p+2n_c};
```

Our GA then performs a local search in the grid, in the geographical sense, to try to improve every individual of the population.

LOCAL-SEARCH$(P, D, N, n_p, n_c, \Delta, y^1, \ldots, y^{n_p+2n_c})$

```
1  for z := 1 to n_p + 2n_c do
2      ℓ := rand({1, ..., N});
3      if (i_ℓ, j_ℓ)^z = (0,0) then
4          continue;
5      end
6      (i*, j*) := (i_ℓ, j_ℓ)^z;
7      b_1 := 0;
8      b_2 := 0;
9      for d_1 := -Δ to Δ do
10         if i* + d_1 < 1 or i* + d_1 > m then
11             continue;
12         end
13         for d_2 := -Δ to Δ do
14             if j* + d_2 < 1 or j* + d_2 > n then
15                 continue;
16             end
17             if p_(i*+d_1, j*+d_2) ≤ p_(i*+b_1, j*+b_2) then
18                 continue;
19             end
20             y^0 := ((i_1, j_1)^z, ..., (i_{ℓ-1}, j_{ℓ-1})^z, (i*+d_1, j*+d_2),
                     (i_{ℓ+1}, j_{ℓ+1})^z, ..., (i_N, j_N)^z);
21             if CHECK-FEASIBILITY(y^0) = 1 then
22                 b_1 := d_1;
23                 b_2 := d_2;
24             end
25         end
26     end
27     (i_ℓ, j_ℓ)^z := (i* + b_1, j* + b_2);
28 end
29 return y^1, ..., y^{n_p+2n_c};
```

In line 2, a coordinate ℓ of individual y^z is chosen at random, and, in case it is not $(0,0)$, a local search in the reservoir grid, within a square of area $(2\Delta+1)^2$ centered at $(i_\ell, j_\ell)^z$, is performed to try to find a location in the search area such that, if substituted in place of $(i_\ell, j_\ell)^z$, will maintain feasibility of the individual and increase its associated p_z value the most. The routine performs this local search for $y^1,\cdots,y^{n_p+2n_c}$, and returns their updated values.

The next genetic operator is mutation, which, again, is performed for every individual of the population $y^1,\cdots,y^{n_p+2n_c}$.

```
MUTATION(P, D, N, n_p, n_c, y^1, ..., y^{n_p+2n_c})

for z := 1 to n_p + 2n_c do
    ℓ := rand({1, ..., N});
    i* := rand({1, ..., m});
    j* := rand({1, ..., n});
    if p_(i*,j*) ≤ p_(i_ℓ,j_ℓ)^z then
        continue;
    end
    y^0 := ((i_1, j_1)^z, ..., (i_{ℓ-1}, j_{ℓ-1})^z, (i*, j*),
           (i_{ℓ+1}, j_{ℓ+1})^z, ..., (i_N, j_N)^z);
    if CHECK-FEASIBILITY(y^0) = 1 then
        (i_ℓ, j_ℓ)^z := (i*, j*);
    end
end
return y^1, ..., y^{n_p+2n_c};
```

This routine picks a coordinate ℓ from individual y^z at random and replaces it with another coordinate picked at random in the entire grid. If the replacement keeps y^z feasible and improves the value of p_z, then y^z is updated; otherwise, y^z remains unchanged.

Our last genetic operator, **INTRUDERS**, creates n_i individuals that enter the population "from outside". The intention of this procedure is to add variability to the population thus far created.

```
INTRUDERS(P, D, N, n_p, n_c, n_i)

1  for z := n_p + 2n_c + 1 to n_p + 2n_c + n_i do
2      u := rand({1, ..., min{10N, mn}});
3      (i_1, j_1)^z := (r_u, s_u);
4      ℓ := 2;
5      for k := 1 to 100N do
6          u := rand({1, ..., min{10N, mn}});
7          i* := r_u;
8          j* := s_u;
9          y^0 := ((i_1, j_1)^z, ..., (i_{ℓ-1}, j_{ℓ-1})^z, (i*, j*),
                  (0,0), ..., (0,0));
10         if CHECK-FEASIBILITY(y^0) = 1 then
11             (i_ℓ, j_ℓ)^z := (i*, j*);
12             ℓ := ℓ + 1;
13         end
14         if ℓ > N then
15             break;
16         end
17     end
18 end
19 return y^{n_p+2n_c+1}, ..., y^{n_p+2n_c+n_i};
```

This routine is similar to **INITIAL-POPULATION**. The main differences are in lines 2 and 6, where the choice of the coordinates of the new individuals is more selective: they are chosen among the top $10N$ values of the P matrix. Thus, this routine brings to the population not only variability, but also members with high fitness.

Our final sub-routine selects and preserves the n_p best fit individuals from the entire population $y^1,\cdots,y^{n_p+2n_c+n_i}$, which will become the initial population of the next generation.

```
SELECTION(P, y^1, ..., y^{n_p+2n_c+n_i})

1  sort y^1, ..., y^{n_p+2n_c+n_i} in decreasing order of
       p_z, z ∈ {1, ..., n_p + 2n_c + n_i}
2  for z := n_p + 1 to n_p + 2n_c + n_i do
3      for k := 1 to N do
4          (i_k, j_k)^z := (0, 0);
5      end
6  end
7  return y^1, ..., y^{n_p+2n_c+n_i};
```

The sorting procedure in line 1 is, in fact, a renumbering of the individuals $y^1,\cdots,y^{n_p+2n_c+n_i}$ based on the values p_z, $z\in\{1,\cdots,n_p+2n_c+n_i\}$. As a clarifying example, suppose we give as input to this line $y^1=((1,2),(3,4))$, $y^2=((5,6),(7,8))$, and $y^3=((9,10),(11,12))$, with $p_1=10$, $p_2=30$, and $p_3=20$. Then the output will be $y^1=((5,6),(7,8))$, $y^2=((9,10),(11,12))$, and $y^3=((1,2),(3,4))$. Lines 2 - 7 simply erase variables $y^{n_p+1},\cdots,y^{n_p+2n_c+n_i}$ and keep $y^1,\cdots,y^{n_p}$ which are the surviving individuals that will start a new generation. (In case $g=n_g$, then there will not be another generation, and **WPP-GA** will return y_1, the individual with the best fitness found after n_g generations.)

A careful analysis of **WPP-GA** and its sub-routines shows that the algorithm terminates, but not necessarily with an optimal solution to the problem, *i.e.*, an individual whose fitness is the highest possible.

3. Mathematical Optimization

Mathematical Optimization approaches quantitative problems using tools such as linear algebra, calculus, and graph theory. In a sense, it is a more sophisticated method than Evolutionary Metaheuristics, and, in practice, usually requires bigger and more structured algorithms to solve a given problem.

The main advantage of using mathematical optimization, and, in our particular case, IP, is that a solution may be *proven* to be optimal, as opposed to GA, which does not guarantee optimality of the best solution found.

It is common practice, in IP, to formulate problems by defining an objective function to be maximized (or minimized), subject to constraints that define the problem in question. Here, the formulation of WPP is as follow:

$$\text{maximize} \sum_{(i,j)\in I\times J} p_{(i,j)} x_{(i,j)},$$

$$\text{subject to } x_{(i_1,j_1)} + x_{(i_1,j_2)} \le 1, \begin{cases} (i_1,j_1),(i_2,j_2) \in I\times J \\ \sqrt{(i_1-i_2)^2+(j_1-j_2)^2} < D \end{cases}$$

$$x_{(i,j)} \in \{0,1\} \qquad (i,j)\in I\times J,$$

where a variable $x_{(i,j)}$ represents the location (i,j) in the oil reservoir grid, and is equal to 1 if the location is chosen, or 0 otherwise. Hence, a feasible solution to this problem is a mn-vector of 0's and 1's that satisfies all constraints defined in the formulation shown. Note that this is a completely different way of representing a solution, in comparison with our GA. Nonetheless, the values of the objective function, here, and the fitness of the individuals, in the GA, are comparable, since they have the same meaning.

To solve our instances with the IP approach, we use a commercial solver based on *branch-and-cut* that can handle large-scale optimization problems. For details on branch-and-cut, the reader is referred to [11].

4. Computational Tests

We ran our computational tests in the Texas Tech High Performance Computing Center's 3.0 GHz CPUs with 16 GB RAM nodes. For the tests using IP, we used CPLEX 12 with default settings, except for the search strategy, which was set to Traditional Branch-and-Cut, and the relative and absolute gap tolerances, which were set to zero. Every test had a time limit of 3600 seconds.

Our test grid has $m = n = 100$, and we set the values of D to $8, 12, 16$, and 20. For each of these values, we tried to solve WPP with N set to $10, 20, \cdots$, up to the point at which either the time limit was reached, or the value of N was *not* reached, *i.e.*, neither method was able to find a solution that represented N well locations. For the GA, we set $n_g = 10000$, $n_p = 1000$, $n_c = 100$, $n_i = 10$, and $\Delta = 10$ for all tests.

The values of the P matrix were obtained with the commercial reservoir simulator Eclipse, using the *Quality Maps* approach (see [12] for details). Some main characteristics of our heterogeneous and anisotropic reservoir are: initial pressure of $4000\, psi$, porosity of 22%, and horizontal permeability average of $175\, mD$ with standard deviation of $91.1\, mD$. The distance between two closest grid points is $300\, ft$, and the thickness of the reservoir is $75\, ft$.

Table 1 shows the results obtained using both GA and IP. The columns "Best Sol' n" displays the value of the best solutions found by both methods. (Note: in GA, this value is called the *fitness* of the best individual, whereas in IP it is called the *objective function val ue* of the best solution. We will use the latter expression in our analysis.) "Time" shows the computational time, in seconds, required for the test to finish. We note that, because of the settings used in CPLEX, all solutions obtained using IP are provably optimal precisely when the time is less than 3600 seconds. The column "Card" shows the cardinality of the solution found, *i.e.*, the number of well locations associated with that solution. Finally, "Gap" refers to the relative percentage difference between the GA and IP best solutions, and is computed using Equation (2).

$$\text{Gap} = \frac{\text{Best Sol'n with IP} - \text{Best Sol'n with GA}}{\text{Best Sol'n with IP}} \cdot 100 \tag{2}$$

From **Table 1**, it is clear that the problem gets harder as the values of D and N increase. On the IP section of the table, this is reflected in the time required to solve the instances. On the GA side, both the computational time and the cardinality of the best solution show the increase in the difficulty of the problem. For instance, for $D = 8$ and $N \ge 120$, the best solution that GA can find has cardinality 117, although solutions with greater cardinality (and better objective function value) do exist, as the IP results show.

Overall, IP was capable of finding better solutions than GA, and, most of the time, was faster. In every test, the best solution found by IP had an objective function value at least as good as the one found by the GA. In 8 cases, both methods found an optimal solution to the problem, and in 11 other cases, the gap between the two methods was less than 1%. This shows that GA can be effective for instances that are less challenging, but, for the harder instances, IP was notably more successful than GA.

The merit of being faster and finding better solutions is due not only to the differences in approaching the problem (IP vs. GA), but also a result of the algorithms used to solve the instances. While our GA has less than 500 lines of code, CPLEX is the product of professional software developers who worked on it for decades. The seemingly small values in the last columns reflect this contrast of algorithms: a gap as small as 0.1% can be very difficult to close, and thus finding a sub-optimal solution can be a much easier task than finding an optimal one. Moreover, the fact that CPLEX can guarantee the optimality of a solution is a feature that a GA does not have. It is only in comparison studies such as the present study, that one can visualize the effectiveness of a GA, in finding optimal solutions. Finally, we note that CPLEX is a multi-purpose solver that can handle a vast number of different problems, while our GA was developed specifically to tackle WPP.

5. Conclusions and Further Research

In this study, we compared the performance of GA and IP to solve instances of the problem of placing vertical wells in an $m \times n$ grid that sits on a reservoir, with the

Table 1. Summary of results using GA and IP.

D	N	GA			IP			Gap
		Best Sol'n	Time	Card	Best Sol'n	Time	Card	
	10	55,187,158	112	10	55,187,158	33	10	0.00
	20	68,445,523	122	20	68,568,163	43	20	0.18
	30	77,462,025	124	30	77,462,025	48	30	0.00
	40	85,106,849	142	40	85,112,969	49	40	0.01
	50	92,306,494	150	50	92,306,494	54	50	0.00
	60	99,004,082	208	60	99,132,004	52	60	0.13
8	70	104,894,614	191	70	105,383,497	57	70	0.46
	80	110,598,373	263	80	111,275,555	54	80	0.61
	90	115,751,327	320	90	116,836,492	74	90	0.93
	100	120,097,642	472	100	121,998,756	76	100	1.56
	110	121,859,490	568	110	126,733,273	97	110	3.85
	120	124,017,004	690	117	130,778,001	149	120	5.17
	130	124,017,004	762	117	134,144,399	284	130	7.55
	140	124,017,004	859	117	136,147,901	3,600	140	8.91
	10	47,495,649	111	10	47,495,649	106	10	0.00
	20	56,566,624	129	20	56,566,624	130	20	0.00
	30	63,899,845	139	30	63,928,663	129	30	0.05
12	40	70,334,844	182	40	70,392,256	141	40	0.08
	50	75,637,819	214	50	76,120,456	181	50	0.63
	60	77,731,339	356	56	80,639,104	263	60	3.61
	70	77,064,512	422	56	82,588,104	3,600	68	6.69
	10	40,836,652	115	10	40,836,652	183	10	0.00
	20	48,708,586	126	20	48,708,586	243	20	0.00
16	30	54,792,273	150	30	54,803,004	322	30	0.02
	40	56,570,909	240	34	58,673,867	1,534	40	3.58
	50	56,758,783	326	35	58,664,640	3,600	40	3.25
	10	36,921,362	113	10	36,921,362	363	10	0.00
20	20	43,793,101	135	20	43,829,509	333	20	0.08
	30	45,037,857	229	23	46,605,956	1,092	28	3.36

objective of extracting the maximum amount of oil from the reservoir. Our results indicate that GA can be effective for easier instances, but it lacks performance for harder ones. In comparison, CPLEX (which takes the IP approach) always found a solution at least as good as our GA, and most of the time faster. Moreover, IP has the advantage of finding provably optimal solutions, while GA is not able to guarantee that a solution is optimal.

The GA presented here can, of course, be modified and have its performance enhanced, either by actually changing the algorithm, or by simply adjusting its parameters. Similarly, the solution via IP can be made faster by considering different formulations of the problem and tuning some parameters on CPLEX. To maintain both approaches simplely, we chose not to vary those parameters, to provide a basic GA, and to use a straightforward formulation for our IP approach.

As further research, we suggest the study of *non-conventional* wells placement, a more challenging problem, since it involves wells that are not necessarily vertical, and thus may require a 3D-grid as the set of possible locations for the wells.

6. Acknowledgements

The authors acknowledge the High Performance Computing Center at Texas Tech University at Lubbock for providing resources that have contributed to the research results reported within this paper.

URL: http://www.hpcc.ttu.edu

We thank Dr. Jerome Onwunalu's for guidance and support, and for providing a GA code used in preliminary tests.

This research was partially supported by the Office of Naval Research (ONR) through grant N000141310041 to de Farias.

REFERENCES

[1] G. AlQahtani, R. Vadapalli, S. Siddiqui and S. Bhattacharya, "Well Optimization Strategies in Conventional Reservoirs," *Proceedings of S PE Saud i Arabi a Sec tion Technical Symp osium and Exhi bition*, Al-Khobar, 8-11 April 2012, 13 Pages.

[2] O. Badru and C. S. Kabir, "Well Placement Optimization in Field Development," *SPE Annual Technical Conference and Exhibition*, Denver, 5-8 October 2003, 9 Pages.

[3] W. Bangerth, H. Klie, M. F. Wheeler, P. L. Stoffa and M. K. Sen, "On Optimization Algorithms for the Reservoir Oil Well Placement Problem," *Computational Geosciences*, Vol. 10, No. 3, 2006, pp. 303-319.

[4] A. S. Cullick, S. Vasantharajan and M. W. Dobin, "Determining Optimal Well Locations from a 3D Reservoir Model," US Patent No. US6549879 B1, 2003.

[5] D. Y. Ding, "Optimization of Well Placement Using Evolutionary Algorithms," *SPE EAGE Annual Technical Conference and Exhib ition*, Rome, 9-12 June 2008, p. 912.

[6] B. Guyaguler, R. N. Horne, L. L. Rogers and J. J. Rosenzweig, "Optimization of Well Placement in a Gulf of Mexico Waterflooding Project," *SPE Annual Technical Conference and Exhibition*, Vol. 5, No. 3, 2000, pp. 110-

118.

[7] J. Onwunalu, M. Litvak, L. J. Durlofsky and K. Aziz, "Application of Statistical Proxies to Speed up Field Development Optimization Procedures," *International Conference and Exhibition*, Abu Dhabi, 3-6 November 2008, 14 Pages.

[8] G. W. Rosenwald and D. W. Green, "A Method for Determining the Optimum Location of Wells in a Reservoir Using Mixed-Integer Programming," *SPE Jour nal*, Vol. 14, No. 1, 1974, 12 Pages.

[9] S. Vasantharajan and A. S. Cullick, "Well Site Selection Using Integer Programming Optimization," *Proceedings of IAMG*'97, 22-26 September 1997, pp. 421-426.

[10] R. Haupt and S. Haupt, "Practical Genetic Algorithms," John Wiley and Sons, Hoboken, 2004.

[11] G. L. Nemhauser and L. A. Wolsey, "Integer and Combinatorial Optimization," John Wiley and Sons, Hoboken, 1988.

[12] P. S. Cruz, R. N. Horne and C. V. Deutsch, "The Quality Map: A Tool for Reservoir Uncertainty Quantification and Decision Making," *SPE Annual Technical Conference and Exhibition*, Vol. 7, No. 1, 2004, 9 Pages.

15

Thermo-Hydrodynamics of Core-Annular Flow of Water, Heavy Oil and Air Using CFX

Antonio José Ferreira Gadelha[1], Severino Rodrigues de Farias Neto[1], Ramdayal Swarnakar[1], Antonio Gilson Barbosa de Lima[2]

[1]Department of Chemical Engineering, Center of Science and Technology, Federal University of Campina Grande, Campina Grande, Brasil

[2]Department of Mechanical Engineering, Center of Science and Technology, Federal University of Campina Grande, Campina Grande, Brasil

ABSTRACT

The transport of heavy and ultra-viscous oil employing the core-flow technique has been increasing recently, because it provides a greater reduction of the pressure drop during the flow. In this context, the effect of temperature and the presence of gas on the thermo-hydrodynamics of a three-phase water-heavy oil-air flow in a horizontal pipe under the influence of gravity and drag forces, using the commercial software ANSYS CFX®, have been evaluated. The standard $\kappa - \varepsilon$ turbulence model, the mixture model for heavy oil-water system and the particle model for heavy oil-gas and water-gas systems, were adopted. Results of velocity, volume fraction, pressure and temperature fields of the phases present along the pipe are presented and discussed. It has been found that the presence of the air phase and the variation in the temperature affect the behavior of annular flow and pressure drop.

Keywords: Heavy Oil; Three-Phase Flow; Heat; Numerical Simulation; CFX

1. Introduction

The worldwide heavy oil reserves are estimated to be 3 trillion barrels [1], while light oil reserves have shown a progressive decline in the last decade. This fact leads to an increased economic interest for the reserves of heavy oil which consequently stimulates further research to make production of heavy oil economically feasible. Currently, Brazil is one of the important worldwide producers of oil, and that most of the Brazilian reserves consist of heavy oil in deep waters, generating technical difficulties in exploitation of such resources. Heavy oil is considered to have API between 10 and 20, a density greater than 0.90 g/ml, a viscosity between 10 cP and 100 cP at reservoir conditions and from 100 cP to 10,000 cP at surface conditions [2]. At present, these oils do not have significant economic value due to the low concentration of smaller chain hydrocarbons. However, with the decline of production of light oil, the importance, and consequently the price of these energy sources are likely to increase. The major obstacle of utilizing heavy oil is its relatively high viscosity, which makes it difficult to transport and higher density which increases the cost of refining. Thus, the transportation of heavy and ultra-viscous oil is a main technological challenge in the petroleum industry. This fact is related to the high pressure drop or friction due to viscous effects of this type of oil during its flow.

According to Trevisan [3], because of unfavorable characteristics of heavy oils, their transport from the production areas to the processing and refining plants is the biggest obstacle encountered for the production of heavy oils. The author also mentions that the alternatives currently used are to transport by truck or heated pipeline. However, these methods are very expensive and are applicable only for short distances. For efficient transportation at considerable distances, it is necessary to use conventional pipelines, but most of these pipelines have viscosity specifications lower than 0.1 Pa·s, which is not true for heavy oils. In order to overcome the difficulties inherent in the production and transport of heavy and ultra-viscous oil, several techniques have been used, so that a decrease in pressure drop during the flow can be provided, thereby a reduction in the viscosity effect of the fluids present can occur. Among the techniques used one can mention: adding heat to the system, diluting the heavy crude oil with a lighter one and forming emulsions using emulsifying agents. However, each of these alternatives has limitations in their use, both technical and economic.

One technique that has higher efficiency compared to other methods is the core-flow technique. This technique consists in injecting a less viscous liquid, usually water, adjacent to the pipeline wall. This prevents the contact of oil with the inner side of the pipeline. It results in a greater reduction in the pressure drop of the flow and consequently in a reduction in the transport cost of such oil. A drawback of this technique is when the oil comes in contact with the inner wall of the pipeline during transport. This may cause a large increase in the system pressure, which can result in a serious damage to the transportation system and the environment [4].

The most important feature of the core-flow technique is that it does not modify the viscosity of the oil but changes the flow pattern and reduces friction during the transport of very viscous products, such as heavy oil. This reduction in friction also causes a reduction in the longitudinal pressure drop and consequently, a reduction in pumping costs. Several papers have reported research works related to the improvement of the core-flow technique [4-9].

However, in oil production, oil and water rarely flow separately and a gas fraction is generally present, which is characterized as a multiphase flow. Such flow can be defined as a system in which fluid components are immiscible and separated by interfaces. The occurrence of multiphase flow in the oil industry is very common in the units of production, transportation and processing of hydrocarbons of an oil field.

Thus, it is important to analyze the influence of the presence of a third phase (gas) in the annular flow (water-oil) with respect to the pressure drop. In order to verify three-phase flow characteristics some experimental studies have been performed [10-13].

Bannwart *et al.* [10] studied the pressure drop and flow patterns of a three phase flow observed in a glass tubing with a diameter of 2.84 cm containing heavy oil (3.4 Pa·s and 970 kg/m^3 at 20°C), water and air under several combinations of individual compositions and tube inclination (horizontal, vertical and inclined). In their study nine flow patterns were verified. According to these authors, when compared to the two-phase flow of heavy oil-water only, the presence of gas increases considerably the mixture velocity and consequently the pressure drop is increased.

Poesio *et al.* [12] made an experimental study related to the core-annular flow with the aim of providing a new database for the three-phase (ultra-viscous oil, water and air) flow and to propose a simple model for the determination of the pressure drop. They observed the effect of the air injection on the pressure drop of the annular liquid-liquid flow and noticed an error less than ±15% of measured value in relation to that of the proposed model.

Strazza *et al.* [13] presented an experimental study of the three-phase flow using water, air and high viscosity oil. The attention was focused on the effect of the gas presence in the core-annular liquid-liquid flow. The experimental flow map, obtained by them, showed that the increase in gas flow breaks up the integrity of the oil core, resulting in a chaotic flow regime. Values for the pressure drop were compared with the proposed theoretical model for the three-phase core flow. The difference between the experimental and the predicted pressure drop was of more or less 20%.

In this background, the objective of this work is to study numerically the three-phase annular flow of heavy oil, water and air (core-flow), at different conditions of temperature and volume fraction of the air.

2. Mathematical Modeling

2.1. Physical Domain of Study

The study domain consists of a 3 meter long and 2.84 cm inner diameter horizontal tube, in which the flow of heavy oil-water-gas takes place. This is shown in **Figure 1**.

2.2. Computational Domain

To study numerically the annular flow behavior of oil-water in the presence of gas, it is necessary to represent the geometry or domain of study in a computational domain or mesh. For this a mesh of hexahedral structured elements was employed. This mesh was made with the ICEM-CFD commercial package available in ANSYS CFX.

To draw up the computational domain, initially, a tube with two inlets, one annular for water injection and another circular for the oil together with gas, was created. This can be seen in **Figure 1**. The ring or the annular space between tube wall and the oil core has a thickness of 1.7 mm. The mesh used in the present work consisted

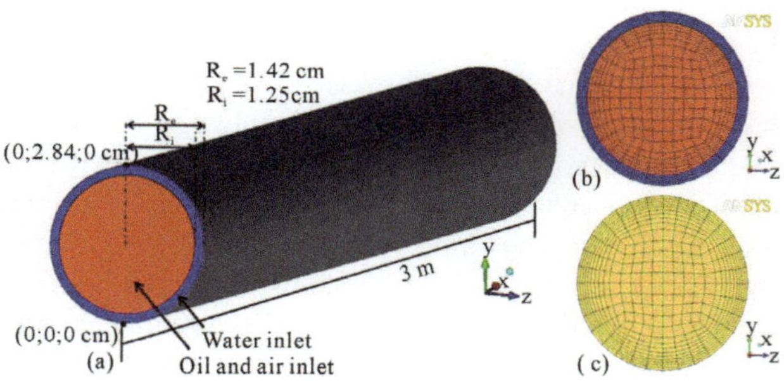

Figure 1. (a) Tube dimensions and details of water and oil inlet sections. (b) Details of the expanded mesh at the inlet region and (c) at the outlet region.

of 464,000 hexahedral elements.

2.3. Mathematical Model

To study the three-phase flow within a horizontal pipe following conditions have been considered:

1) Incompressible and steady state flow;

2) No chemical reactions;

3) Existence of gravitational and drag effects;

4) The viscosities of water, gas and ultra-viscous heavy oil are functions of temperature;

5) There is no interfacial mass transfer between water, oil and gas phases;

Thus, the conservation equations of mass, momentum and energy applied to the multiphase flow are reduced to:

2.3.1. Conservation Equations

1) Mass Conservation Equation

$$\nabla \bullet \left(r_\alpha \rho_\alpha \vec{U}_\alpha \right) = 0 \tag{1}$$

where r_α, ρ_α e $\vec{U}_\alpha$ correspond to the volume fraction, density and velocity vector of phase α, respectively.

2) Momentum Conservation Equation

$$\begin{aligned} &\nabla \bullet \left[r_\alpha \left(\rho_\alpha \vec{U}_\alpha \otimes \vec{U}_\alpha \right) \right] = -r_\alpha \nabla p_\alpha + \\ &\nabla \bullet \left\{ r_\alpha \mu_\alpha \left[\nabla \vec{U}_\alpha + \left(\nabla \vec{U}_\alpha \right)^\Gamma \right] \right\} + \vec{S}_{M_\alpha} + \vec{M}_\alpha \end{aligned} \tag{2}$$

where ρ is pressure, $\vec{S}_{M_\alpha}$ represents a term for the external forces acting on the system per unit volume and $\vec{M}_\alpha$ describes the total forces per unit volume (interfacial drag force).

In the mixture model, available in ANSYS CFX, the total forces per unit volume only consider the interfacial forces (drag) and are given by:

$$\vec{M}_\alpha = \vec{M}_{\alpha\beta} = C_D A_{\alpha\beta} \rho_{\alpha\beta} \left| \vec{U}_\beta - \vec{U}_\alpha \right| \left(\vec{U}_\beta - \vec{U}_\alpha \right) \tag{3}$$

where C_D is the drag coefficient and the sub-indexes α and β correspond to the phases α and β present in the flow.

In Equation (3), $A_{\alpha\beta}$ corresponds to the interfacial contact area between the phases α and β which is given by:

$$A_{\alpha\beta} = \frac{6 r_\beta}{d_\beta} \tag{4}$$

where in r_β and d_β represent the volume fraction and the diameter of the particle of the phase β, respectively. Adopting the particle model, α represents the continuous phase (heavy oil or water) and β being the dispersed phase (gas).

For the mixture model the interfacial contact area between phases α and β is given by:

$$A_{\alpha\beta} = \frac{r_\alpha \cdot r_\beta}{d_{\alpha\beta}} \tag{5}$$

where $d_{\alpha\beta}$ is the length of the mixture.

3) Energy Conservation Equation

$$\nabla \bullet \left[r_\alpha \left(\rho_\alpha \vec{U}_\alpha h_\alpha - \lambda_\alpha \nabla T_\alpha \right) \right] = Q_\alpha \tag{6}$$

where h_α, λ_α and T_α describe static enthalpy, thermal conductivity and temperature respectively of α phase and Q_α describes heat transfer to the α phase through the interfaces with the other phases, which is given by:

$$Q_\alpha = \sum_{\beta \neq \alpha} Q_{\alpha\beta} \tag{7}$$

where:

$$Q_{\alpha\beta} = -Q_{\beta\alpha} \Rightarrow \sum_\alpha Q_\alpha = 0 \tag{8}$$

The heat transfer through the boundary is usually described in terms of a coefficient of overall heat transfer, $h_{\alpha\beta}$, which is the amount of heat energy through a unit area per unit time per unit of temperature difference between the phases.

Thus, the rate of heat transfer, $Q_{\alpha\beta}$, per unit of time through the phase interfacial boundary area, per unit volume $A_{\alpha\beta}$, of phase β to phase α is given by:

$$Q_{\alpha\beta} = h_{\alpha\beta} A_{\alpha\beta} \left(T_\beta - T_\alpha \right) \tag{9}$$

Many times it is convenient to express the coefficient of heat transfer in terms of the dimensionless Nusselt number, defined by Equation (10):

$$h = \frac{\lambda Nu}{d} \tag{10}$$

In the particle model the thermal conductivity (λ) is considered as being the thermal conductivity of the continuous phase, and the length d is considered to be the diameter of the dispersed phase. So it can be written as:

$$h_{\alpha\beta} = \frac{\lambda_\alpha Nu_{\alpha\beta}}{d_\beta} \tag{11}$$

2.3.2. Turbulence Model

As turbulence model, the standard $\kappa - \varepsilon$ model was used, where it is assumed that the Reynold's tensors are proportional to the mean velocity gradients, with the proportionality constant being characterized by turbulent viscosity (idealization known as Boussinesq hypothesis).

The characteristic of this type of model is that two transport equations modeled separately are solved for the turbulent length and the time scale or solved for any two linearly independent combinations of them. The transport equations for the turbulent kinetic energy, κ, and turbulent dissipation rate, ε, respectively are:

$$\nabla \bullet \left\{ r_\alpha \left[\rho_\alpha \vec{U}_\alpha \kappa_\alpha - \left(\mu + \frac{\mu_{t\alpha}}{\sigma_k} \right) \nabla \kappa_\alpha \right] \right\} = r_\alpha \left(G_\alpha - \rho_\alpha \varepsilon_\alpha \right) \quad (12)$$

$$\nabla \bullet \left\{ r_\alpha \rho_\alpha \vec{U}_\alpha \varepsilon_\alpha - \left(\mu + \frac{\mu_{t\alpha}}{\sigma_\varepsilon} \right) \nabla \varepsilon_\alpha \right\} = r_\alpha \frac{\varepsilon_\alpha}{\kappa_\alpha} \left(C_1 G_\alpha - C_2 \rho_\alpha \varepsilon_\alpha \right) \quad (13)$$

where G_α is turbulent kinetic energy generated in the phase α, C_1 and C_2 are empirical constants. Also, in this equation, ε_α the dissipation rate of turbulent kinetic energy of phase α and κ_α the turbulent kinetic energy for phase α, are defined by:

$$\varepsilon_\alpha = \frac{c_\mu q_\alpha^3}{l_\alpha} \quad (14)$$

$$\kappa_\alpha = \frac{q_\alpha^2}{2} \quad (15)$$

where l_α is the length of spatial scale, q_α is the velocity scale, $c_\mu = 0.09$ is an empirical constant.

The variable $\mu_{t\alpha}$ is the turbulent viscosity and is given by the equation:

$$\mu_{t\alpha} = c_\mu \rho_\alpha \frac{\kappa_\alpha^2}{\varepsilon_\alpha} \quad (16)$$

where, the constants used in the above equations are: $C_1 = 1.44$; $C_2 = 1.92$; $\sigma_k = 1.0$ and $\sigma_\varepsilon = 1.3$.

2.3.3. Boundary Conditions

1) In the annular section referring to water inlet a prescribed and not null value, for the axial velocity component and volume fraction of water in the x direction, was adopted such that:

$$R_i < y < R_e, \text{ at } x = 0; \begin{cases} u_w \neq 0 \\ v_w = w_w = 0 \\ u_o = v_o = w_o = 0 \\ u_g = v_g = w_g = 0 \\ r_o = r_g = 0 \\ r_w = 1 \\ T = T_w \end{cases}$$

where u, v e v correspond to the velocity vector components in the x, y and z directions, in this order, the sub-indices w, o and g represent water, oil and gas phases respectively, and T the temperature.

2) In the core section referring to oil inlet a prescribed and not null value, for the axial velocity component and volume fraction of oil and air in the x direction, was adopted such that:

$$0 < y < R_i, \text{ at } x = 0; \begin{cases} u_o = u_g \neq 0 \\ v_o = w_o = 0 \\ v_g = w_g = 0 \\ u_w = v_w = w_w = 0 \\ r_o = 0.95 \\ r_g = 0.05 \\ r_w = 0 \\ T = T_o = T_g \end{cases}$$

3) At the borders referring to the tube wall the condition of non-slip was considered, namely:

$$y = R_e, \text{ at } 0 \le x \le L; \begin{cases} u_w = v_w = w_w = 0 \\ u_o = v_o = w_o = 0 \\ u_g = v_g = w_g = 0 \\ T = T_P = 288K \end{cases}$$

4) At the outlet section ($x = L$), a constant mean pressure p_{est} = 101,325 Pa, was prescribed, where L is the length of the tube.

In the present work a root mean square (RMS) residue equal to 10^{-7} kg/s was considered as the convergence criterion. The thermo-physical properties of the fluids used in the simulation are presented in **Tables 1** and **2**. **Table 3** summarizes all the cases of this study.

3. Results and Discussion

3.1. Influence of the Air Phase on Flow Hydrodynamics

Figures 2 and **3** depict the superficial velocity profiles for oil and water, respectively. The two-phase flow (oil-water) refers to the case 01, and the three-phase (oil-water-air), refers to the case 04 (**Table 3**), at four positions of the axial direction, Z = 0 m and were obtained under the same conditions.

By observing the velocity profiles in **Figures 2** and **3**, it can be noted that the introduction of the air phase in two-phase flow (water-oil), causes a significant change

Table 1. Thermo-physical properties of the fluids (25°C) used in the simulations.

Properties	Water	Heavy Oil	Air	Source
Density (kg/m^3)	997,200	971,000	0.778	[14]
Specific heat (J/kg·K)	4181,700	1800,000	1025,766	[15]
	Water/Oil	Water/Air	Oil/Air	
Surface tension (N/m)	0.067	0.0725	0.026	[14]

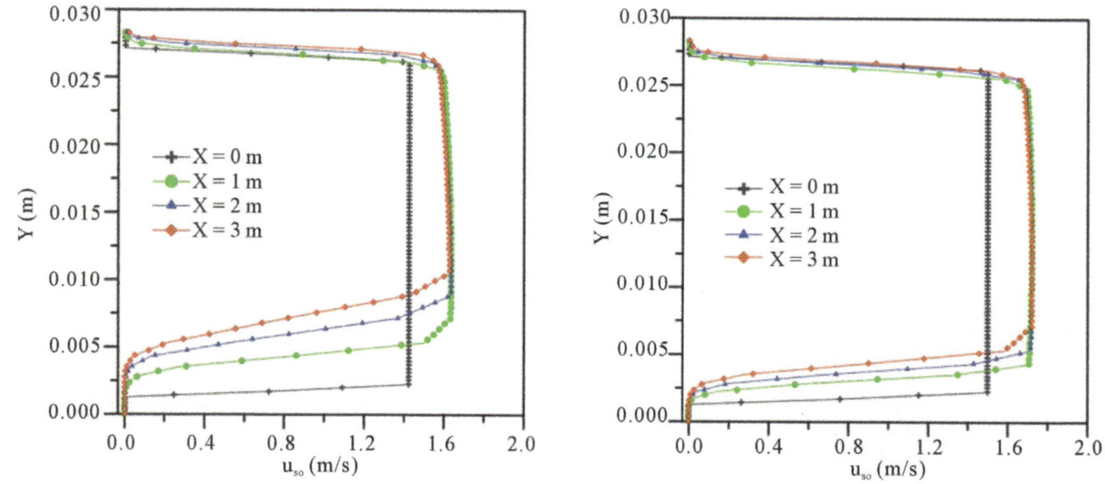

(a) (b)

Figure 2. Superficial velocity profiles of the oil: (a) three-phase (water-oil-air) flow and (b) two-phase (water-oil) flow.

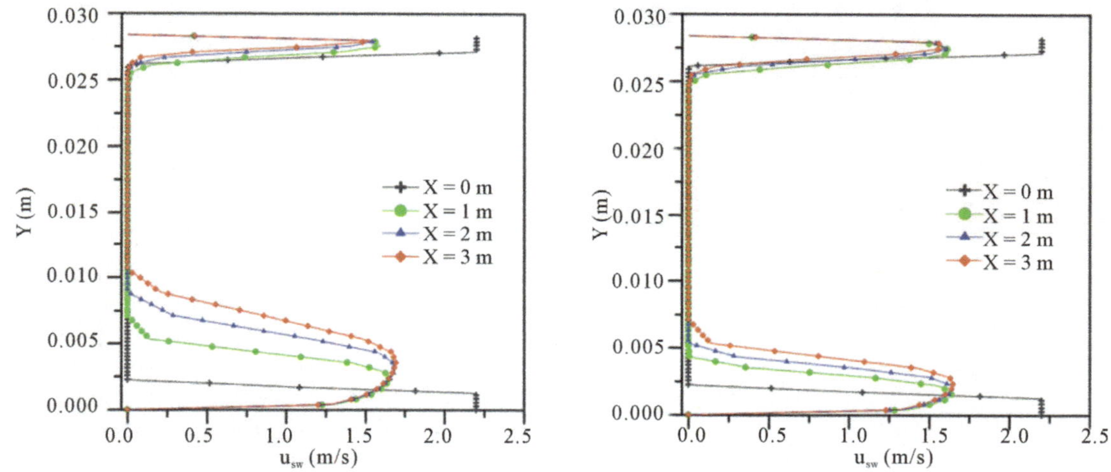

(a) (b)

Figure 3. Superficial velocity profiles of the water: (a) Three-phase (water-oil-air) flow and (b) Two-phase (water-oil) flow.

Table 2. Dynamic viscosity of the fluid as a function of temperature (in °C).

Fluid	Dynamic Viscosity (Pa.s)**	Source
Water	$\mu_W = \left(\frac{997.2}{2.443299\times10^{-2}\times T - 6.153676}\right)$	[16]
Oil	$\mu_O = 0.6402 + 18.9612\times e^{(-0.074\times T)}$	[3]
Air	$\mu_g = 2.8\times10^{-7}\times T^{0.735476}$	[17]*

*Equation fitted to the data provided by [17]; **Applied in the range of 0°C ≤ T ≤ 100°C.

in fluid-dynamics of water and oil phases. It affects the velocity gradient in the lower region of the pipe due to the increase in water flow in this region. A similar behavior was observed independently by [10-13]. This effect can be better seen in **Figures 4(a)** and **(b)**, where the behavior of the volumetric fraction field of oil in the pipe, at different axial positions, is illustrated. Here it is clearly perceived that the presence of air creates a greater elevation of the oil core and, thus increases the area occupied by annular section of water in the lower region of the pipe, close to the wall. Therefore, it can be said that the divergence between buoyancy and lubrication forces are more intense in three phase flow than in two-phase flow.

3.2. Effect of Temperature on Flow Hydrodynamics

Figure 5 depicts the superficial velocity profiles, at different temperatures (288.15 K, 303.15 K and 323.15 K) for the cases: 02, 03 and 05, on the axial position equal to 1.0 m and direction Z = 0 m. It is noted that an increase in the temperature of the phases, at the entrance of the pipe, causes a small variation in the oil phase superficial velocity due to the change in the viscosity.

Figure 6 shows the volumetric fraction field of the oil phase at a distance X = 1 m, for different temperatures of

Table 3. Conditions used in the simulations.

Case	u_w (m/s)	u_o (m/s)	u_g (m/s)	r_w	r_o *in mixture*	r_g *in mixture*	T_w, T_o *and* T_g (K)
01 (Two-phase)	2.20	1.50	-	1.00	1.00	-	313.15*
02 (Three-phase)	2.20	1.50	1.50	1.00	0.95	0.05	288.15
03 (Three-phase)	2.20	1.50	1.50	1.00	0.95	0.05	303.15
04 (Three-phase)	2.20	1.50	1.50	1.00	0.95	0.05	313.15
05 (Three-phase)	2.20	1.50	1.50	1.00	0.95	0.05	323.15

*T_w and T_o only.

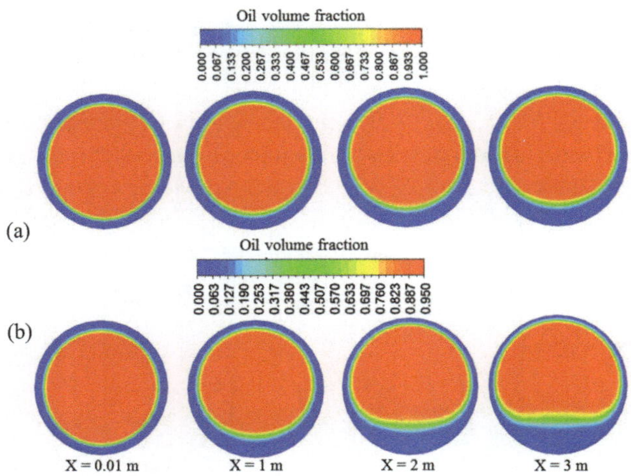

Figure 4. Oil volume fraction field in two-phase flow (a) and three-phase flow (b), at different YZ cross sections along the pipe.

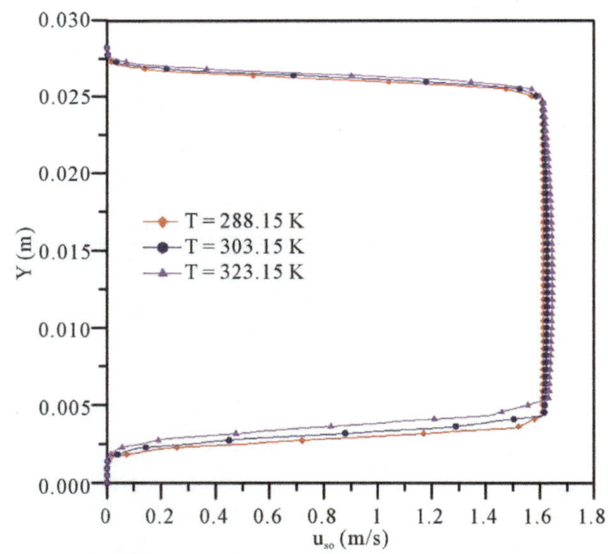

Figure 5. Superficial velocity profiles of the oil at different temperatures (X = 1 m and Z = 0 m).

the phases at the pipe entrance. It is noticed that as the temperature increases, the oil core level undergoes a small increase. This effect of temperature can be explained in terms of the reduction in the oil viscosity, whereby the resistance to the flow of this fluid, caused by viscous forces, is reduced.

3.3 Effect of Temperature and the Presence of the Gas on Pressure Drop

Table 4 presents the values for the pressure drop as a function of temperature for Cases 01 to 05 (two and three phases). It can be seen that for the three-phase flow (oil-water-gas) an increase in temperature results in a reduction of pressure drop, $\boldsymbol{\Delta P}$, which is more pronounced in the temperature range 288 to 313 K. This fact can be explained due to the decrease in viscosity of oil and water with increase in temperature, which reduces the resistance to the flow in the pipe and thereby causing a decrease in pressure drop. The viscosity of the air increases with increasing temperature, but as it is present in a lesser volumetric fraction, this effect is small compared to other phases (oil-water). Comparing the two-phase flow (Case 01) with the three-phase flow (case 04), it can be noted that the presence of air causes an increase in pressure drop of the flow. Trevisan [3] and Bannwart *et al.* [10] have verified a similar behavior for the simultaneous flow of heavy oil, water and gas. According to the latter, this is due to the fact that the gas increases the velocity of the fluid. The increase in the velocity also increases the friction factor and hence the pressure drop of the three phase flow.

3.4. Profiles and Temperature Fields of the Phases

Figures 7 and **8** show the temperature profiles for water and oil respectively (Case 04) at four axial positions (0, 1, 2 and 3 m) along the pipe It can be seen that the water temperature at the pipe entrance (0 m) is uniform, due to the boundary conditions assumed. One can perceive a temperature decrease, as the fluids move away from the entrance. A strong temperature gradient near the pipe wall, due to the boundary condition adopted, is also observed.

By observing **Figure 7**, which illustrates the water temperature profile along the tube, it is verified that the

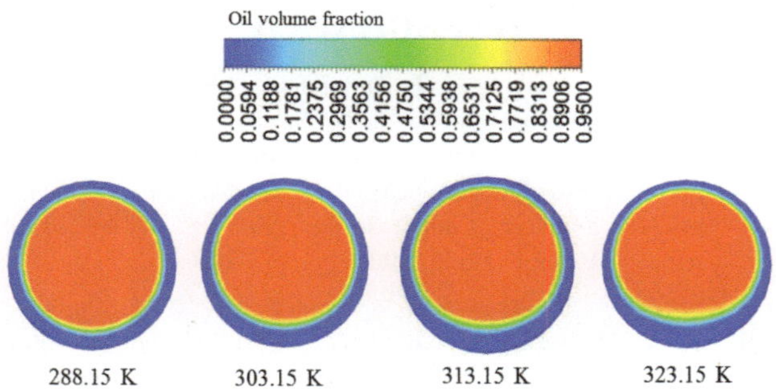

Figure 6. Volumetric fraction field of oil for different temperatures, on YZ plane (transversal section), at 1 m distance from the entrance.

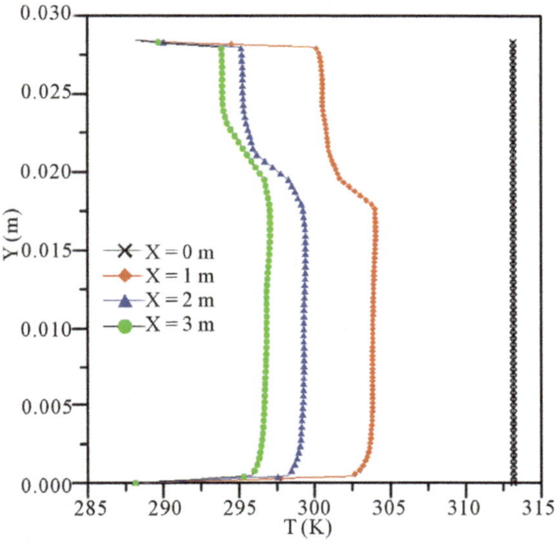

Figure 7. Water temperature profiles at four axial positions (X) along the pipe, at Z = 0 m (Case 04).

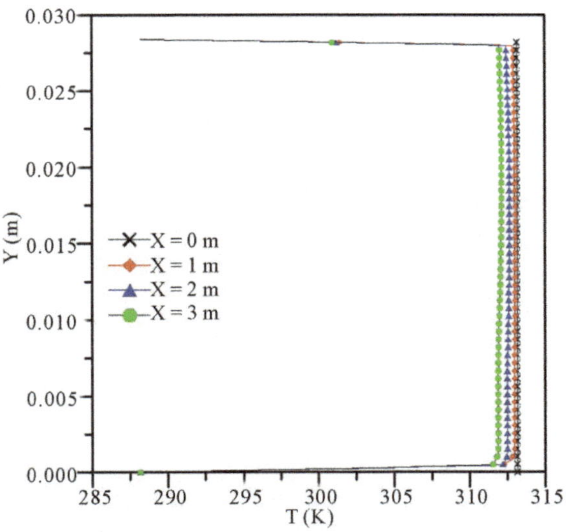

Figure 8. Oil temperature profiles at four axial positions (X) along the pipe, at Z = 0 m (Case 04).

water has a greater temperature reduction at the upper region of the pipe. This behavior is due to the fact that, as the water tends to accumulate in the lower region, the film of water formed on top of the flow is relatively thinner, and this film thus undergoes greater influence of the low temperature adopted for the pipe wall.

With respect to the oil temperature profile (**Figure 8**), there is a uniform behavior in the central region of the pipe. However, as one moves away from the inlet section, a small temperature decrease can still be noted. This can be attributed to the heat transfer, since the pipe wall is at a lower temperature than the oil. By comparing **Figures 7** and **8**, it can be seen that the oil at the pipe exit has a temperature higher than the temperature of the water, which is due to the fact that the water flows next to the wall preventing contact between the oil and the wall, thus water behaves as a thermal insulator.

Figures 9 and **10** present the details of temperature fields of water and oil, at the inlet (**Figure 9(a)**) and at the outlet (**Figure 9(b)**) sections near the pipe wall, respectively. It is in this region where the main temperature changes occur. It is possible to visualize the formation of the thermal boundary layer, near the pipe entrance, (**Figures 9(c)** and **(d)**), which is due to the temperature difference between the wall and the adjacent fluid. **Figures 9** and **10**, therefore, corroborate that the temperature distributions in oil and water phases differ in the pipe and that water undergoes a higher decrease in its temperature.

Table 4. Pressure drop per unit length as a function of the temperature of the mixture at the entrance of the pipe.

Case	T_w, T_o, T_g (K)	ΔP (Pa/m)
01 (Two-Phase)	313.15*	1267.50
02 (Three-Phase)	288.15	1561.25
03 (Three-Phase)	303.15	1483.83
04 (Three-Phase)	313.15	1465.88
05 (Three-Phase)	323.15	1461.25

*T_w and T_o only.

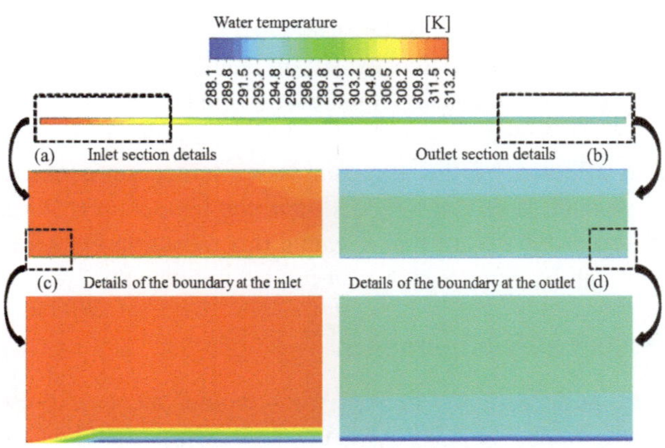

Figure 9. Water temperature field in the XY plane along the pipe (Case 04).

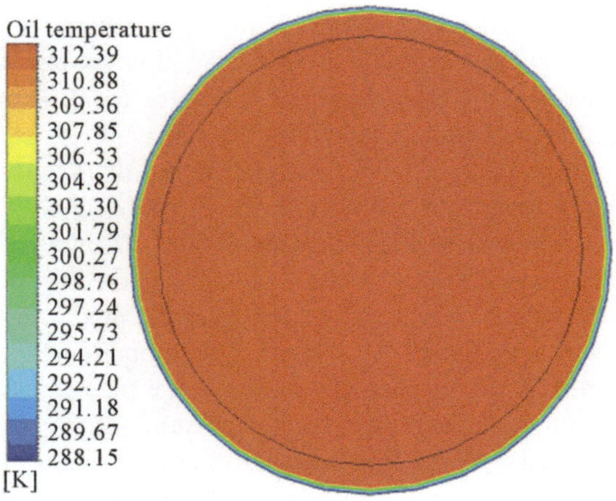

Figura 10. Oil temperature field in the YZ plane at X = 1 m, along the pipe (Case 04).

With respect to the temperature of the air phase, as it is dispersed in the oil core, its temperature profile is similar to the temperature profile of the continuous phase (heavy oil), presenting a uniform behavior in the vicinity of the center of the pipe.

4. Conclusions

Based on the results obtained it can be concluded that:

1) The utilized mathematical model was able to predict the behavior of a non-isothermal three-phase oil-water-air flow in a horizontal pipe;

2) It is found that the position of the oil core in the pipe and the velocity profiles of the phases are affected by the presence of air in the heavy oil-water two-phase flow. However, the core-annular flow pattern maintained its integrity;

3) The core-annular flow pattern is maintained even with temperature variation, and that the core of oil and air mixture tends to stratification (to be eccentric) while keeping away from the wall by a formed thin film of water;

4) Increasing the temperature of the fluids injected causes a reduction in the flow pressure drop, due to the decrease in the viscosity of oil and water, while the presence of air causes an increase in the pressure drop;

5) Temperature profiles of the phases along the pipe are affected by the assumed lower temperature of the pipe wall. It causes a reduction in the temperature of the fluids at the exit of the pipe, and this reduction is greater for annular water which is in contact with the pipe wall.

5. Acknowledgements

The authors of the present work would like to express their thanks to CNPQ, CAPES, FINEP, ANP/UFCG/PRH-25, PETROBRAS and JBR Engenharia Ltda. (Brazil), for the technical and financial support.

REFERENCES

[1] K. C. O. Crivelaro, Y. T. Damacena, T. H. F. Andrade, A. G. B. de Lima and S. R. de Farias Neto, "Numerical Simulation of Heavy oil Flows in Pipes Using the Core-Annular Flow Technique," *WIT Transactions on Engineering Sciences, Computational Methods in Multiphase Flow V*, Vol. 63, No. 5, 2009, pp. 193-203.

[2] R. M. O. Vara, A. C. Bannwart and C. H. M. Carvalho, "Production and Transportation of Heavy Oil by Water Injection," *1st Brazilian Congress R & D in Petroleum and Gas—1st PDPETRO*, UFRN-SQB Regional, Natal, 25-28 November 2001 (In Portuguese).

[3] F. E. Trevisan, "Flow Patterns and Pressure Drop in Three Phase Horizontal Flow of Heavy Oil, Water and Air," Master's Thesis, Petroleum Science and Engineering, Faculty of Mechanical Engineering, State University of Campinas (UNICAMP), Campinas, 2003 (In Portuguese).

[4] A. C. Bannwart, "Modeling Aspects of Oil-Water Core-Annular Flows," *Journal of Petroleum Science and Engineering*, Vol. 32, No. 2-4, 2001, pp. 127-143.

[5] G. Ooms and P. Poesio, "Stationary Core-Annular Flow through a Horizontal Pipe," *Physical Review E*, Vol. 68, No. 066301, 2003, pp. 1-7.

[6] A. Bensakhria, Y. Peysson and G. Antonini, "Experimental Study of the Pipeline Lubrication for Heavy Oil Transport," *Oil & Gas Science and Technology*, Vol. 59, No 5, 2004, pp. 523-533.

[7] S. Ghosh, T. K. Mandal, G. Das and P. K. Das, "Review of Oil Water Core Annular Flow," *Renewable and Sustainable Energy Reviews*, Vol. 13, No. 8, 2009, pp. 1957-1965.

[8] J. S. de S. Santos, "Numerical Study of Submerged Risers Lubrication to Transport of Heavy Oil," Master's Thesis, Chemical Engineering, Center of Science and Technology, Federal University of Campina Grande (UFCG), Campina Grande, 2009 (In Portuguese).

[9] T. H. F. Andrade, K. C. O. Crivelaro, S. R. de Farias Neto and A. G. B. de Lima, "Numerical Study of Heavy Oil Flow on Horizontal Pipe Lubricated by Water," In: A. Öchsner, L. F. M. da Silva and H. Altenbach, Eds., *Materials with Complex Behaviour II: Advanced Structured Materials*, Springer-Verlag, Heidelberg, Berlin Heidelberg, Vol. 16, 2012, pp. 99-118.

[10] A. C. Bannwart, O. M. H. Rodriguez, F. E. Trevisan, F. F. Vieira and C. H. M. de Carvalho, "Experimental Investigation on Liquid-Liquid-Gas Flow: Flow Patterns and Pressure-Gradient," *Journal of Petroleum Science and Engineering*. Vol. 65, No. 1-2, 2009, pp. 1-13.

[11] P. Poesio, G. Sotgia and D. Strazza, "Experimental Investigation of Three-Phase Oil-Water-Air Flow through a Pipeline," *Multiphase Science and Technology*, Vol. 21, No. 1-2, 2009, pp. 107-122.

[12] P. Poesio, D. Strazza and G. Sotgia, "Very-Viscous-Oil/Water/Air Flow through Horizontal Pipes: Pressure Drop Measurement and Prediction," *Chemical Engineering Science*, Vol. 64, No. 6, 2009, pp 1136-1142.

[13] D. Strazza, D. Chiecchi and P. Poesio, "High Viscosity Oil-Water-Air Three Phase Flows: Flow Maps, Pressure Drops and Bubble Dynamics," *7th International Conference on Multiphase Flow—ICMF*, Tampa, 30 May-4 June 2010, pp. 1-7.

[14] F. N. Silva, T. H. F. Andrade, S. R. de Farias Neto and A. G. B. de Lima, "Numerical Study of Three Phase Flow (Water-Heavy Oil-Gas) Type Core Flow in a Connection 'T'," *6th Brazilian Congress of Research and Development at Petroleum and Gas—6th PDPETRO*, Florianópolis, 9-13 October 2011 (In Portuguese).

[15] F. P. Incropera, A. S. Lavine and D. P. DeWitt, "Fundamentals of Heat and Mass Transfer," 6th Edition, LTC, Rio de Janeiro, 2008 (In Portuguese).

[16] C. W. S. Santana, E. G. Tôrres and I. de S. Lacerda, "Adjustment Equations for Kinematic Viscosity of Petroleum Products Depending on the Temperature," *Proceedings of the 3rd Brazilian Congress of R & D in Petroleum and Gas—3rd PDPETRO*, Rio de Janeiro, 2-5 October 2004 (In Portuguese).

[17] F. Kreith and M. S. Bohn, "Principles of Heat Transfer," Editora Edgard Blücher, São Paulo, 1977 (In Portuguese).

Removal of Carbon Monoxide from Hydrogen-rich Fuels over CeO_2-promoted Pt/Al_2O_3

Akkarat Wongkaew[1*], Pichet Limsuwan[2]
[1]Department of Chemical Engineering, Faculty of Engineering, Burapha University, Chonburi, Thailand
[2]Department of Physics, Faculty of Science, King Mongkut's University of Technology Thonburi, Bangkok, Thailand

ABSTRACT

A comparative study of catalytic CO oxidation and selective CO oxidation over Pt/Al_2O_3 and CeO_2-promoted Pt/Al_2O_3 catalysts has been investigated for the removal of a trace amount of CO from the reformed gas. The catalysts were prepared by sol gel and incipient wetness impregnation. CO oxidation and selective CO oxidation were carried out with a 5%Pt/Al_2O_3 and a 5%Pt/15%CeO_2/Al_2O_3. The presence of 15%CeO_2 in the 5%Pt/Al_2O_3 dramatically improves the activities to CO oxidation and selective CO oxidation at low temperature (<180℃). FTIR results indicate that CO could react with lattice oxygen from ceria and release CO_2 as a product. Low space velocity would obtain high CO conversion at low temperatures while high space velocity would obtain high CO conversion at high temperatures. The results also show that a 5%Pt/15%CeO_2/Al_2O_3 can completely oxidize 1% CO at 180℃ with selectivity of 52% and space velocity of 70,000 $cm^3 \cdot g^{-1} \cdot h^{-1}$. Under the realistic gas feed with 1%O_2, this catalyst is very stable and retains its activity and selectivity at 180℃ during 72 h.

Keywords: CO Oxidation; CeO_2-Pt/alumina; CO Adsorption; Selective CO Oxidation; Fuel Processing

1. Introduction

The polymer-electrolyte-membrane fuel cell (PEMFC) has been attracting significant attention in several applications including electric vehicles and residential power-generations. This is because of its many attractive features such as high power density, rapid start-up, and high efficiency [1, 2]. As the PEMFCs utilize hydrogen gas as a fuel and since hydrogen can be produced by means of a fuel reformer followed by water gas shift reaction for further conversion of CO to H_2, carbon monoxide is always present in the hydrogen stream. Generally, catalytic steam reforming of methanol or partial oxidation of gasoline followed by water gas shift reaction will produce a gas stream with 40%-75% H_2, 15%-20% CO_2, ~10% H_2O, 0-25% N_2 and 0.5%-1.0% CO. This amount of CO contained in the reformed gas is high enough to poison the Pt anode of PEM fuel cells and in turn dramatically degrades the fuel cell potential and energy conversion efficiency [3, 4]. Experimentally, it has been found that the tolerable level of CO without harmful effects is about 10 ppm [5]. This means the 0.5%-1.0% CO needs to be reduced to 10 ppm or less in order to increase the use of proton exchange membrane (PEM) fuel cells running with on-board generated hydrogen. The most economical and straightforward technique for this purpose is selective catalytic oxidation of CO in the H_2-rich reformed gas using O_2 or preferential oxidation (PROX). This method needs a suitable catalyst to enhance the CO oxidation reaction with minimal oxidation of hydrogen which is the desired fuel. The crucial requirement for the PROX re- actor is a high CO conversion with high selectivity. A number of catalysts have been investigated for the PROX reaction [6-10]. Noble metals supported on alumina such as Pt, Au, Ru and Pd, have been proposed as ideal catalysts for PROX reaction, especially Pt/Al_2O_3, Pt/ Fe_2O_3/ Al_2O_3, Pt/CeO_2, Pt/CeO_2_ZrO_2, Pt/zeolite [11-13]. An improvement in the selectivity at low temperatures is needed. Oxidation of CO on alumina-supported Pt catalysts is known to take place via the Langmuir-Hinshel- wood mechanism. Kahlich et al. [14] studied the kinetics of selective CO oxidation in H_2-rich gas on Pt/Al_2O_3 and observed that CO conversion never reached 100% for a 0.5%Pt/Al_2O_3. The maximum CO conversion was ~80% at temperatures as high as 250 ℃. Other studies reported that CO conversion occurred in the reaction temperature range of 200℃-250℃ [15] or else needed high oxygen concentration for complete elimination of CO, corresponding to lower selectivity. We have previously [16] investigated the catalytic activity of 2%Pt/Al_2O_3, which was prepared by the sol-gel method, in selective CO oxidation reaction under the excess H_2 gas stream. We found that 2% Pt was well dispersed in alumina supports. Therefore, it can selectively oxidize CO down to ppm level with constant selectivity and high space velocity. The performance of Pt catalysts can be improved by modifying supports such as adding alkali metals into supports [17] or by promoting with other metal oxides such as Fe_2O_3 or CeO_2 [18]. Serre, et al. [19] found that the presence of CeO_2 in a 2%Pt/Al_2O_3 after a reductive pretreatment drastically enhanced the activity of the catalyst to CO oxidation. The promoting effect of ceria was attributed to the enhancement of the metal dispersion and the stabilization of Al_2O_3 support toward thermal sintering. Moreover, ceria can be a chemically active component, working as an oxygen store that releases lattice oxygen in the presence of reductive gases and re- placement of the lattice oxygen with oxygen gas when oxygen gas is present in excess [20]. Parinyaswan et al. [21] investigated the performance of Pt-Pd/CeO_2 catalysts for selective CO oxidation. This catalyst could maximally convert 83% CO to CO_2 with selectivity of 60% at 90℃with a gas feed containing 1% CO, 1% O_2, 25% CO_2 and 10% H_2O. This means there would still be 1,700 ppm of CO left in the gas feed and this gas could dramatically deplete the efficiency of a PEM fuel cell in a very short time. The authors suggest a multi-stage reactor to reduce CO to below 10 - 100 ppm for the use of this catalyst with PEM fuel cells. Silva et al. [22] studied the effect of the presence of ceria with Pt over alumina catalysts for the partial oxidation of methane reaction. They reported that the addition of ceria in alumina led to the formation of a homogenous solid solution, which exhibited a high-oxygen storage capacity. Brown et al. [23] compared the activities of Pt over alumina with ceria-promoted Pt over alumina in the production of hydrogen from methanol decomposition. They reported that promoting with ceria had a positive effect on activity and selectivity. Indeed, the use of ceria coupled with alumina as a support for Pt might enhance the catalytic selective CO oxidation of platinum over alumina catalysts. Son et al. [24,25] investigated the performance of Ce-Pt/γ-Al_2O_3 for selective oxidation of CO in H_2 for PEFCs. They found that the addition of 5% Ce in the Pt over alumina dramatically enhanced CO conversion and selectivity at low temperatures. The catalyst completely converted 1%CO to CO_2 at 200℃ with 50% selectivity. Although the effect of other gases such as CO_2 and H_2O was stud- ied, their gas compositions containing 1% CO, 2.3% H_2O, 10.09% CO_2 and H_2 as balance were far from a realistic gas composition containing 40%-75% H_2, 0.5%-2.0% CO, 15%-20% CO_2, 10% H_2O and 0-25% N_2 by volume. Therefore, the activity of promoted platinum over alu- mina with ceria for preferential CO oxidation still needs to be investigated.

In order to obtain a better understanding of the activity of the ceria promoted platinum alumina catalyst to the selective CO oxidation, 5%Pt/15%CeO_2/Al_2O_3 and 5% Pt/Al_2O_3 catalysts were tested for their activities to both CO oxidation in H_2 free-stream and CO oxidation in the presence of excess H_2-containing feed stream and the obtained results were compared with others reported in literatures. All supports in this work were prepared by sol gel method. The CO coverage of the catalysts was also studied using FTIR. The FTIR results were used to explain the enhancement of catalyst activity in the presence of ceria. It should be pointed out that all gases containing in reformed gas affect to the selectivity to CO oxidation of the catalysts as reported in literatures. Therefore, the effect of space velocity on the activity of the catalyst was investigated. These results will lead us to the proper operating conditions in order to obtain high selectivity and high CO conversion of ceria promoted Pt alumina catalyst.

2. Experimental

2.1. Catalyst Preparation

A cerium aluminum oxide supported platinum (Pt) catalyst with 5.0 wt% Pt loading was prepared by sol gel technique [26] and incipient wetness impregnation. Aluminum is protoxide; cerium (III) acetate and hydrogen hex anchor oplatinate (IV) hydrate were obtained from Aldrich. Preparation of supports containing cerium aluminum oxide began with dissolving the desired amount of aluminum is protoxide in hot demonized water at 80℃. After 30 min of aging with continuous stirring, nitric acid (HNO_3) was added to start the hydrolysis reaction resulting in a fibrillar sol. Then, the known amount of cerium (III) acetate was incorporated into the solution at room temperature. The solution was stirred overnight to obtain uniformity. The obtained solution was heated to 60℃ and kept at this temperature until gelation occurred. The gel was dried in air at 110℃ overnight, and then calcined at 500℃ for 13 hours. After calcination, the resulting powder was ground and sieved to obtain a 100-140 mesh powder. Incipient wetness impregnation was used to deposit platinum into the support. With this method, a desired amount of solution of hydrogen hexachloroplatinate (IV) hydrate was added into a cerium alumina support and then mixed together until the mixture uniformly. The obtained solid was dried in air at 110°C overnight, and then calcined at 500℃ for 13 hours. The final powder was 5% Pt/15% CeO_2/Al_2O_3. In this work, the activity of this catalyst was compared with activity of a 5% Pt/ Al_2O_3. For a 5% Pt/Al_2O_3 analogous preparation techniques were used such that the Al_2O_3 support was prepared by sol gel and Pt was impregnated in the support by incipient wetness impregnation. Before testing the activities of these catalysts, the catalysts were purged with H_2 at 400℃ for 5 hrs.

2.2. Characterization

The BET surface area and average pore radius of catalysts were determined with adsorption-desorption isotherms of N_2 at 77 K using a MicroMeritics ASAP 2010 instrument. Average crystalline sizes of oxides were determined by Scherrer's equation using the X-ray line broadening from X-ray diffraction, Bruker AXS model D 8 Discover equipped with a CuKα radiation with a nickel filter. Diffraction intensity was measured in the 2 theta ranges between 20° and 85°, with a step of 0.02° for 8 s per point.

2.3. Catalytic Activity

CO oxidation and CO oxidation in the H_2-rich stream was performed in a fixed-bed reactor. The reaction temperatures inside the reactor were measured with a K-type thermocouple placed on the top of the catalyst bed and were controlled by a temperature controller (OMEGA: CN3251). The amount of catalyst used in each run was 68 mg. The total gas flow rates of the reaction mixture were 40 $cm^3 \cdot min^{-1}$ and 80 $cm^3 \cdot min^{-1}$, corresponding to the space velocity of 70,000 $cm^3 \cdot g^{-1} \cdot h^{-1}$ and of 35,000 $cm^3 \cdot g^{-1} \cdot h^{-1}$, respectively. For CO oxidation, the activity tests were conducted with a feed mixture of 1% CO, 1% O_2 and He as balance. For selective CO oxidation, a feed mixture contained 1% CO, 0.5%-1% O_2, 0-10% H_2O, 0-20% CO_2, 55% H_2 and He as balance. A volumetric flow controller with an accuracy of 0.5 $cm^3.min^{-1}$ was used for measuring the total gas flow rate at the bypass (for calibration purposes) and at the outlet of the reactor.

A Varian CP-4900 micro gas chromatograph (micro GC) equipped with 2 channels (A and B) was used for analysis of the outlet gas compositions from the reactor. Channel A was used to detect H_2, O_2, CO and CH_4 by a Molsieve 5A PLOT column. Channel B was used to detect CO_2 by a PoraPLOT Q column. Along with GC, FTIR was used to detect CO at low concentrations (ppm level) in the outlet gas from the reactor. Because water deteriorates the performance of these columns, an ice cooled water condenser was used to remove water from the gas streams before entering the GC and FTIR.

The CO conversion was obtained by comparing the CO concentration in the feed measured at the bypass line and the CO concentration in the outlet stream from the reactor. Selectivity was defined as the ratio of oxygen consumed by CO oxidation to the total oxygen consumption (obtained by subtracting the O_2 concentration at the reactor outlet from the O_2 concentration in the feed). The amount of O_2 not used in the CO oxidation reaction was assumed to oxidize H_2 in the H_2 oxidation reaction. Importantly, there was no methane formation observed under reaction conditions performed in this study.

3. Results and Discussion

3.1. Characterization of the Catalysts

The alumina support prepared via the sol gel method yielded a BET area of 227.8 m^2/g with an average pore size of 5.3 nm while the sol gel made 15%CeO_2/Al_2O_3 had a BET area of 230.0 m^2/g with an average pore size of 5.2 nm. Pt crystalline sizes were estimated from the line broadening of Pt (111) peaks. For both a 5%Pt/Al_2O_3 catalyst and a 5%Pt/15%CeO_2/Al_2O_3 catalyst, no Pt (111) peaks were observed indicating that Pt metal was well impregnated and dispersed on the supports for both catalysts. For ceria structure in a 5%Pt/15%CeO_2/Al_2O_3 fresh catalyst reduced under hydrogen, XRD results matched with those for CeO_2 with average CeO_2 crystallite sizes of ~6.8 nm. Dispersions measured by CO chemisorption were 48% for the5%Pt/Al_2O_3 and 54% for the 5%Pt/ 15%CeO_2/Al_2O_3.

3.2. Activity Tests for CO Oxidation with Free-H2 Gas Stream

The two catalysts were tested for their activities in CO oxidation as a function of temperature under gas feed containing 1%CO, 1%O_2 and He as balance as seen in **Figure 1**.

At 100℃ the integral conversions were 2% and 10% over the 5%Pt/Al_2O_3 and the 5%Pt/15%CeO_2/Al_2O_3 catalysts. Increasing the temperature to 150℃ increased the CO conversion from 2% to 27% for the 5%Pt/Al_2O_3 and from 10% to 60% for the 5%Pt/15%CeO_2/Al_2O_3. Further increasing reaction temperature to 170℃ increased the CO conversion to 70% for the 5%Pt/Al_2O_3 and to 98.9% for the 5%Pt/15%CeO_2/Al_2O_3. Finally, CO completely converted to CO_2 at 180℃ for both catalysts. Clearly, the catalyst containing ceria showed better activity in CO oxidation, especially at low temperatures. For the 5%Pt/Al_2O_3, it has been known that CO strongly chemisorbs over Pt sites from room temperatures to 150℃

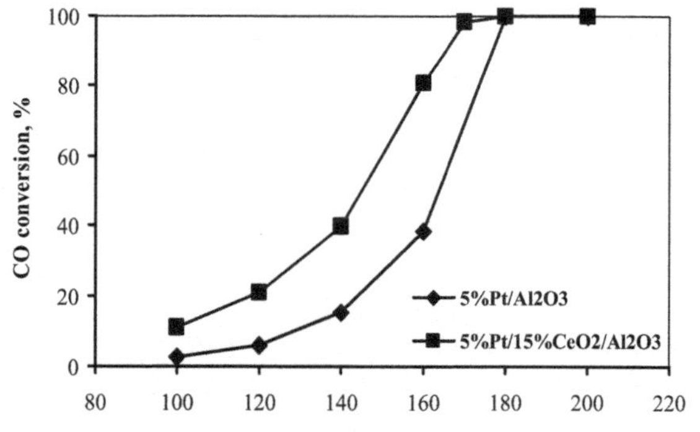

Figure 1. Activities of platinum over alumina and promoted platinum over alumina catalysts to CO oxidation. Gas composition: 1%CO, 1%O_2 and He as balance.

[14,27]. At these temperatures, the competition between CO molecules and O_2 molecules over the active sites is crucial. The reaction occurs when O_2 molecules adsorb and dissociate next to CO molecules. At tem- peratures less than 150℃, Pt sites are occupied by CO molecules. This leads to low activity of platinum over alumina catalysts in the CO oxidation reaction. Increas- ing reaction temperature above 150℃ dramatically in- creases the rate of reaction due to desorption of CO molecules from the active sites and leaving available sites for O_2 molecules to be adsorbed. The addition of ceria enhances the rate of reaction of the catalyst. This is due to the oxygen storage property proposed as following [28]:

$$CO_{ads} + 2CeO_2 \rightarrow CO_2 + * + Ce_2O_3$$

$$\frac{1}{2}O_2 + * \rightarrow O_{ads}$$

$$O_{ads} + Ce_2O_3 \rightarrow 2CeO_2 + *$$

where "*" stands for an adsorption site on platinum and "ads" for an adsorbed species. From this model, the key point of high activity of the $5\%Pt/15\%CeO_2/Al_2O_3$ catalyst at low reaction temperatures (<150℃) results from a capacity of ceria to switch between the two-oxidation states Ce^{4+} and Ce^{3+}. Ce^{3+} returns back to Ce^{4+} by gaseous oxygen molecules. This phenomenon occurs even at room temperature. However, 170℃ is high enough for CO molecules to desorbs from active sites and leave the active sites available for other reactant molecules to be adsorbed and reacted [29]. Therefore, both catalysts performed comparably above 170℃. FTIR was used to check the chemisorbed CO species. The results are shown in **Figure 2**. Before the test, the catalyst pellet diluted with KBr was purged with helium at 200℃ until catalyst surface was clean. Then, the sample was cooled to 100℃ under a helium purge. After the temperature of the sample reached 100℃, the sample was purged with gas stream containing 1% CO balance with helium. The adsorption of CO over the catalyst sample was recorded.

Figure 2(a) shows the CO adsorption over the $5\%Pt/15\%CeO_2/Al_2O_3$ pellet. A strong peak at 2062 cm^{-1} was observed. This peak corresponds to CO adsorbed over $Pt/CeO_2/Al_2O_3$ [30]. Other peaks at 2395-2312 cm^{-1} were also observed and these peaks indicate the presence of gas phase CO_2. The intensity of the CO_2 band slowly decreased with time and finally disappeared. The formation of CO_2 was due to CO oxidation reaction and the O_2 reactant must have come from lattice of CeO_2. This result was in agreement with a model of oxygen transport in Pt/ceria catalyst [31]. Next, the same experiment was carried out with the $5\%Pt/Al_2O_3$. The results are shown in **Figure 2(b)**. A strong peak at 2085 cm^{-1} was observed. This peak corresponds to CO adsorbed on Pt with neighboring oxidized Pt [32]. Unlike platinum over ceria-promoted alumina catalyst, no peak of CO_2 gas phase was observed. This means that in the absence of gas phase O_2 no CO oxidation reaction occurs with this catalyst. The CO adsorbed on Pt peak for promoted Pt catalyst appeared at 2062 cm^{-1} while that of Pt/Al_2O_3 appeared at 2085 cm^{-1} [29,32]. The downward shift of CO adsorbed wave number may be due to inducing of C-O bond weakening for CO adsorbed on Pt by Ce [19]. The reduction of the bond strength of adsorbed CO makes it more reactive with an oxygen atom from ceria lattice into CO adsorbed near Pt-CeO_2 interface followed by desorption of CO_2 leaving Pt site available for other gas molecules. Therefore, the presence of a small amount of ceria could enhance the activity of platinum over alumina due to its oxygen storage property.

3.3. Effect of Ceria on the Activity of Catalysts

The effect of ceria on the activity of catalysts was shown in **Figure 3**.

Figure 3(a) shows the activities of the two catalysts as function of reaction temperature for a dry gas composition of 1%CO, 1%O_2, 55%H_2 and He as balance. At 110 ℃, CO conversions were 46% and 60% over the 5% Pt/Al_2O_3 and the $5\%Pt/15\%CeO_2/Al_2O_3$ catalysts. Increasing reaction temperatures further to 180°C led to dramatic increases in CO conversion from 46% to 90% for the $5\%Pt/Al_2O_3$ and from 60% to 99.3% for the $5\%Pt/15\%CeO_2/Al_2O_3$. At 190℃, CO conversion reached a maximum of approximately 99.97% for the 5%Pt/15% CeO_2/Al_2O_3 while CO conversion reached a maximum of approximately 93.70% for the $5\%Pt/Al_2O_3$. Further increasing reaction temperature to 230℃ resulted in decreases in CO conversion to 57% for the $5\%Pt/Al_2O_3$ and 89% for the $5\%Pt/15\%CeO_2/Al_2O_3$. **Figure 3(b)** shows oxygen consumption as a function of reaction temperature. For the $5\%Pt/Al_2O_3$, O_2 was quickly consumed by the reactions. O_2 conversion increased from 79% to 100% when reaction temperature increased from 110℃

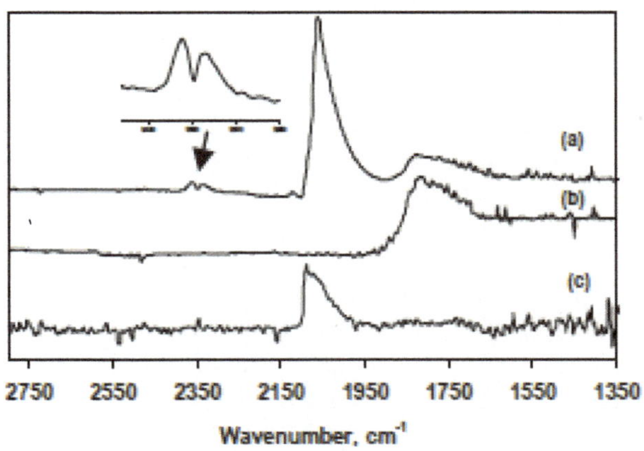

Figure 2. FTIR study for CO adsorbed over catalysts under 1%CO in He at 100° C: (a) $5\%Pt/15CeO_2/Al_2O_3$, (b) 15% CeO_2/Al_2O_3 and (c) $5\%Pt/Al_2O_3$.

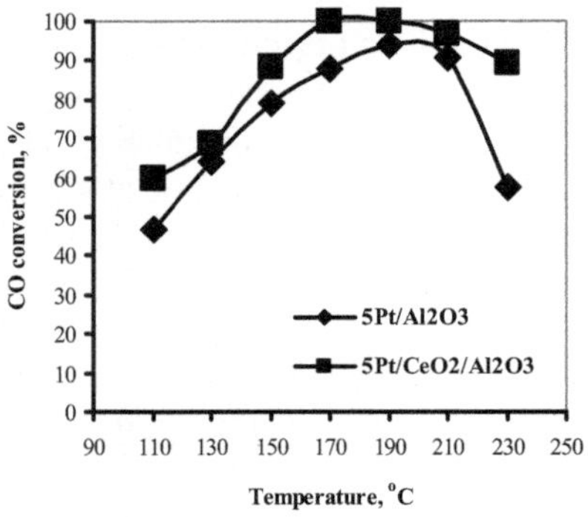

(a)

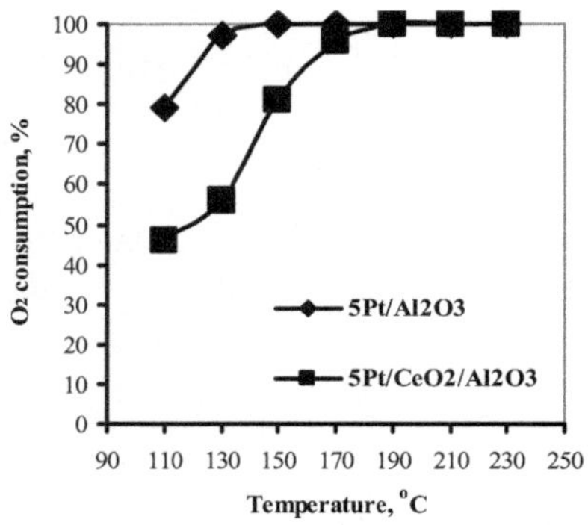

(b)

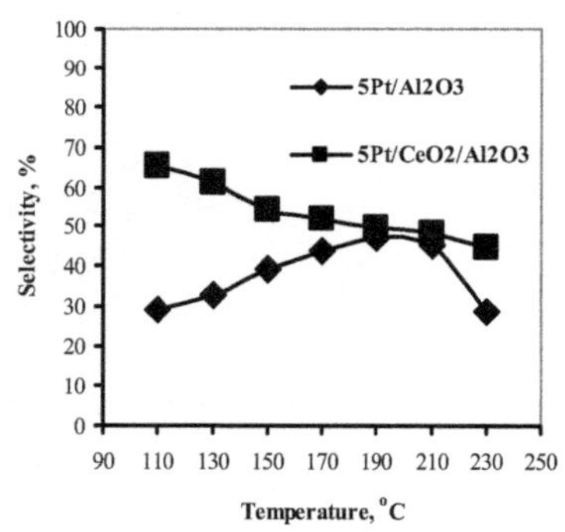

(c)

Figure 3. Comparison of activities to selective CO oxidation of the 5%Pt/Al_2O_3 and the 5%Pt/ 15%CeO_2/Al_2O_3 catalysts as a function of temperature. Gas composition: 1%CO, 1%O_2, 55%H_2 and He as balance.

to 150℃. Unlike the 5%Pt/Al_2O_3, O_2conversion for the 5%Pt/15%CeO_2/Al_2O_3 slowly increased from 46% to 56%, 81%, 86% and 100% when reaction temperature increased from 110℃ to 130℃, 150℃, 170℃, and 190 ℃, respectively. These results led to differences in selectivity for CO oxidation as shown in **Figure 3(c)**. As mentioned previously, no methane formation was observed under these operating conditions. Consequently, selectivity was defined as the ratio of oxygen used for CO oxidation to total oxygen consumed by the reactions. As shown in **Figure 3(c)**, the selectivity's of the two catalysts are different. At 110℃, selectivity was 29% for the 5%Pt/Al_2O_3. This means that oxygen consumed by the reactions mostly goes to H_2 oxidation. At the same temperature, selectivity was 65% for the 5%Pt/15% CeO_2/Al_2O_3. This means that the presence of ceria in the catalyst enhanced the rate of CO oxidation. Further increases reaction temperature to 180℃ increased selectiveity to 44% for the 5%Pt/Al_2O_3 but decreased selectivity to 50% for the 5%Pt/15%CeO_2/Al_2O_3. At 230℃, selectivity of both catalysts dropped to 28% for the 5%Pt/ Al_2O_3 and 45% for the 5%Pt/15%CeO_2/Al_2O_3. The decrease of selectivity for both catalysts at high temperatures is due to the competition between CO oxidation and H_2 oxidation reactions. At high temperatures, the H_2 oxidation reaction occurs faster than the CO oxidation reaction [32]. Further investigation of the 5%Pt/15% CeO_2/Al_2O_3 catalyst was conducted to understand its behavior under different reaction conditions.

3.4. Space Velocity Effect

The dependence of CO conversion and selectivity on flow rate is shown in **Figures 4(a)-(c)**.

Space velocity was increased from 35,000 to 70,000 $cm^3 \cdot g^{-1} \cdot h^{-1}$. Considering at the same temperatures, increasing the space velocity decreased the CO conversion and decreased the O_2 consumption. With 1% CO, 1% O_2, 20% CO_2, 10% H_2O, 55% H_2 and He as balance, maximum CO conversion for the low space velocity run was 100% at 150°C while maximum CO conversion for the high space velocity run was 100% at 180-190℃. Selectivity did not change with space velocity at temperatures less than 150℃. At higher temperatures, selectivity for the low space velocity run was lower than that for the high space velocity run. This result indicates that the use of this catalyst depends on the reaction condition. Low space velocity would obtain high CO conversion at low temperatures while high space velocity would obtain high CO conversion at high temperatures.

Activities of 1%Pt-1%CeO_2 over activated carbon [33] were compared with those of 5%Pt/15%CeO_2/Al_2O_3 in selective CO oxidation under realistic gas compositions. Although the authors reported that their catalyst is very active to CO oxidation in H_2-excess stream and CO conversion drastically increased with the presence of CO_2, combining both CO_2 and H_2O in the gas feed stream resulted in CO conversion of 100% and selectivity to CO oxidation of 50% at 150℃ with space velocity of 24,000 $cm^3 \cdot g^{-1} \cdot h^{-1}$. Unlike this catalyst, CO conversion of the 5% Pt/15%CeO_2/Al_2O_3 decreased with the presence of CO_2 in the gas stream. This is due to the carbonate formation blocking the available active sites. However, combining both CO_2 and H_2O in the gas feed enhanced the activity of the catalyst especially at low temperatures (<160℃). Under the realistic gas composition, our catalyst obtained 100% CO conversion at 140-150℃ with selectivity in CO oxidation of 50%-52% and space velocity of 35,000 $cm^3 \cdot g^{-1} \cdot h^{-1}$.

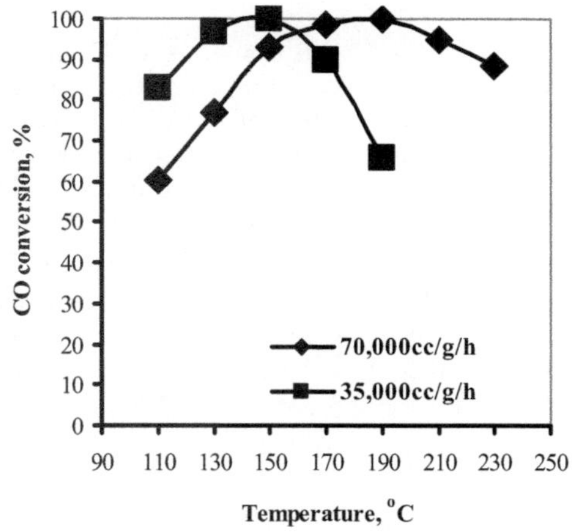

(a)

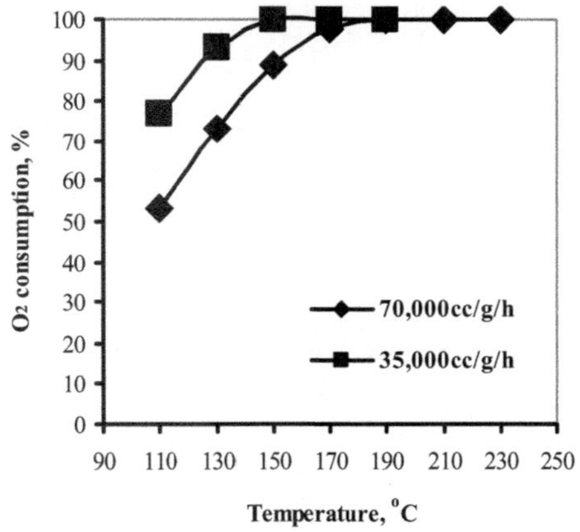

(b)

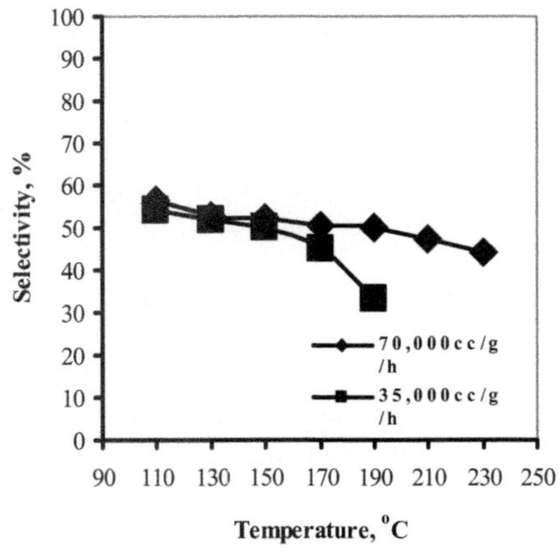

(c)

Figure 4. Dependence of CO conversion and selectivity of CO oxidation on space velocity and temperature for a 5%Pt /15%CeO_2/Al_2O_3. Gas composition: 1%CO, 1%O_2, 10%H_2O, 20%CO_2, 55%H_2 and He as balance.

In our work, 5%Pt/15%CeO_2/Al_2O_3 demonstrated excellent performance in preferential CO oxidation. This catalyst is very stable under the realistic gas conditions during 72 hr of use. The higher selectivity in CO oxidation could obtain under the low O_2 concentration.

4. Conclusions

The CO poisoning of PEMFCs is a major problem to deplete the efficiency and energy conversion of PEMFCs. To remove a trace amount of CO in the reformed gas is essential. In this work, the catalytic performance of promoted platinum over alumina with ceria in selective CO oxidation in the presence of excess hydrogen has been studied. The addition of ceria improved the catalytic activity in oxidation reactions. It resulted from the oxygen storage property of ceria. Space velocity also affected the CO conversion. High CO conversion at low temperature was obtained when low space velocity was chosen. Finally, the 5%Pt/15%CeO_2/Al_2O_3 is an excellent catalyst for removal of trace CO in reformed gas. It reduces 1% CO in the realistic reformed gas to less than 10 ppm with selectivity of 52% at 180℃ with space velocity of 70,000 $cm^3 \cdot g^{-1} \cdot h^{-1}$.

5. Acknowledgements

The authors gratefully acknowledge the financial support from Thailand Research Fund (TRF) under the contract # MRG 4680140.

REFERENCES

[1] K. Jost, "Gasoline-reforming Fuel Cell," *Automotive Engineering* , 1997, pp. 151-152.

[2] X. Cheng, Z. Shi, N. Glass, L. Zhang, J. Zhang, D. Song, A.-S. Liu, H. Wang and J. Shen, "A Review of PEM Hydrogen Fuel Cell Contamination: Impacts, Mechanisms, and Mitigation," *Journal of Power Sources,* Vol. 165, 2007, pp. 739-756.

[3] S. J. C. Cleghorn, X. Ren, T. E. Springer, M. S. Wilson, C. Zawodzinski, T. A. Zawodzinski and S. Gottesfeld, "PEM fuel Cells for Transportation and Stationary Power Generation Applications," *International Journal of Hydrogen Energy*, Vol. 22, 1997, pp. 1137-1144.

[4] W. A. Adams, J. Blair, K. R. Bullock and C. L. Gardner, "Enhancement of the Performance and Reliability of CO Poisoned PEM Fuel Cells," *Journal of Power Sources* Vol. 145, No. 1, 2005, pp. 55-61.

[5] S. Gottesfeld, US patent 4,910,099 Preventing CO poisoning in fuel cells (Mar 20, 1990).

[6] Y. H. Kim, S.-D. Yim and E. D. Park, "Selective CO Oxidation in a Hydrogen-rich Stream Over Ru/SiO_2, Catal," *Today*, Vol. 185, 2012, pp. 143-150.

[7] J. Li, P. Zhu, S. Zuo, Q. Huang and R. Zhou, "Influence of Mn Doping on the Performance of CuO-CeO Catalysts for Selective Oxidation of CO in Hydrogen-rich Streams," *Applied Catalysis A: General*, Vol. 381, 2010, pp. 261-266.

[8] J. W. Park, J. H. Jeong, W. L. Yoon, C. S. Kim, D. K. Lee, Y.-K. Park and Y. W. Rhee, "Selective Oxidation of CO in Hydrogen-rich Stream over Cu–Ce Catalyst Promoted with Transition Metals," *International Journal of Hydrogen Energy*, Vol. 30, No. 2, 2005, pp. 209-220.

[9] Y. -F. Han, M. Kinne, R. J. Behm, "Selective Oxidation of CO on Ru/γ-Al_2O_3 in Methanol Reformate at Low Temperatures, *Appl. Catal. B: Environ,* Vol. 52, No. 2, 2004, pp. 123-134.

[10] C.-T. Chang, B.-J. Liaw, Y.-P. Chen, Yin-Zu Chen, "Characteristics of Au/Mg_xAlO Hydrotalcite Catalysts in CO Selective Oxidation, *Journal of Molecular Catalysis A: Chemical*, Vol. 300, No. 1-2, 2009, pp. 80-88.

[11] M. Kotobuki, A. Watanabe, H. Uchida, H. Yamashita, M. Watanabe, "Development of Pt/ZSM-5 Catalyst with High CO Selectivity for Preferential Oxidation of Carbon Monoxide in a Reformed Gas," *Chem. Lett.* Vol. 34, 2005, pp. 866-867.

[12] J. L. Ayastuy, A. Gil-Rodríguez, M. P. González-Marcos, M. A. Gutiérrez-Ortiz, Effect of Process Variables on Pt/CeO_2 Catalyst Behavior for the PROX Reaction, *International Journal of Hydrogen Energy*, Vol. 31, 2006, pp. 2231-2242.

[13] M. Kotobuki, A. Watanabe, H. Uchida, H. Yamashita, M. Watanabe, "High Catalytic Performance of Pt-Fe Alloy Nanoparticles Supported in Mordenite Pores for Preferential CO Oxidation in H_2-rich Gas, *Applied Catalysis A: General*, Vol. 307, 2006, pp. 275-283.

[14] M. J. Kahlich, H. A. Gasteiger and R. J. Behm, "Kinetics of the Selective CO Oxidation in H_2-Rich Gas on Pt/Al_2O_3," *Journal of Catalysis*, Vol. 171, 1997, pp. 93-105.

[15] I. H. Son, M. Shamsuzzoha and A. M. Lane, "Promotion of Pt/γ-Al_2O_3 by New Pretreatment for Low-Temperature Preferential Oxidation of CO in H_2 for PEM Fuel Cells," *Journal of Catalysis*, Vol. 210, 2002, pp. 460-465.

[16] A. Manasilp and E. Gulari, "Selective CO Oxidation over Pt/alumina Catalysts for Fuel Cell Applications, *Applied Catalysis B: environmental*, Vol. 37, 2002, pp. 17-25.

[17] H. Tanaka, M. Kuriyama, Y. Ishida, S.-I. Ito and K. Tomishige, "Preferential CO Oxidation in Hydrogen-rich Stream over Pt Catalysts Modified with Alkali Metals: Part II. Catalyst Characterization and Role of Alkali Metals," *Applied Catalysis A: Generl*, Vol. 343, No. 1-2, 2008, pp. 125-133

[18] Y. Li, Q. Fu, M. F., Stephanopoulos, Low-temperature water-gas shift reaction over Cu- and Ni-loaded cerium oxide catalysts, *Applied Catalysis B: Environmental*, Vol. 27, No. 3, 2000, pp. 179-191.

[19] C. Serre, F. Garin, G. Belot and G. Maire, "Reactivity of Pt/Al_2O_3 and Pt-$CeO_2Al_2O_3$ Catalysts for the Oxidation of Carbon Monoxide by Oxygen: I. Catalyst Characterization by TPR Using CO as Reducing Agent," *Journal of Catalysis*, Vol. 141, No. 1, 1993, pp. 1-8.

[20] A. Martînez-Arias, J. M. Coronado, R. Cataluña, J. C. Conesa and J. Soria, "Influence of Mutual Platinum-Dispersed Ceria Interactions on the Promoting Effect of Ceria for the CO Oxidation Reaction in a Pt/CeO_2/Al_2O_3 Catalyst," *The Journal of Physical Chemistry Letters B*, Vol. 102, 1998, pp. 4357-4365.

[21] A. Parinyaswan, S. Pongstabodee and A. Luengnaruemitchai, "Catalytic Performances of Pt–Pd/CeO_2 Catalysts for Selective CO Oxidation," *International Journal of Hydrogen Energy*, Vol. 31, No. 13, 2006, pp. 1942-1949.

[22] F. A. Silva, D. S. Martinez, J. A. C. Ruiz, L. V. Mattos, C. E. Hori, G. B. Noronha, The Effect of the Use of Cerium-doped Alumina on the Performance of Pt/CeO_2/Al_2O_3 and Pt/$CeZrO_2$/Al_2O_3 Catalysts on the Partial Oxidation of Methane, *Applied Catalysis A: General*, Vol. 335, 2008, pp. 145-152.

[23] J. C. Brown and E. Gulari, "Hydrogen Production from Methanol Decomposition over Pt/Al_2O_3 and Ceria Promoted Pt/Al_2O_3 Catalysts, *Catalysis Communications*, Vol. 5, No. 8, 2004, pp. 431-436.

[24] I. H. Son and A. M. Lane, Promotion of Pt/γ-Al_2O_3 by Ce for Preferential Oxidation of CO in H_2, *Catalysis Letters*, Vol. 76, No. 3-4, 2001, pp. 151-154.

[25] I. H. Son, Study of Ce-Pt/γ-Al_2O_3 for the Selective Oxidation of CO in H_2 for Application to PEFCs: Effect of Gases, *Journal of Power Sources*, Vol. 159, 2006, pp. 1266-1273.

[26] B. E. Yoldas, "Alumina Sol Preparation from Alkoxides," *Ceram. Bull,* Vol. 54, 1975, pp. 289-290.

[27] J. H. B. J. Hoebink, J. P. Huinink and G. B. Marin, "A Quantitative Analysis of Transient Kinetic Experiments: The Oxidation of CO by O_2 over Pt," *Applied Catalysis A: General*, Vol. 160, 1997, pp. 139-151.

[28] C. Serre, F. Garin, G. Belot and G. Maire, "Reactivity of Pt/Al_2O_3 and Pt-$CeO_2Al_2O_3$ Catalysts for the Oxidation of Carbon Monoxide by Oxygen : II. Influence of the Pretreatment Step on the Oxidation Mechanism," *Journal of Catalysis*, Vol. 141, 1993, pp. 9-20.

[29] D. Liu, G.-H. Que, Z.-X. Wang and Z.-F. Yan, "In Situ FT-IR Study of CO and H_2 Adsorption on a Pt/Al_2O_3 Catalyst," *Catalysis Today*, Vol. 68, No. 1-3, 2001, pp. 155-160.

[30] A. Holmgren, G. Andersson and D. Duprez, "Interactions of CO with Pt/ceria Catalysts," *Applied Catalysis B: En-*

vironmental, Vol. 22, No. 3, 1999, pp. 215-230.

[31] A. Holmgren, D. Duprez, B. Andersson, "A Model of Oxygen Transport in Pt/Ceria Catalysts from Isotope Exchange," *Journal of Catalysis*, Vol. 182, No. 2, 1999, pp. 441-448.

[32] J. L. Ayastuy, M. P. González-Marcos, A. Gil-Rodríguez, J. R. González-Velasco, M. A. Gutiérrez-Ortiz, "Selective CO Oxidation over $Ce_XZr_{1-X}O_2$-supported Pt Catalysts, *Catalysis Today*, Vol. 116, 2006, pp. 391-399.

[33] E. Simsek, S. Ozkara, A. E. Aksoylu and Z. I. Onsan, "Preferential CO Oxidation over Activated Carbon Supported Catalysts in H_2-rich Gas Streams Containing CO_2 and H_2O, *Applied Catalysis A: General*, Vol. 316, No. 2, 2007, pp. 169-174

17

Correlation and Prediction of the Solubility of Solid Solutes in Chemically Diverse Supercritical Fluids Based on the Expanded Liquid Theory

Loubna Nasri[1*], Salima Bensaad[2], Zouhir Bensetiti[3]
[1]Department of Pharmaceutical Engineering, University Constantine 3, Constantine, Algeria
[2]Department of Chemistry, University Constantine 1, Constantine, Algeria
[3]Unisignal Inc., Brossard, Canada

ABSTRACT

For the proper design of any extraction procedure based on supercritical solvents, it is essential to have a sound knowledge of the solubility data of different compounds and the accurate way to represent it. The solute's solubility in a supercritical solvent is dependent on the solute, the solvent, and the operating conditions (temperature and pressure). Developing a comprehensive experimental data set is an onerous task and time consuming and, thus, the incentive to develop predictive tools is substantial. In this paper, a technique is presented and tested to correlate and predict solute's solubility in different supercritical fluids with a methodology based on the expanded liquid theory, in which the solid-fluid equilibrium is modeled using the local composition model of UNIQUAC in which the interaction parameters are related to the solvent reduced density with empiric equations. The most advantages of this model include: it does not require the knowledge of critical properties and sublimation pressure of solid solutes and does take into account the binary interaction between solid solute and solvent. The evaluation of the proposed model capabilities is done by testing it on a large data base consisting of experimental solubility data taken from literature of 33 binary systems solid-SC fluid. The results obtained for both correlation and prediction show good agreement with the experimental data used. For the comparison we have considered some literature models that account for effect of the system conditions (temperature and pressure) in addition to the sublimation pressure of the solute through their introduction of the enhancement factor and a model based on a modified Peng-Robinson equation of state.

Keywords: Solubility Modeling; Supercritical Fluids; Correlation; Prediction; Expanded Liquid; UNIQUAC

1. Introduction

During the past few years, widespread attention has been focused on supercritical fluids due to their potential application in extraction processes in foods processing, pharmaceuticals, flavors, chemicals and petroleum industries. The main advantages of supercritical fluid extraction over conventional extraction methods include increased speed, easy solvent separation and better recovery, and reduction in both solvent usage and solvent waste generation. The solubility of a solute in a supercritical fluid is perhaps the most important thermophysical property that must be determined and modeled for an efficient design of any extraction procedure based on supercritical solvents. The determination of solubilities of a wide variety of solids and liquids of low volatility in supercritical fluids has received considerable attention in recent years. However, despite the vital importance of the solubility data of different compounds from chemical, biochemical, pharmaceutical and industrial points of view, there is still a lack of fundamental solubility and mass-transfer data available in the literature to facilitate the development of commercial-scale processes. Since the experimental determination of the solubilities of various solutes in supercritical fluids at each operating condition is tedious, time-consuming and not reported in literatures, there is a considerable interest in mathematical models that can accurately predict the solubilities of solid solutes in supercritical fluids [1]. Therefore it is essential to have a model that not only can accurately correlate but also predict phase equilibrium properties.

*Corresponding author.

Some of the models that have been used for correlating solubility data can be classified in two classes, equations of state based models (EOS) [2] and empirical models [3,4]. EOS based models require the prior knowledge of a certain number of parameters such as the critical properties (temperature and pressure), acentric factor and the sublimation pressure of the solid solute. These parameters are not available and specifically for many high molecular weight compounds and are calculated using group contribution methods, which could lead to solubility error prediction. Due to the lack of information on these properties, empirical models are often used for the correlation of experimental solubility data. These models are known as density-based models and consist of equations that contain constants that are empirically adjusted for each compound. Although simple, these models rely much on the knowledge of the thermodynamic behaviour of the supercritical solvent rather than of the solute, and are mostly capable of correlating rather than predicting the solubility. They are used for quantitative determination of the solute solubility in supercritical phase at equilibrium, and do not provide qualitative information about the solute-solvent interaction.

In a previous presentation [5], we have adopted a methodology for the correlation and the prediction of the solubility of 10 aromatic pollutants in the supercritical carbon dioxide. In this work we present an extension of the methodology to the solubility of other solids with different functional groups in different supercritical fluids namely carbon dioxide, ethylene and ethane. The methodology is based on the expanded liquid model theory [6,7] which does not require the knowledge of the solute critical properties and sublimation pressure. In this case the supercritical phase is considered as an expanded liquid and is modeled using excess Gibbs energy models such as Margules, Van Laar, and local composition based models *i.e.* Wilson, NRTL and UNIQUAC. In this study we focus on the use of the UNIQUAC model that has been widely used in modeling vapour-liquid and liquid-liquid equilibria data. This model does not only take the size and nature of the molecules into consideration, but also accounts for the strength of solute-solvent intermolecular forces. And because the primary concentration variable is the surface fraction rather than mole fraction, the UNIQUAC model is applicable to solutions containing small or large molecules, including polymers. To assess the correlative and predictive capabilities of this model, a database is built by consisting of the experimental solubility data of 33 binary systems solid-super-critical fluid where solid solutes have different functional groups.

2. Model Development

In the supercritical state, a fluid has a high density when compared with a gas. In fact, the density of a supercritical fluid is closer to that of a liquid than that of a gas. Consequently, in theoretical treatments the supercritical fluid phase can be treated approximately as an expanded liquid. This allows the phase equilibria between the solute and the supercritical fluid to be represented thermodynamically by solid-liquid equilibrium relations and conventional activity coefficients. To estimate the solid solubility in the supercritical phase, the knowledge of the activity coefficients are required. These coefficients are determined from the knowledge of the component fugacities, thus when the equilibrium of the pure solid and the supercritical phase is reached, we have:

$$f_2^s = f_2^L \tag{1}$$

f_2^s is the fugacity of the solute in the solid phase considered as pure solid and equal to f_2^{os} because the solubility of supercritical fluid (SCF) in the solid phase is considered to be negligibly small. The f_2^L is the fugacity of the solid solute in the supercritical phase and is equal to:

$$f_2^L = \gamma_2 y_2 f_2^{oL} \tag{2}$$

Equation (1) could be written as follows:

$$f_2^{os} = \gamma_2 y_2 f_2^{oL} \tag{3}$$

where γ_2, y_2 and f_2^{oL} are the activity coefficient, the solid solubility represented in mole fraction and the fugacity of the pure solid solute in the expanded liquid phase respectively. According to Prausnitz *et al.* [8], we have:

$$\ln\left(\frac{f_2^{os}}{f_2^{ol}}\right) = \frac{-\Delta H_2^f}{R}\left(\frac{1}{T}-\frac{1}{T_m}\right) - \frac{\Delta C_P}{RT}\left(\frac{T-T_m}{T}\right) + \frac{\Delta C_P}{R}\ln\left(\frac{T}{T_m}\right) \tag{4}$$

Prausnitz *et al.* [8] stated that, to a fair approximation, the heat capacity terms can be neglected. Equations (3) and (4) then combined to yield an expression for the solute solubility:

$$y_2 = \frac{1}{\gamma_2}\exp\left[\frac{-\Delta H_2^f}{R}\left(\frac{1}{T}-\frac{1}{T_m}\right)\right] \tag{5}$$

ΔH_2^f is the enthalpy of fusion, T_m is the melting point temperature of the solid solute. Since the solid solubility in the supercritical phase is very small, we can assume to be at infinite dilution condition. Consequently, the activity coefficient of the solid solute is the one at infinite dilution and the density of the solution is that of the pure solvent. Thus Equation (5) becomes:

$$y_2 = \frac{1}{\gamma_2^\infty}\exp\left[\frac{-\Delta H_2^f}{R}\left(\frac{1}{T}-\frac{1}{T_m}\right)\right] \tag{6}$$

The activity coefficient of the solid solute at infinite dilution γ_2^{∞} was calculated using the UNIQUAC model which consists of two parts, a combinatorial part $\gamma_2^{C,\infty}$ that attempts to describe the dominant entropic contribution, and a residual part $\gamma_2^{R,\infty}$ that is due primarily to intermolecular forces that are responsible for the enthalpy of mixing. The combinatorial part is determined only by the composition and by the sizes and shapes of the molecules; it requires only pure-component data. The residual part, however, depends also on intermolecular forces; the two adjustable binary parameters a_{12} and a_{21}, therefore, appear only in the residual part [8]:

$$\ln \gamma_2^{\infty} = \ln \gamma_2^{C,\infty} + \ln \gamma_2^{R,\infty} \tag{7}$$

$$\ln \gamma_2^{C,\infty} = 1 - \frac{r_2}{r_1} + \ln \frac{r_2}{r_1} - q_2 \frac{z}{2}\left(1 - \frac{r_2 q_1}{r_1 q_2} + \ln \frac{r_2 q_1}{r_1 q_2}\right) \tag{8}$$

Here q and r are the surface area and volume parameters; z is the coordination number that is usually taken equal to 10. The residual part at infinite dilution is given by the following equation [8]:

$$\ln \gamma_2^{R,\infty} = q_2 \left(1 - \ln \tau_{12} - \tau_{21}\right) \tag{9}$$

where

$$\tau_{12} = \exp\left(-\Delta u_{12}/RT\right) = \exp\left(-a_{12}/T\right) \tag{10a}$$

and

$$\tau_{21} = \exp\left(-\Delta u_{21}/RT\right) = \exp\left(-a_{21}/T\right) \tag{10b}$$

Δu_{12} and Δu_{21} are characteristic energies and are related to the interaction parameters a_{12} and a_{21} through Equation (10). Finally combining Equations (9) and (10) leads to:

$$\ln \gamma_2^{R,\infty} = q_2 \frac{a_{12}}{T} + q_2 \left(1 - e^{\frac{-a_{21}}{T}}\right) \tag{11}$$

Equation (11) could be written in reduced form by introducing the reduced temperature, thus we obtain:

$$\ln \gamma_2^{R,\infty} = q_2 \frac{a'_{12}}{T_r} + q_2 \left(1 - e^{\frac{-a'_{21}}{T_r}}\right) \tag{12}$$

With $a'_{12} = \frac{a_{12}}{T_c}$ and $a'_{21} = \frac{a_{21}}{T_c}$, T_c is the solvent critical temperature.

The binary interaction parameters a'_{12} and a'_{21} are related to the energy of interaction between the solid solute and the solvent in the supercritical phase, and cannot be kept constant and specifically at high pressure conditions. Therefore to take into account the pressure and temperature effects, these parameters are assumed to be density dependent and were fitted to the following equations:

$$a'_{12} = \alpha_{12} \cdot \rho_r^{\beta_{12}} \tag{13a}$$

$$a'_{21} = \alpha_{21} \cdot \rho_r^{\beta_{21}} \tag{13b}$$

ρ_r is the reduced density of the solvent equal to ρ/ρ_{c1} where ρ_{c1} is its critical density, α_{12}, β_{12}, α_{21} and β_{21} are the regressed parameters of the model.

3. Database Compilation

By considering 33 system solid-SC fluid, an exhaustive solutes solubility database consisting of more than 2218 solubility data in supercritical fluids is built-up for the elaboration and validation of the proposed model. It is acknowledged that the systems studied do not include all the data available but should be sufficient to provide a thorough testing of the potential of Equations (6) to 13(b). The density of supercritical fluid solvents used in this work, is estimated using the Span and Wagner equation of state [10] when they are not reported in the solubility data sources, the physical properties of the solvents are given in **Table 1**. **Table 2** show the solutes and their thermodynamic properties obtained from literature used in this study. Description of the database is given in **Table 3** which lists the binary systems used together with the number of solubility data points, and the lower and upper limits of the operating conditions, solubility and the references. Detailed information about all the complete references from which experimental solubility data are taken are provided in **Table 4**.

4. Solutes Solubility Correlation

The surface area and volume parameters are calculated as the sum of the group volume and area parameters (R and Q) given by the UNIFAC group specifications [8]. These parameters and properties listed in **Table 2**, together with those of different fluids listed in **Table 1** are used to calculate the combinatorial part of the activity coefficient from Equation (8). In other hand, Equation (12) is used to calculate the residual part of the solid solute activity coefficient. Thermodynamic properties of the solid solute listed in **Table 2** are used together with Equations (7), (8), and (12) to estimate the solubility y_2 using Equation (6). The interaction parameters a'_{12} and a'_{21} are then regressed according to Equations 13(a) and 13(b). The best regression is based on minimizing the error between the regressed and experimental solubility data. The objective function used minimizes the sum of average absolute relative deviation (AARD) according to Equation (14):

$$\text{FOBJ} = \text{AARD}(\%) = \frac{100}{N}\sum_{1}^{N} \frac{\left|y_{2(\text{exp})} - y_{2(\text{regr})}\right|}{y_{2(\text{exp})}} \tag{14}$$

where N is the number of experimental solubility data of each solute.

Table 1. Solvents physical properties.

Solvent	T_{c1} (K)	P_{c1} (bar)	ω_1	ρ_{c1} (mol/cm^3) ×100	r_1	q_1
CO_2	304.2[a]	73.83[a]	0.239[a]	1.063[a]	1.296[b]	1.261[b]
ethane	305.33[a]	48.8[a]	0.09[a]	0.687[a]	1.802	1.696
ethylene	382.35[a]	50.4[a]	0.089[a]	0.764[a]	1.488	1.574

[a]From reference [8], [b]From reference [9].

Table 2. Solute species properties.

Solute	Formula	M (g/mol)	T_m (K)	ΔH_2^f (J/mol)	V_2^S (cm^3/mol)	r_2	q_2
anthracene	$C_{14}H_{10}$	178.23	492.5 [10]	28829.0 [11]	138.9 [11]	6.77	4.48
biphenyl	$C_{12}H_{10}$	154.21	342.1 [10]	18601.0 [11]	132.0 [11]	6.04	4.24
6-caprolactam	$C_6H_{11}NO$	113.16	342.3 [12]	16096.0 [12]	162.0 [13]	4.67 [14]	3.736 [14]
1,10-decanediol	$C_{10}H_{22}O_2$	174.3	347.1 [15]	43505.3 [16]	158.4 [17]	8.74	7.8
2,3-dimethylnaphtalene	$C_{12}H_{12}$	156.23	377.8 [11]	25101.0 [11]	156.0 [11]	6.45	4.57
2,6-dimethylnaphtalene	$C_{12}H_{12}$	156.23	383.2 [11]	25055.0 [11]	199.0 [11]	6.45	4.57
2,7-dimethylnaphtalene	$C_{12}H_{12}$	156.23	369.0 [18]	23349.0 [18]	154.0 [18]	6.45	4.57
fluoranthene	$C_{16}H_{10}$	202.2	383.3 [15]	18728.0 [12]	163.0 [11]	7.5	4.72
fluorene	$C_{13}H_{10}$	116.23	387.9 [19]	19580.0 [19]	139.3 [19]	6.39	4.22
hexamethylbenzene	$C_{12}H_{18}$	228.38	438.7 [19]	20640.0 [19]	152.7 [13]	7.59	5.8
lauric acid	$C_{12}H_{24}O_2$	200.3	316.9 [12]	36650.0 [12]	229.0 [20]	8.94	7.47
myristic acid	$C_{14}H_{28}O_2$	228.38	327.1 [15]	45362.8 [15]	257.5 [13]	10.29	8.55
naphtalene	$C_{10}H_8$	128.17	353.2 [11]	19123.0 [11]	125.0 [11]	4.98	3.44
1-naphtol	$C_{10}H_8O$	144.2	368.1 [15]	23477.5 [15]	117.8 [21]	5.34	3.72
2-naphtol	$C_{10}H_8O$	144.2	396.0 [11]	17511.0 [11]	118.0 [11]	5.34	3.72
phenanthrene	$C_{14}H_{10}$	178.23	372.2 [11]	16465.0 [11]	182.0 [11]	6.77	4.48
phenol	C_6H_6O	94.10	313.9 [11]	11289.0 [11]	89.0 [13]	3.55	2.68
pyrene	$C_{16}H_{10}$	202.26	424.2 [19]	17111.0 [19]	158.5 [19]	7.5	4.72
resorcinol	$C_6H_6O_2$	110.11	383.1 [15]	21291.6 [15]	86.7 [21]	3.91	2.96
stearic acid	$C_{18}H_{36}O_2$	284.5	341.9 [15]	56569.2 [15]	302.4 [13]	12.99	10.71
triphenylene	$C_{18}H_{12}$	228.29	470.9 [15]	24190.0 [18]	175.0 [20]	8.57	5.52
triphenylmethane	$C_{19}H_{16}$	244.3	365.6 [19]	20920.0 [19]	240.9 [19]	9.512	6.588

5. Solutes Solubility Prediction

In order to evaluate the predictive ability of the proposed model, the solubility data for each considered component are split with specific tool in two sets. The first data set is the training set on which the minimization routine is performed and the interaction parameters are regressed. This data set contains 70% of the data randomly picked up from the experimental solubility data for each component. The second data set, namely the test set, contains the remaining 30% data and is intended for testing the generalized capabilities of the UNIQUAC model. The interaction parameters have been regressed using the training solubility data set, and then used directly to predict the solid solubility y_2 using the test data set. The predictive ability of the model is then assessed by comparing the obtained AARD values for each data set and so for each component.

Table 3. Database description.

Binary systems	*N*	*T*-range (K)	*P*-range (bar)	ρ_r range	Solubility order	References
naphtalene-CO_2	242	308 - 338.05	75.4 - 400	0.51 - 1.98	7E-4 - 6E-2	L1 - L17
naphtalene-ethylene	166	285 - 323	54.5 - 303.9	0.52 - 1.99	3E-4 - 8E-2	L1, L6, L61, L62
naphtalene-ethane	48	308 - 328.2	48.1 - 250	0.56 - 1.99	8E-4 - 5E-2	L24, L59
anthracene-CO_2	206	298 - 343	92.6 - 470	0.66 - 2.0	3E-6 - 3E-4	L20, L23, L28, L29, L30, L31, L32, L33, L34
anthracene-ethane	13	308 - 343	104.3 - 345.6	1.49 - 1.99	7E-5 - 6E-4	L24
anthracene-ethylene	27	323 - 358	104.4 - 414.7	0.75-1.97	1E-5 - 9E-4	L59
biphenyl-CO_2	57	308 - 330.6	80 - 379.46	0.55 - 1.99	4E-4 - 6E-2	L9, L5, L12
biphenyl-ethane	8	308 - 318.3	70.5 - 180	1.32 - 1.94	1E-2 -4E-2	L24
6-caplolactam-CO_2	32	307 - 324	101 - 208	1.5 - 1.81	2E-2 - 7E-2	L46
2,3-dimethylnaphtalene-CO_2	25	308 - 328	99 - 280	0.68 - 1.96	3E-4 -9E-3	L21, L25
2,3-dimethylnaphtalene-ethylene	18	308 - 328	77 - 280	0.64 - 1.93	3E-4 - 5E-2	L21
2,6-dimethylnaphtalene-CO_2	23	308 - 328.2	79 - 280	0.63 - 1.96	3E-4 - 9E-3	L8, L21
2,6-dimethylnaphtalene-ethylene	18	308 - 328	78 - 280	0.62 - 1.94	2E-4 - 2E-3	L21
2,7-dimethylnaphtalene-CO_2	10	308.2 - 328.2	88 - 249	0.69 - 1.91	7E-4 - 1E-2	L8
fluoranthene-CO_2	68	308 - 348	86 - 354.6	0.62 - 1.99	9E-6 - 1E-3	L35, L26
fluorene-CO_2	146	308 - 343	80.9 - 483.4	0.6 - 1.98	6E-5 - 9E-3	L18, L19, L23
phenanthrene-CO_2	281	308 - 343	82.2 - 414.5	0.56 - 2.0	3E-5 - 4E-3	L12, L17, L18, L19, L20, L21, L22, L24, L25, L26, L63
phenanthrene-ethylene	42	298 - 343	69.9 - 280	0.45 - 1.49	3E-5 - 1E-2	L59, L21, L19
phenanthrene-ethane	17	313 - 333	69.9 - 264	0.76 - 1.95	2E-4 - 6E-3	L19
pyrene-CO_2	235	308 - 343	83.4 - 483.4	0.78 - 1.99	1E-5 - 9E-4	L18, L19, L20, L27
pyrene-ethane	15	333 - 333	100.2 - 314.5	1.37 - 1.98	2E-4 - 1E-3	L20
triphenylene-CO_2	53	308 - 328	85 - 355.6	0.83 - 1.99	1E-6 - 5E-5	L35, L27
triphenylmethane-CO_2	111	308 - 328	73.7 - 414.5	0.507 - 1.99	3E-5 - 4E-3	L19, L38
lauric acid-CO_2	24	308 - 318	77 - 260	0.53 - 1.94	2E-5 - 5E-2	L43, L44, L45
palmitic acid-CO_2	27	308 - 328	91 - 248	1.1 - 1.9	3E-5 - 2E-3	L4, L52, L53
myristic acid-CO_2	30	308 - 318	128.5 - 226.5	1.47 - 1.89	1E-3 - 1E-2	L43, L44
stearic acid-CO_2	28	308 - 338	90 - 237	0.94 - 1.91	2E-5 - 1E-3	L56, L60, L57, L58
Resorcinol	26	308 - 338	121.6 - 405.3	0.84 - 1.97	1E-4 -9E-4	L55
phenol-CO_2	73	309 - 333.2	79.3 - 249.43	0.51 - 1.91	1E-3 - 6E-3	L40, L41,
1,10-decanediol	15	318 - 328	133.7 - 307.3	1.26 - 1.91	3E-4 - 8E-4	L47
1-naphtol-CO_2	64	308 - 328	88.7 - 296.5	0.54 - 1.98	6E-5 - 2E-3	L33, L51
2-naphtol-CO_2	48	308.05 - 343.2	100 - 363.6	0.65 - 1.99	8E-5 - 2E-3	L24, L37, L29
2-naphtol-ethane	22	308 - 343.2	61 - 364	0.56 - 1.98	2E-5 - 1E-3	L24

Table 4. References for solubility data.

L1: Russian journal of physical chemistry 1964; 38: 9	L30: Ind. Eng. Chem. Res. 1987, 26, 7, 1476-1482
L2: J. Supercritical Fluids. 1996; 9: 3	L31: J. Chem. Eng. Data 1996, 41, 97-100
L3: J. Chem. Eng. Data 1998; 43: 400-402	L32: J. Chem. Eng. Data 1997, 42, 636-640
L4: J. Chem. Eng. Data 1999; 48: 951-957	L33: J. Chem. Eng. Data 1995, 40, 953-858
L5: J. Chem. Eng. Data 1980; 25, 4, 326-329	L34: Utilisation des fluides supercritiques pour l'extraction des fullerènes, thesis presented by Valérie Quillet,1996: Bordeaux 1, France
L6: J. Supercritical. Fluids 1988; 1: 1	L35: J. Chem. Eng. Data 2000, 45, 53
L7: J. Chem. Eng. Data 1989, 36: 4, 430-432	L37: Ind. Eng. Chem. Res 26, 1, 1987, 56-65
L8: J. Chem. Eng. Data 1993, 38, 3,	L38: J. Supercritical Fluids 2004, 32, 115-121
L9: Fluid Phase Equilibria 1992; 81:321-341	L39: Fluid Phase Equilibria 2004, 226, 9-13
L10: Fluid Phase Equilibria 1995, 107, 189-200	L40: J. Chem. Eng. Data 1980, 25, 257-259
L11: Ind. Eng. Chem. Res. 2000, 39, 4609-4614	L41: Hwahak Konghak Journal 1993, 31, 6, 637
L12: J. Supercritical. Fluids 1995, 8, 1, 15-19	L43: J. Chem. Eng. Data 1988, 33, 3, 327-333
L13: J. chem. Eng. Data 1988, 33, 1, 35-37	L44: J. Chem. Eng. Data 2008, 53, vol. 11
L14: J. Chem. Eng. Data 1985, 30, 1	L45: J. Am. Oil. Chem. Soc. 1992, 69, p. 1069
L15: J. Physical Chemistry 1986, 90, 17	L46: J. Chem. Eng. Data 1996, 41, 1418-1420
L16: J. Chem. Eng. Data 2000, 45: 464-466	L47: J. Chem. Eng. Data 1986, 31, 285-288
L17: J. Chem. Eng. Data 2000, 45: 358-361	L51: Fluid Phase Equilibria 1987, 34, 37-47
L18: J. Chem. Eng. Data 1990, 35: 355-360	L52: J. Chem. Eng. Data 1991, 36, 4, 430-432
L19: Ind. Eng. Chem. Fundamentals 1982, 21, 3	L53: J. Chem. Eng. Data 1988, 33, 230-234
L20: J. Supercritical Fluids 1997, 10, 175-189	L55: Fluid Phase Equilibria 1998, 152, 299-305
L21: J. Chem. Eng. Data 1981, 26, 1, 47-51	L56: J. Chem. Eng. Data 1993, 38, 506-508
L22: Ber. Bunsen. Phys. Chem. 1984, 88, 865-869	L57: J. Chem. Eng. Data 2008, 53, 2913-2917
L23: Phys. Chem. 1984, 88, 865-869	L58: J. Chem. Thermodynamics 2010, 42, 193-197
L24: J. Chem. Eng. Data 1986, 31, 2, 204-212	L59: AIChE journal 1981, 27, No 5, 773-779
L25: J. Chem. Eng. Data 2001, 46, 5, 1156-1159	L60: J. Chem. Eng. Data 1989, 34, 184-187
L26: J. Chem. Eng. Data 1996, 41, 1466-1469	L61: J. Amer. Chemic. Socie 1953, 57, 575-578
L27: Ind. Eng. Chem. Res 1995, 34, 340-346	L62: J. Amer. Chemic. Socie 1948,70, 4085-4089
L28: J. Chem. Eng. Data 1987, 32, 148-150	L63: J. Chem. Eng. Data 1986, 31, 3, 303-308
L29: Fluid Phase Equilibria 2003, 207, 183-192	

6. Results and Discussion

6.1. Solubility Data Correlation

The analysis of the correlation model results is done through statistical calculations. **Table 5** provides the quantitative results of the regression for the UNIQUAC model. The AARD is included for each binary system and is listed together with the adjustable parameters values. Each component parameters are obtained by fitting them on its own solubility data available in the database. For all investigated systems the resultant AARD values are lower than 20% (except for the system phenol-CO_2). The model achieves an overall AARD of 12.16 on a range of 2.8 to 29.7. These results are indicative of a good correlation performance of the proposed model and also show that the calculated solubility data are in good agreement with the experimental ones.

In order to compare more clearly, the experimental data and correlated data in this work are compared in **Figures 1-3** with naphthalene in SC CO_2 at T = 308 K,

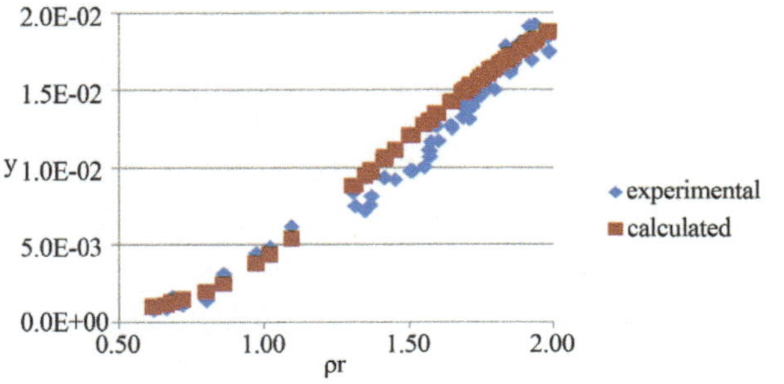

Figure 1. Experimental and correlated solubility versus reduced density of naphthalene-CO_2 system at 308 K.

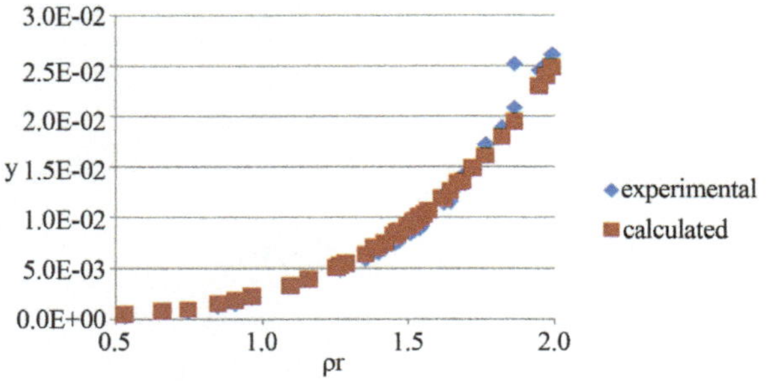

Figure 2. Experimental and correlated solubility versus reduced density of naphthalene-ethylene system at 298 K.

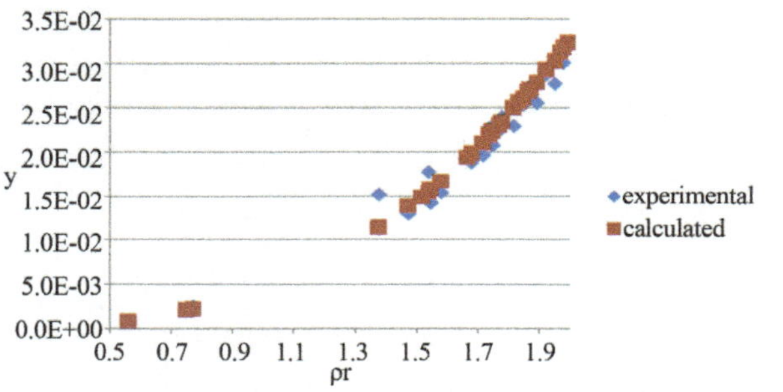

Figure 3. Experimental and correlated solubility versus reduced density of naphthalene-ethane system at 308 K.

naphthalene in SC ethylene at T = 298 K, and naphthalene in SC ethane at T = 308 K respectively. These systems are arbitrarily taken from the database for illustration. From the figures, we see that it is good accordance with experimental and calculated ones, and the precision of the model in this work is very acceptable.

6.2. Comparison with Literature

We have considered for the comparison some literature models that account for effect of the system conditions (temperature and pressure) in addition to the physical properties as sublimation pressure of the solute through their introduction of the enhancement factor and a model based on a modified equation of state.

6.2.1. Wang and Tavlarides Model

Based on a dilute solution theory, Wang and Tavlarides [22] proposed the following model:

$$\frac{1}{T\left[\ln(E)-\ln(Z)\right]} = R\left(\frac{c}{b}+\frac{V_1}{b}\right),\ \left(V_1=\frac{1}{\rho_1}\right) \qquad (15)$$

where Z is the compressibility factor, E is the *enhancement factor* that is defined as the enhancement of actual mole fraction solubility of the solid solute y_2 over the solubility in an ideal gas P_2^S/P, *i.e.*, $y_2 = \frac{EP_2^S}{P}$. P_2^S and P are the sublimation vapor pressure and total pressure respectively. By introducing reduced variables, the dimensionless form of Equation (15) is given by:

$$\frac{1}{T_r\left[\ln(E)-\ln(Z)\right]} = \alpha\left(\beta+\frac{1}{\rho_r}\right) \qquad (16)$$

where $\alpha = \frac{T_{c_1}R}{\rho_{c_1}b}$; $\beta = c\rho_{c_1}$.

6.2.2. Méndez-Santiago and Teja Models

Based on the theory of dilute solutions, Méndez-Santiago and Teja [23] began with the Henry's constant and presented a simple linear relationship for the solubility of solids in supercritical fluids as given below:

$$T\ln E = A_1 + B_1\rho \qquad (17)$$

To apply the model to compounds whose sublimation vapor pressure is unknown, a Clausius-Clapeyron-type expression for the sublimation pressure $\ln P_2^s = A+\frac{B}{T}$ and $E=\frac{y_2P_2^s}{P}$ are substituted into Equation (17) resulting in the following expression:

$$T\ln(y_2P) = A' + B'\rho + C'T \qquad (18)$$

The dimensionless form of Equations (17) and (18) are given bellow [4]:

$$T_r\ln E = T_r\ln\frac{y_2P}{P_2^s} = \alpha+\beta\rho_r \qquad (19)$$

where:

$$\alpha = \frac{A_1}{T_{c_1}};\ \beta = \frac{B_1\rho_{c_1}}{T_{c_1}}$$

$$T_r\ln(y_2P_r) = \alpha' + \beta'\rho_r + \gamma'T_r \qquad (20)$$

where:

$$P_r = \frac{P}{P_{c_1}};\ \alpha' = \frac{A'}{T_{c_1}};\ \beta' = \frac{B'\rho_{c_1}}{T_{c_1}};\ \gamma' = C' - T_r\ln P_{c_1}$$

The linear behavior of Equation (17) is well observed in the range (0.5 - 2) of reduced density of the super-

Table 5. Regression parameters and average deviation of the model.

Binary systems	N	α_{12}	β_{12}	α_{21}	β_{21}	AARD (%)
anthracene-CO_2	206	2.380	−0.210	−0.360	0.598	14.58
anthracene-ethylene	27	2.284	−0.356	−0.277	0.841	8.06
anthracene-ethane	13	1.840	−0.711	−0.053	0.844	6.87
biphenyl-CO_2	57	1.620	−0.450	−0.107	0.930	16.39
biphenyl-ethane	8	1.303	−0.403	−0.066	1.942	8.99
6-caprolactam	32	1.630	0.851	−0.330	1.940	5.66
1,10-decanediol-CO_2	15	1.803	−0.357	−0.090	−0.370	3.94
2,3-dimethylnaphtalene-CO_2	25	0.680	−0.370	2.424	−2.160	19.71
2,3-dimethylnaphtalene-ethylene	18	0.525	−0.827	3.659	−2.390	13.04
2,6-dimethylnaphtalene-CO_2	23	0.900	−0.560	0.802	−0.702	9.68
2,6-dimethylnaphtalene-ethylene	18	0.513	−0.890	3.240	−2.440	15.97
2,7-dimethylnaphtalene-CO_2	10	1.192	−0.332	0.271	−1.579	2.78
fluoranthene-CO_2	68	1.310	−0.150	1.209	−2.010	12.55
fluorene-CO_2	146	1.040	−0.370	1.174	−1.300	6.87
lauric acid-CO_2	24	0.680	−0.600	2.480	−2.090	11.48
myristic acid-CO_2	30	1.260	−0.470	0.377	−2.150	15.42
naphthalene-CO_2	242	0.920	−0.090	0.757	−2.450	16.57
naphthalene-ethylene	166	0.741	−0.490	1.327	−1.779	19.26
naphthalene-ethane	48	0.718	−0.705	0.675	−1.194	12.01
1-naphtol-CO_2	64	1.350	−0.280	2.695	−1.960	16.05
2-naphtol-CO_2	48	1.530	−0.114	1.735	−2.260	12.31
2-naphtol-ethane	22	1.660	−0.115	1.550	−1.916	17.46
palmitic acid-CO_2	27	1.530	−0.380	6.295	−42.14	10.42
phenanthrene-CO_2	281	1.970	−0.350	0.053	1.950	13.15
phenanthrene-ethylene	42	2.060	−0.370	−0.120	1.810	16.31
phenanthrene-ethane	17	2.040	−0.180	−0.230	1.470	14.67
phenol-CO_2	73	1.460	−0.200	0.990	−1.530	29.70
pyrene-CO_2	235	2.190	−0.370	−0.061	1.890	4.78
pyrene-ethane	15	1.870	−0.460	−0.060	1.750	1.16
resorcinol-CO_2	26	3.120	−0.290	−0.020	3.250	6.71
stearic acid-CO_2	28	1.120	−0.320	0.461	−0.500	8.49
triphenylene-CO_2	53	3.370	−0.300	−0.800	−1E-4	4.90
triphenylmethane-CO_2	111	1.800	−0.170	−0.270	0.690	15.08

critical solvent [23], for this reason we have omitted from the database all data points over this range of density (see **Table 3**).

6.2.3. Schmitt and Reid Model

By the assumption that: the system pressure is much greater than the sublimation pressure of the solute, that the solute is incompressible, and that no solvent is dissolving in the solid phase, the solubilities of solids in supercritical solvents are usually correlated with the equation bellow [24]:

$$y_2 = \left(\frac{P_2^S}{P}\right)\frac{\exp\left(\frac{v_2^S}{RT}\left(P-P_2^S\right)\right)}{\Phi_2^{sf}} \quad (21)$$

v_2^S is the molar volume of the solute (**Table 2**), the fugacity coefficient of the solute in the supercritical phase $\left(\Phi_2^{sf}\right)$ is determined from an equation of state applicable to the solute-solvent mixture. The Peng-Robinson equation of state, *i.e.* PR-EOS is a commonly used approach for correlating solubility in supercritical fluids.

Schmitt and Reid [24] used this equation of state for modeling solid solubilities in supercritical fluids but not in the traditional manner due to the lack of the critical properties values and the low accuracy of their estimation. In fact they eliminated the binary interaction parameter and they assumed the solid solute parameters "a_2" and "b_2" independent of temperature and they eliminated the terms with y_2 in the combining and mixing rules since the values of y_2 were sufficiently small. Thus they propose the simplified equation bellow for the fugacity coefficient:

$$\begin{aligned}\ln \Phi_2^{sf} &= (b_2/b_1)(Z-1) - \ln\left[P(V-b_1)/RT\right] \\ &-\left(a_1/8^{0.5}RTb_1\right)\left[2(a_2/a_1)^{0.5} - (b_2/b_1)\right] \\ &\times \ln\left[(V+2.414b_1)/\left(V-\sqrt{2}b_1\right)\right]\end{aligned} \quad (22)$$

For the supercritical solvent, the parameters a_1 and b_1 are given by:

$$a_1 = 0.4572R^2\, T_{c_1}^2/P_{c_1}$$
$$\times\left[1+\left(0.3746+1.5423\omega_1-0.2699\omega_1^2\right)\left(1-\left(T/T_{c_1}\right)^{0.5}\right)\right]^2$$

$$b_1 = 0.0778RT_{c_1}/P_{c_1}$$

Equation (21) becomes,

$$\begin{aligned}\ln y_2 &= \ln\left[P_2^s/P\right] + \frac{\left(P-P_2^s\right)v_2^s}{RT} \\ &-\Big[(b_2/b_1)(Z-1) - \ln\left[P(V-b_1)/RT\right] \\ &-\left(a_1/8^{0.5}RTb_1\right)\left[2(a_2/a_1)^{0.5}-(b_2/b_1)\right] \\ &\times\ln\left[(V+2.414b_1)/\left(V-\sqrt{2}b_1\right)\right]\Big]\end{aligned} \quad (23)$$

The experimental solubility data are regressed by using Equation (23) to determine the parameters of the solute a_2 and b_2 for each binary system considered. These parameters are expressed in Pa $(m^3/mol)^2$ and m^3/mol respectively. The Equation (23) is dimensionally consistent and already dimensionless, but the parameters "a_2" and "b_2" are not. For comparison on the same basis the dimensionless parameters are as follow:

$$a_2' = \left(a_2\rho_{c_1}^2/P_{c_1}\right) \text{ and } b_2' = \left(b_1\rho_{c_1}\right)$$

6.3. Analysis and Discussion

As mentioned in the previous section, the three first correlations used for comparison require the knowledge of the sublimation pressure. **Table 6** gives the coefficients for the estimation of the sublimation pressure of the different solutes taken from literature and the temperature range of applicability.

To compare all the correlations on the same basis, the average absolute relative deviation (AARD) is determined for each system's solute using each model. **Table 7** display the parameters of the comparison models and shows in the last column the AARD produced by each equation for the different fluid-solute systems considered. The lowest error fit is marked with (*), the proposed model gives the lowest AARD for 11 binary systems and provides a better fit than the modified EOS for 24 binary systems. From this comparison we want to emphasize that the proposed UNIQUAC model provides better quantitative capabilities with one common EOS which does not require the knowledge of solute critical properties.

Judging from the mean AARD values given in **Table 8**, we can see clearly that the UNIQUAC model gives good results compared to the other models. The best fit is achieved by the Mendez-Teja model with three parameters given by Equation (20). Whereas, the largest values of AARD and the highest mean AARD are achieved by the model of Wang-Tavlarides (Equation (16)). However, as acknowledged by the authors [22], this model somewhat oversimplified a real system since the interaction between solute and solvent molecules is assumed to follow an interaction behavior similar to the potential well model. In this theory, the system consists of a free volume and a constant solvent cluster volume (*i.e.*, they assume that this volume does not vary with temperature and pressures); the solute thus is either a quasi-gas type (moving in the cluster) or an ideal gas type. As a result, the model developed produces a large error for the supercritical fluid solubility.

6.4. Solubility Prediction

6.4.1. Binary Systems

In order to evaluate the predictive capabilities of the UNIQUAC model and to overcome the over-fitting problem which may alter the model generalization capabilities, solubility data for each considered component

Table 6. Coefficients for sublimation pressure estimation.

$$P^S(Pa) = 10^{\left(A - \frac{B}{T}\right)}$$

Solute	A	B	T-range (K)	Reference
anthracene	14.755	5313.7	308 - 343	[24]
biphenyl	14.804	4367.4	308 - 343	[24]
6-caprolactam	15.480	4811.1	307 - 324	[25][a]
1,10-decanediol	20.901	7217.0	308 - 328	[26][a]
2,3-dimethylnaphtalene	14.065	4302.5	308 - 328	[27]
2,6-dimethylnaphtalene	14.417	4415.9	308 - 328	[27]
2,7-dimethylnaphtalene	14.464	4386.7	308 - 328	[28]
fluoranthene	14.795	5357.0	298 - 358	[29]
fluorene	14.205	4561.8	308 - 343	[27]
lauric acid	22.022	7322.0	295 - 314	[15]
myristic acid	20.861	7291.0	311 - 325	[15]
naphthalene	13.865	3823.1	250 - 340	[15][a]
1-naphtol	10.683	3148.9	308 - 328	[24]
2-naphtol	14.815	4923.9	308 - 343	[24]
palmitic acid	22.341	8069.0	308 - 328	[20]
phenanthrene	14.343	4776.7	300 - 360	[15][a]
phenol	13.689	3586.4	309 - 333	[20]
pyrene	13.395	4904.0	308 - 398	[30]
stearic acid	21.021	7957.0	308 - 338	[31]
triphenylene	14.462	5804.1	300 - 340	[20]
triphenylmethane	14.781	5228.0	308 - 328	[19]

[a]Data interpolated in this work.

were split into two sets. The first set of solubility data obtained from randomly sampling 70% of the experimental data, served for the optimization of the model adjustable parameters and for training the model. The second set of solubility data was then used to test its predictive capabilities. In this step we have considered only solutes for which solubility data points are more than 20. **Table 9** summarizes the key information on the two data sets, namely, the training and the test data sets together with the correlation and prediction results. This table lists the number of solubility data points used for each component and deviations in term of the AARD values obtained for both the training and the test data sets. To assess the predictive ability of the model, the two AARD values are compared. To be predictive, the model should respect the following rule: the AARD values for both data sets should be of the same order of magnitude for each component. **Table 9** shows that the AARD values obtained for the test data set are in accordance with those of the training data set and are generally of the same order of magnitude. This results implies that the proposed model do not show any over-fitting problem or over prediction of the experimental solubility data. Therefore we can conclude that the predictive ability of the model is well demonstrated.

6.4.2. Ternary Systems

The study of mixed-solute systems is important because most potential applications of supercritical fluid extraction involve the removal of a desired compound from a matrix of components. However, in this section an attempt is made to predict the solubilities of mixed compounds in supercritical carbon dioxide. Experimental data provided by: Kosal and Holder [32] for mixed an

Table 7. Comparison with the literature models considered.

System	N	α	β	α'	β'	γ'	a'_2	b'_2	AARD%
anthracene-CO_2	206								
Equation (19)		4.10	4.73						14.88
Equation (20)				−33.09	4.58	15.57			14.41*
Equation (16)		16.17	0.65						21.44
Equation (23)							8.7E-5	1.24E-4	16.74
UNIQUAC model									14.58
anthracene-ethylene	27								
Equation (19)		6.33	5.22						14.30
Equation (20)				−32.12	5.33	12.80			7.60*
Equation (16)		18.05	0.776						21.80
Equation (23)							1.26E-4	1.56E-4	17.05
UNIQUAC model									8.06
anthracene-ethane	13								
Equation (19)		3.42	5.05						10.38
Equation (20)				−33.06	4.77	15.76			5.66*
Equation (16)		15.00	0.602						22.64
Equation (23)							1.19E-4	1.54E-4	10.96
UNIQUAC model									6.87
biphenyl-CO_2	57								
Equation (19)		3.87	4.04						17.69
Equation (20)				−35.57	4.19	24.15			14.48
Equation (16)		17.58	0.910						13.11*
Equation (23)							8.39E-5	1.4E-4	19.14
UNIQUAC model									16.39
biphenyl-ethane	8								
Equation (19)		3.11	4.42						9.81
Equation (20)				−40.01	4.97	27.71			5.76
Equation (16)		15.06	0.705						3.77*
Equation (23)							1.0E-4	1.41E-4	6.59
UNIQUAC model									8.99
6-caprolactam-CO_2	32								
Equation (19)		−0.08	7.96						6.95
Equation (20)				−29.20	6.89	14.50			7.51
Equation (16)		9.71	0.08						12.71
Equation (23)							6.36E-5	4.33E-5	16.40
UNIQUAC model									5.66*

Continued

1,10-decanediol-CO_2	15								
Equation (19)		5.40	4.55						4.22
Equation (20)				−45.09	4.37	28.71			3.03*
Equation (16)		20.29	0.859						6.60
Equation (23)							1.08E-4	1.47E-4	6.07
UNIQUAC model									3.94
2,3-dimethylnaphtalène-CO_2	25								
Equation (19)		2.68	4.74						20.05
Equation (20)				−27.72	4.56	14.76			19.74
Equation (16)		12.57	0.493						26.07
Equation (23)							7.19E-5	1.14E-4	32.97
UNIQUAC model									19.71*
2,3-dimethylnaphtalene-ethylene	18								
Equation (19)		2.82	5.72						16.78
Equation (20)				−42.37	6.48	24.85			14.62
Equation (16)		12.67	0.373						16.37
Equation (23)							9.26E-5	1.10E-4	17.80
UNIQUAC model									13.04*
2,6-dimethylnaphtalène-CO_2	23								
Equation (19)		4.23	3.77						12.75
Equation (20)				−31.54	3.90	19.42			11.78
Equation (16)		17.16	0.885						11.85
Equation (23)							8.66E-5	1.57E-4	13.56
UNIQUAC model									9.68*
2,6-dimethylnaphtalene-ethylene	18								
Equation (19)		2.73	5.59						11.77
Equation (20)				−38.84	6.13	21.93			10.31*
Equation (16)		14.28	0.521						13.81
Equation (23)							8.64E-5	1.12E-4	10.41
UNIQUAC model									15.97
2,7-dimethylnaphtalène-CO_2	10								
Equation (19)		4.87	3.44						8.52
Equation (20)				−30.31	3.63	19.05			6.73
Equation (16)		18.66	1.004						9.95
Equation (23)							9.79E-5	1.74E-4	12.62
UNIQUAC model									2.78*
fluoranthene-CO_2	68								

Continued

Equation (19)		6.73	4.42						18.38
Equation (20)				−29.06	4.28	13.99			11.21*
Equation (16)		24.59	1.06						35.14
Equation (23)							1.49E-4	1.96E-4	24.30
UNIQUAC model									12.35
fluorene-CO_2	146								
Equation (19)		4.74	3.97						7.59
Equation (20)				−29.77	3.97	16.87			7.59
Equation (16)		19.63	0.99						15.23
Equation (23)							9.59E-5	1.53E4	13.80
UNIQUAC model									6.87*
lauric acid–CO_2	24								
Equation (19)		4.39	6.83						13.56
Equation (20)				−37.51	6.66	21.85			11.06*
Equation (16)		19.63	0.55						14.77
Equation (23)							1.39E-4	1.54E-4	14.66
UNIQUAC model									11.48
myristic acid-CO_2	30								
Equation (19)		7.41	5.72						14.66
Equation (20)				−48.06	5.72	32.48			14.65*
Equation (16)		25.65	0.82						17.20
Equation (23)							1.88E-4	2.13E-4	15.12
UNIQUAC model									15.45
naphtalene-CO_2	242								
Equation (19)		3.39	3.31						19.45
Equation (20)				−34.37	3.68	24.12			11.79
Equation (16)		15.55	0.948						11.13*
Equation (23)							6.77E-5	1.34E-4	14.37
UNIQUAC model									16.57
naphthalene-ethylene	167								
Equation (19)		3.24	4.18						16.02
Equation (20)				−32.94	4.54	20.65			1.28
Equation (16)		14.00	0.655						10.53*
Equation (23)							8.43E-5	1.37E-4	16.46
UNIQUAC model									19.26
naphthalene-ethane	48								
Equation (19)		2.65	3.70						15.19

Continued

Equation (20)				−32.25	3.85	22.23			12.47
Equation (16)		14.22	0.833						10.15*
Equation (23)							8.03E-5	1.37E-4	14.12
UNIQUAC model									12.01
1-naphtol-CO_2	64								
Equation (19)		4.02	3.83						20.41
Equation (20)				−35.10	3.86	21.15			17.99
Equation (16)		18.48	1.015						18.27
Equation (23)							7.32E-5	1.18E-4	23.33
UNIQUAC model									16.05*
2-naphtol-CO_2	48								
Equation (19)		3.71	4.53						14.08
Equation (20)				−30.50	4.30	15.73			11.91*
Equation (16)		15.55	0.948						84.51
Equation (23)							8.28E-5	1.24E-4	18.93
UNIQUAC model									12.31
2-naphtol-ethane	22								
Equation (19)		2.68	4.13						8.74
Equation (20)				−33.93	4.10	18.27			8.68*
Equation (16)		13.55	0.704						16.68
Equation (23)							8.4E-5	1.26E-4	10.92
UNIQUAC model									17.46
Palmitic acid-CO_2	27								
Equation (19)		8.26	5.34						7.85
Equation (20)				−52.37	5.32	35.22			7.70*
Equation (16)		26.65	0.86						14.10
Equation (23)							2.03E-4	2.34E-4	13.15
UNIQUAC model									10.42
phenanthrene-CO_2	281								
Equation (19)		4.92	4.37						13.57
Equation (20)				−32.20	4.40	18.06			13.37
Equation (16)		20.59	0.96						17.34
Equation (23)							1.07E-4	1.64E-4	17.27
UNIQUAC model									13.15*
phenanthrene-ethylene	42								
Equation (19)		4.08	5.83						12.55
Equation (20)				−31.68	5.64	15.03			9.55*

Continued

Equation (16)		16.54	0.55						21.72
Equation (23)							1.35E-4	1.66E-4	16.80
UNIQUAC model									16.31
phenanthrene-ethane	17								
Equation (19)		2.21	5.81						16.13
Equation (20)				−30.77	5.83	14.73			15.11
Equation (16)		13.80	0.49						19.45
Equation (23)							1.09E-4	1.31E-4	17.85
UNIQUAC model									14.67*
phenol-CO_2	73								
Equation (19)		2.12	3.23						28.34
Equation (20)				−36.15	3.65	25.92			24.84*
Equation (16)		12.32	0.82						26.54
Equation (23)							4.66E-5	1.01E-4	26.38
UNIQUAC model									29.70
pyrene-CO_2	235								
Equation (19)		5.66	4.89						9.60
Equation (20)				−31.67	4.79	14.92			6.41
Equation (16)		20.75	0.85						14.44
Equation (23)							1.14E-4	1.52E-4	11.24
UNIQUAC model									4.78*
pyrene-ethane	15								
Equation (19)		5.34	4.95						2.89
Equation (20)				−30.24	4.95	13.83			2.89
Equation (16)		20.55	0.87						1.99
Equation (23)							1.49E-4	1.7E-4	6.07
UNIQUAC model									1.16*
resorcinol-CO_2	26								
Equation (19)		2.83	4.38						6.02
Equation (20)				−33.73	4.42	17.67			5.05*
Equation (16)		11.67	0.61						66.21
Equation (23)							4.58E-5	7.38E-5	65.60
UNIQUAC model									6.71
stearic acid-CO_2	28								
Equation (19)		10.55	4.59						8.80
Equation (20)				−53.95	5.16	35.78			7.53*
Equation (16)		24.25	0.65						30.96

Continued

Equation (23)							1.98E-4	2.08E-4	31.30
UNIQUAC model									8.49
triphenylene-CO_2	53								
Equation (19)		8.15	5.04						12.84
Equation (20)				−32.91	5.03	17.78			4.77*
Equation (16)		26.18	1.01						20.40
Equation (23)							1.59E-4	1.92E-4	14.19
UNIQUAC model									4.90
triphenylmethane-CO_2	111								
Equation (19)		7.46	4.42						13.18
Equation (20)				−32.29	4.42	22.69			13.17
Equation (16)		35.63	1.68						10.90*
Equation (23)							1.65E-4	2.22E-4	24.71
UNIQUAC model									15.09

Table 8. AARD range and mean AARD.

	Solvent CO_2		Solvent ethane		Solvent ethylene		Mean AARD	Mean AARD	Mean AARD
	minima	maxima	minima	maxima	minima	maxima	(solvent CO_2)	(solvent ethylene)	(solvent ethane)
Equation (19)	4.20	28.34	2.9	16.3	11.7	16.8	13.34	14.3	10.5
Equation (20)	3.03	24.84	2.9	15.1	7.6	14.6	11.21	10.7	8.40
Equation (16)	6.60	84.51	1.99	22.6	10.5	21.8	22.68	16.8	12.4
Equation (23)	6.07	65.60	6.07	17.85	10.4	17.8	20.27	15.7	11.1
UNIQUAC model	2.78	29.70	1.16	17.46	8.06	19.2	11.70	14.5	10.19

thracene and phenanthrene, Pennisi and Chimowitz [26] for mixed 1,10-decanediol and benzoic acid, Iwai *el al.* [33] for mixed 2,6- and 2,7-dimethylnaphtalenes are used. The solubility data of the considered compounds are very small and have an order of 10^{-6} - 10^{-3}. As a consequence, we can assume that the density of the supercritical phase is that of the pure solvent, and the activity coefficient of each solute is the one at infinite dilution. In this case the only interaction parameters that are taken into account are those of solutes-CO_2. Therefore predicted solubilities are estimated using Equations (6) to 13(b) and interaction parameters listed in **Table 5** are directly implemented to estimate the solubility of each component in the mixture.

The interpolation results are given in **Table 10**, the absolute average relative deviation (AARD) obtained are generally low confirming the predictive ability of the model but have an order of magnitude considerably higher than those for the solutes in binary systems. This can be attributed to the fact that effect of solute-solute interactions is not always negligible and assuming an infinite dilution of all solutes can be inaccurate in some systems. However, there is significant experimental evidence to suggest that solute-solute interactions are important even in dilute solutions [34-36].

7. Conclusion

In this work, we have proposed the correlation and prediction of the solubility of solid solutes with different functional groups in chemically diverse supercritical fluids with a methodology based on the expanded liquid theory, in which the solid-fluid equilibrium is modeled using the local composition model of UNIQUAC. For the elaboration of the model we have considered 33 systems solid-SC fluid to built-up an exhaustive solutes solubility database consisting of more than 2218 experimental solubility data in supercritical fluids taken from literature. The proposed model achieves an overall AARD of 11.85 on a whole database and on a range of 1.16 to 29.7, these results show that the calculated solubility data are in

Table 9. Prediction capabilities of the model.

	Training set		Test set	
Systhem	*N*	AARD%	*N*	AARD%
anthracene-CO_2	144	14.51	62	14.70
anthracene-ethylene	18	8.36	9	7.57
biphenyl-CO_2	39	15.15	18	20.27
6-caprolactam-CO_2	22	6.18	10	4.76
2,3-dimethylnaphtalene-CO_2	17	16.93	8	21.42
2,6-dimethylnaphtalene-CO_2	16	8.10	7	15.24
fluoranthene-CO_2	47	11.33	20	16.22
fluorene-CO_2	102	6.48	44	7.86
lauric acid-CO_2	16	10.06	8	20.48
myristic acid-CO_2	20	12.98	10	18.79
naphthalene-CO_2	169	16.36	73	17.28
naphthalene-ethylene	116	18.50	50	21.20
naphthalene-ethane	33	10.16	15	16.42
1-naphtol-CO_2	44	15.76	20	17.05
2-naphtol-CO_2	33	8.32	15	21.22
2-naphtol-ethane	15	18.24	7	17.20
palmitic acid-CO_2	18	10.54	9	12.23
phenanthrene-CO_2	196	13.44	85	12.54
phenanthrene-ethylene	29	16.90	13	15.68
phenol-CO_2	51	26.18	22	47.01
pyrene-CO_2	164	4.71	71	5.04
Resorcinol-CO_2	18	6.51	8	7.17
stearic acid-CO_2	19	7.87	9	11.15
triphenylene-CO_2	37	5.05	16	5.66
triphenylmethane-CO_2	77	16.73	34	11.90

Table 10. Prediction results for solutes considered in ternary systems.

Solute	*N*	*T*-range (K)	*P*-range (bar)	ρ_r range	AARD (%)
anthracene	10	308 - 318	104 - 242	1.19 - 1.92	27.0
phenanthrene	10	308 - 318	104 - 242	1.19 - 1.92	15.5
2,6-dimethyl naphtalene	12	308 - 318	90 - 247	0.72 - 1.92	18.1
2,7-dimethyl naphtalene	12	308 - 318	90 - 247	0.72 - 1.92	16.9
1,10-decanediol	10	308 - 318	164 - 307	1.64 - 1.99	26.2

good agreement with the experimental ones. Comparison between the considered literature correlations and the proposed model was justified using comparative criteria. The model's performance is similar and sometimes superior to the literature models considered. The advantages of this model include the following: it does not require the knowledge of critical properties of the solutes and does take into account the binary interaction between solid solute and solvent. Moreover the predictive capabilities of the proposed model were well demonstrated both for solid-solvent and mixed solids-solvent systems.

REFERENCES

[1] G. Madras, C. Kulkarni and J. Modak, "Modeling the Solubilities of Fatty Acids in Supercritical Carbon Dioxide," *Fluid Phase Equilibria*, Vol. 209, No. 2, 2003, pp. 207-213.

[2] N. R. Foster, G. S. Gurdial, J. S. L. Yun, K. K. Liong, K. D. Tilly, S. S. T. Ting, H. Singh and J. H. Lee, "Significance of the Crossover Pressure in Solid Supercritical Fluid Phase-Equilibria," *Industrial and Engineering Chemistry Research*, Vol. 30, No. 8, 1991, pp. 1955-1964.

[3] J. Chrastil, "Solubility of Solids and Liquids in Supercritical Gases," *Journal of Physical Chemistry*, Vol. 86, No. 15, 1982, pp. 3016-3021.

[4] D. L. Sparks, R. Hernandez and L. A. Estévez, "Evaluation of Density-Based Models for the Solubility of Solids in Supercritical Carbon Dioxide and Formulation of a New Model," *Chemical Engineering Science*, Vol. 63, No. 17, 2008, pp. 4292-4301.

[5] L. Nasri, Z. Bensetiti and S. Bensaad, "Correlation of the Solubility of Some Organic Aromatic Pollutants in Supercritical Carbon Dioxide Based on the UNIQUAC Equation," *Energy Procedia*, Vol. 18, 2012, pp. 1261-1270.

[6] J. W. Lee, J. M. Min and H. K. Bae, "Solubility Measurement of Disperse Dyes in Supercritical Carbon Dioxide," *Journal of Chemical and Engineering Data*, Vol. 44, No. 4, 1999, pp. 684-687.

[7] J. W. Lee, M. W. Park and H. K. Bae, "Measurement and Correlation of Dye Solubility in Supercritical Carbon Dioxide," *Fluid Phase Equilibria*, Vol. 173, No. 2, 2000, pp. 277-284.

[8] J. M. Prausnitz, R. N. Lichtenthaler and G. de Azevedo, "Molecular Thermodynamics of Fluid-Phase Equilibria," 3rd Edition, Prentice Hall Inc., Engelwood Cliffs, 1999.

[9] M. Vázquez da Silva and D. Barbosa, "Prediction of the Solubility of Aromatic Components of Wine in Carbon Dioxide," *Journal of Supercritical Fluids*, Vol. 31, No. 1, 2004, pp. 9-25.

[10] R. Span and W. Wagner, "A New Equation of State for Carbon Dioxide Covering the Fluid Region from the Triple-Point Temperature to 1100 K at Pressures up to 800 MPa," *Journal of Physical and Chemical Reference Data*, Vol. 25, No. 6, 1996, pp. 1509-1596.

[11] S. E. Guigard and W. H. Stiver, "A Density-Dependant Solute Parameter for Correlating Solubilities in Supercritical Fluids," *Industrial and Engineering Chemistry Research*, Vol. 37, No. 9, 1998, pp. 3786-3792.

[12] NIST database. http://webbook.nist.gov/chemistry/name-ser.html

[13] K.-W. Cheng, M. Tang and Y.-P. Chen, "Calculations of Solid Solubility in Supercritical Fluids Using a Simplified Cluster Solvation Model," *Fluid Phase Equilibria*, Vol. 214, No. 2, 2003, pp. 169-186.

[14] C. L. Laws, "Thermophysical Properties of Chemicals and Hydrocarbons," Wiliam Andrew Inc., 2008.

[15] D. R. Lide, "CRC Handbook of Chemistry and Physics," 84th Edition, 2003-2004.

[16] L. Li, Z.-C. Tan, S.-H. Meng and Y.-J. Song, "A Thermochemical Study of 1,10-Decanediol," *Thermochim Acta*, Vol. 342, No. 1-2, 1999, pp. 53-57.

[17] M. Mukhopaday and G. V. Raghuram Rao, "Thermodynamique Modeling for Supercritical Fluid Process Design," *Industrial & Engineering Chemistry Research*, Vol. 32, No. 5, 1993, pp. 922-930.

[18] W. H. Stiver, S. E. Guigard and Beausoleil, "Predicting Ternary Solubilities Using a Solubility Parameter Approach," *Proceedings of the 5th International Symposium on Supercritical Fluids*, Atlanta, April 8-12, 2000.

[19] S. Garnier, E. Neau, P. Alessi, A. Cortesi and I. Kikic, "Modelling Solubility of Solids in Supercritical Fluids Using Fusion Properties," *Fluid Phase Equilibria*, Vol. 158-160, 1999, pp. 491-500.

[20] C.-C. Huang, M. Tang, W.-H. Tao and Y.-P. Chen, "Calculation of the Solid Solubilities in Supercritical Carbon Dioxide Using a Modified Mixing Model," *Fluid Phase Equilibria*, Vol. 179, No. 1-2, 2001, pp. 67-84.

[21] P. Coutsikos, K. Magoulas and D. Tassios, "Solubilities of Phenols in Supercritical Carbon Dioxide," *Journal of Chemical and Engineering Data*, Vol. 40, No. 4, 1995, pp. 953-958.

[22] X. Wang and L. Tavlarides, "Solubility of Solutes in Compressed Gases: Dilute Solution Theory," *Industrial and Engineering Chemistry Research*, Vol. 33, No. 3, 1994, pp. 724-729.

[23] J. Méndez-Santiago and A. Teja, "The Solubility of Solids in Supercritical Fluids," *Fluid Phase Equilibria*, Vol. 158-160, 1999, pp. 501-510.

[24] W. J. Schmitt and R. C. Reid, "Solubitity of Monofunctional Organic Solids in Chemically Diverse Supercritical Fluids," *Journal of Chemical and Engineering Data*, Vol. 31, No. 2, 1986, pp. 204-212.

[25] A. Laitinen and M. Jaentti, "Solubility of 6-Caprolactam in Supercritical Carbon Dioxide," *Journal of Chemical and Engineering Data*, Vol. 41, No. 6, 1996, pp. 1418-1420.

[26] K. J. Pennisi and E. H. Chimowitz, "Solubilities of Solid 1,10-Decanediol and a Solid Mixture of 1,10-Decanediol and Benzoic Acid in Supercritical Carbon Dioxide," *Journal of Chemical and Engineering Data*, Vol. 31, No. 3, 1986, pp. 285-288.

[27] Z. Eckert, "Correlation and Prediction of Solid-Supercritical Fluid Phase Equilibria," *Industrial & Engineering Chemistry Process Design and Development*, Vol. 22, No. 4, 1983, pp. 582-588.

[28] Z. Huang, S. Kawi and Y. C. Chiew, "Application of the Perturbed Lennard-Jones Chain Equation of State to Solute Solubility in Supercritical Carbon Dioxide," *Fluid Phase Equilibria*, Vol. 216, No. 1, 2004, pp. 111-122.

[29] A. Delle Site, "The Vapor Pressure of Environmentally Significant Organic Chemicals: A Review of Methods and Data at Ambient Temperature," *Journal of Physical and Chemical Reference Data*, Vol. 26, No. 1, 1997, p. 157.

[30] V. Oja and E. M. Suuberg, "Vapor Pressures and Enthalpies of Sublimation of Polycyclic Aromatic Hydrocarbons and Their Derivatives," *Journal of Chemical and Engineering Data*, Vol. 43, No. 3, 1998, pp. 486-492.

[31] K. Tochigi, T. Iizumi and K. Kojima, "High Pressure Vapor-Liquid and Solid-Gas Equilibria," *Industrial and Engineering Chemistry Research*, Vol. 37, No. 9, 1998, pp. 3731-3740.

[32] E. Kosal and G. D. Holder, "Solubility of Anthracene and Phenanthrene Mixtures in Supercritical Carbon Dioxide," *Journal of Chemical and Engineering Data*, Vol. 32, No. 2, 1987, pp. 148-150.

[33] Y. Iwai, Y. Mori, N. Hosotani, H. Higashi, T. Furuya, Y. Arai, K. Yamamoto and Y. A Mito, "Solubilities of 2,6-and 2,7-Dimethylnaphthalenes in Supercritical Carbon Dioxide," *Journal of Chemical and Engineering Data*, Vol. 38, No. 4, 1993, pp. 509-511.

[34] J. F. Brenneck and C. A. Eckert, "Phase Equilibria for Supercritical Fluid Process Design," *AIChE Journal*, Vol. 35, No. 9, 1989, pp.1409-1427.

[35] R. T. Kurnik and R. C. Reid, "Solubility of Solid Mixtures in Supercritical Fluids," *Fluid Phase Equilibria*, Vol. 8, No. 1, 1982, pp. 93-105.

[36] J. Kwiatkowski, Z. Lisicki and W. Majewski, "An Ex-

perimental Method for Measuring Solubilities of Solids in Supercritical Fluids," *Berichte der Bunsengesellschaft für physikalische Chemie*, Vol. 88, No. 9, 1984, pp. 865-869.

The Mechanical Properties of the System and Training $Zr_{59}Nb_5Cu_{18}Ni_8AL_{10}$ Bulk Metallic Glasses

Nasr-Eddine Chakri[1,2*], Badis Bendjemil[2,3], M. Baricco[4]
[1]LASEA, Department of Physic, University of Badji Mokhtar, Annaba, Algeria
[2]LASEA, Department of Chemistry, University of Badji Mokhtar, Annaba, Algeria
[3]University of Guelma, Guelma, Algeria
[4]Università di Torino, Torino, Italy

ABSTRACT

In the present work, the glass formation of $Zr_{59}Nb_5Cu_{18}Ni_8Al_{10}$ (numbers indicate at %) alloy with diameter of 2 mm was prepared through water-cooled copper mold casting and in a ribbon form by the single roller melt-spinning method. This study is primarily devoted to evaluating the results obtained with the two methods of the development. The thermal stability was evaluated by differential scanning calorimetry (DSC) at a heating rate of 10°C/mn. The characteristic data of the bulk metallic glass are presented, including glass transition temperature (T_g) and crystallization temperature (T_x). The microstructure and constituent phases of the alloy composite have been analyzed by using X-ray diffraction, and observed by Scanning Electron Microscopy (SEM). The mechanical properties of bulk $Zr_{59}Nb_5Cu_{18}Ni_8Al_{10}$ were alloy measured by compression tests at room temperature.

Keywords: Ribbon and Bulk Metallic Glasses; Amorphous; Fracture Morphology; Mechanical Properties

1. Introduction

Amorphous alloys, also known as metallic glasses, have attracted increasingly intense scientific and technological attention since amorphous metal was first discovered [1]. Due to the lack of defects such as grain boundaries and dislocations, amorphous alloys exhibit extraordinary properties including superior mechanical properties and lower thermal conductivity as well as extreme wear and corrosion resistance. Amorphous alloys are therefore considered to be promising structural as well as functional materials for practical applications. However, the critical cooling rate typically required is above 10^6 K/s, for retaining the metastable amorphous phase during cooling process. Consequently, the size of the amorphous alloys obtained was limited to thin ribbons, tiny wires and small powders. This strongly restricts the further applications of the amorphization technique.

Since the late 1980s, investigations of multi-component liquid alloys with very deep eutectics have led to the development of metallic glasses with high glass-forming ability (GFA). The development of these metal alloys enables the fabrication of glassy samples with a thickness greater than 1 mm, which is known as bulk metallic glasses (BMGs). This development makes the application of BMGs a step closer to reality. Recently, many kinds of BMGs have been discovered and investigated in various alloy systems such as Fe- [2], Zr- [3,4], Ni- [5,6], Mg- [7], Pd- [8], Ti- [9-11], Cu- [12] based systems. The cooling rate for these BMGs varies in the range of 10^{-1} to 10^3 K/s. With a range of differences between the quenching rates of the bulk and ribbon metal glasses, a window of opportunity becomes available for the investigation of glasses with the same composition but produced in various structural states.

The aim of the present work is to study the glass formation behaviour of the $Zr_{59}Nb_5Cu_{18}Ni_8Al_{10}$ prepared by the melt-spinning process, in the ribbon shape, and then injected into a copper mould to prepare alloy rods of about 2 mm in diameter. The mechanical properties of the composite have been examined through compression tests and indentation fracture of amorphous alloy is presented.

2. Experimental

An ingot of the $Zr_{59}Nb_5Cu_{18}Ni_8Al_{10}$ alloy (composition is

*Corresponding author.

given in nominal atomic percentages) was prepared by arc-melting mixtures of Zr 99.99 mass % purity, Ti 99.8 mass % purity, Cu 99.9 mass % purity, Ni 99.9 mass % purity and Al 99.9 mass % purity in an argon atmosphere purified using Nb-gettering. In order to obtain a master alloy, followed by induction melting the master alloy and injection casting into a water-cooled copper mold to form 50 mm long cylinders with diameters of 2 mm. For comparison, thin ribbons of 5 mm width and about 30 μm thickness was prepared using a single-roller, melt spinning technique under a vacuum atmosphere.

All the ribbons and cylinders were investigated by differential scanning calorimetry (DSC) using a constant heating rate of 10˚C/s. The structure of the samples was examined by X-ray diffraction (XRD) with Cu Kα (λ = 1.54056 Å) radiation. Scanning electron microscopy ((JEOL JSM6400F) was employed for the analysis of the fracture features. The compression properties were tested by using an Instron testing machine at a strain rate of 5×10^{-4} s^{-1} at room temperature.

3. Results and Discussion

Figure 1 shows the constant-rate heating (10˚C/mn) DSC curves of $Zr_{59}Nb_5Cu_{18}Ni_8Al_{10}$ analysis was carried out for melt-spun ribbons and as-cast cylinders. In the temperature range investigated, all the curves are characterized by two exothermic peaks, revealing a multi-step crystallization path that does not depend on the way of preparation. The analysis of the thermal stability data, summarized in **Table 1**, reveals striking similarities between the as-cast cylinders and the melt-spun ribbon.

The DSC curves indicate a small difference in glass transition temperature (T_g), crystallization temperature (T_x), super cooled liquid region ($\Delta T_x = T_x - T_g$) between the ribbon and rod samples, for the as cast 2 mm-diameter.

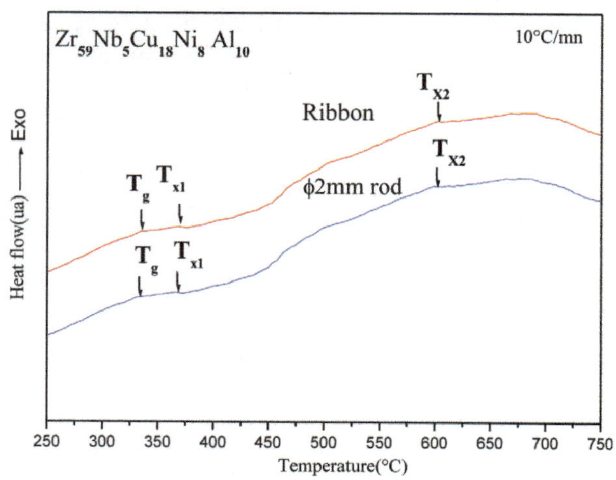

Figure 1. DSC curves of $Zr_{59}Nb_5Cu_{18}Ni_8Al_{10}$ alloy ribbon and rods with diameters of 2 mm.

Table 1. Temperature of the glass transition (T_g), onset crystallization peak of the first (T_{x1}) and the second (T_{x2}), extension of the supercooled liquid region ($\Delta T_x = T_{x1} - T_g$) for the melt-spun ribbons and as-cast rod 2 mm respectively.

Temperature	Melt-spun ribbon	As-cast rod 2 mm
T_g (˚C)	335	333
T_{x1} (˚C)	369	367
T_{x2} (˚C)	602	601
ΔT_x (˚C)	34	34

Cylinder shows that alloy presents two over-lapping exothermic peak around 367 - 601 (˚C). The glass transition temperature, T_g, is 333 (˚C). The melt-spun ribbon for the same alloy presents two exothermic peaks corresponding to two distinct stages of phase transformations T_{x1} and T_{x2} 369, 602 (˚C), respectively and an obvious T_g is 335 (˚C).

Figure 2 shows XRD pattern of the cast $Zr_{59}Nb_5$ $Cu_{18}Ni_8$ Al_{10} rod with a diameter of 2 mm, together with the XRD pattern of the melt-spun glassy alloy ribbon. Only a broad peak is seen around a diffraction angle of 40˚ for the bulk sample and ribbon, indicating the formation of a glassy phase.

The SEM observation on the fracture surface of broken samples provides insight into the fracture mechanism of the in situ composite. **Figure 3** shows the secondary electron images taken for the fractured surface. River patterns predominate in the fracture surface (**Figure 3(a)**).

The fracture path is basically intergranular (**Figure 3(b)**), which implies that the remaining glassy phase in between the amorphous grains plays a role in the deformation process. Milman *et al.* [13] have proposed a parameter defining a plasticity factor for qualifying the brittleness of a material.

Figure 4 shows the compressive stress-strain curve at a strain rate of 5×10^{-4} s^{-1}.The fracture of the samples occurs at a fracture stress σ_f of 2044 MPa and elastic a deformation up to a strain of about range [1.18%, 1.28%]. Young's modulus is determined as 165 GPa forthe2 mm diameter sample.

4. Conclusions

In conclusion, the comparison between the ribbon and massive alloys was obtained directly from the melt-spinning technique. The same sample alloy produced by injection casting of the molten alloy into copper molds with cylindrical cavities reveals that both samples are characterized by the formation phase amorphous obtained in X-ray diffraction.

The bulk glassy rods possess good mechanical proper-

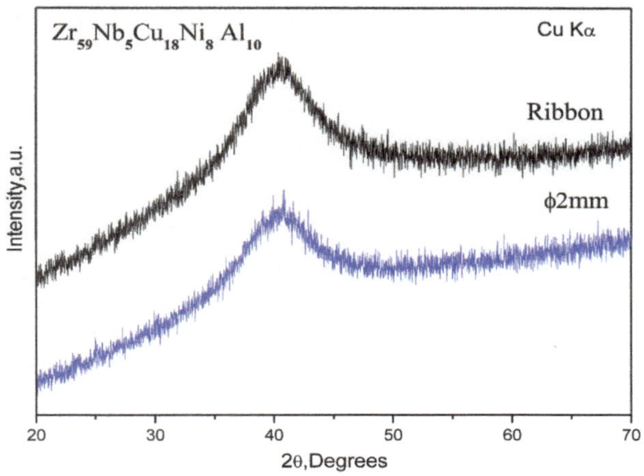

Figure 2. XRD pattern of the cast $Zr_{59}Nb_5Cu_{18}Ni_8Al_{10}$ rod with a diameter of 2 mm, together with the XRD pattern of the melt-spun glassy alloy ribbon.

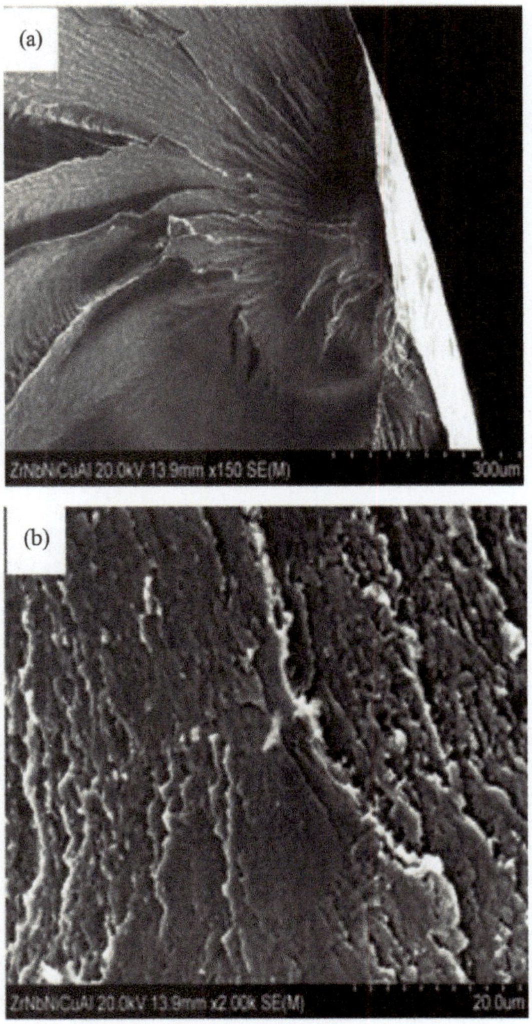

Figure 3. SEM micrographs of the fracture morphology of $Zr_{59}Nb_5Cu_{18}Ni_8Al_{10}$ amorphous rod in as-cast state with diameter of 2 mm (a) low magnification exhibiting the river patterns; (b) the higher magnification images showing the typical characteristics of an intergranular fracture.

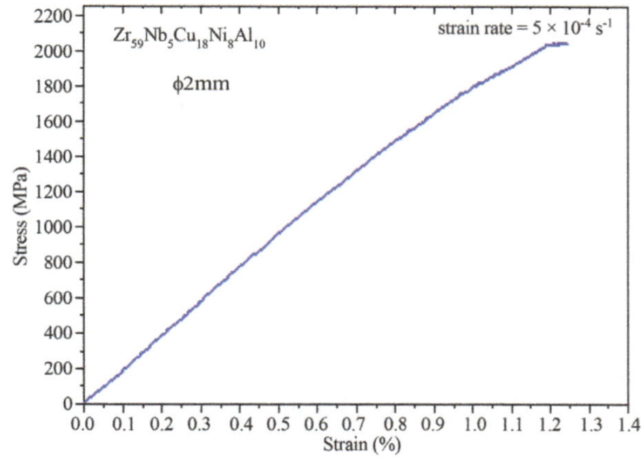

Figure 4. Compressive stress-strain curves of cast $Zr_{59}Nb_5Cu_{18}Ni_8Al_{10}$ glassy alloy rods under anniaxial compression testing at room temperature.

ties, the compressive fracture strength and elastic strain to fracture of the amorphous alloy with 1, 24 at % also exhibit ultra high fracture strength of2044 MPa and a Young's modulus of 165 GPa.

5. Acknowledgements

We acknowledge the group of Prof. M. Barrico for the synthesis of the samples and SEM pictures, XRD analysis.

REFERENCES

[1] P. Duwez, R. H. Willens and W. Klement, "Metastable Electron Compound in Ag-Ge Alloys," *Journal of Applied Physics*, Vol. 31, No. 6, 1960, p. 1137.

[2] J. Shen, Q. J. Chen, J. F. Sun, H. B. Fan and G. Wang, "Exceptionally High Glass-Forming Ability of an FeCoCrMoCBY Alloy," *Applied Physics Letters*, Vol. 86, No. 15, 2005, Article ID: 151907.

[3] Q. K. Jiang, X. P. Nie, Y. G. Li, Y. Jin, Z. Y. Chang, X. M. Huang and J. Z. Jiang, "Ni-Free Zr-Based Bulk Metallic Glasses with critical Diameter above 20 mm," *Journal of Alloys and Compounds*, Vol. 443, No. 1-2, 2008, pp. 191-194.

[4] A. Peker and W. L. Johnson, "A Highly Processable Metallic Glass: $Zr_{41.2}Ti1_{3.8}Cu_{12.5}Ni_{10.0}Be_{22.5}$," *Applied Physics Letters*, Vol. 63, No. 17, 1993, p. 2342.

[5] H. Choi-Yim, D. H. Xu and W. L. Johnson, "Ni-Based Bulk Metallic Glass Formation in the Ni-Nb-Sn and Ni-Nb-Sn-X (X = B,Fe,Cu) Alloy Systems," *Applied Physics Letters*, Vol. 82, No. 7, 2003, pp. 1030-1031.

[6] S. J. Pang, C. H. Shek, K. Asami, A. Inoue and T. Zhang, "Formation and Corrosion Behaviour of Glassy Ni-Nb-Ti-Zr-Co(Cu)," *Proceeding of the 12th International Symposium on Metastable and Nano-Materials*, Original

Reseach Article J. Alloys Compd., Vol. 434-435, 2007, pp. 240-243.

[7] X. K. Xi, D. Q. Zhao, M. X. Pan, W. H. Wang, Y. Wu and J. J. Lewandowski, "Fracture of Brittle Metallic Glasses: Brittleness or Plasticity," *Physical Review Letters*, Vol. 94, No. 12, 2005, Article ID: 125510.

[8] N. Nishiyama and A. Inoue, "Direct Comparison between Critical Cooling Rate and Some Quantitative Parameters for Evaluation of Glass-Forming Ability in Pd-Cu-Ni-P Alloys," *Materials Transactions*, Vol. 43, No. 8, 2002, pp. 1913-1917.

[9] F. Q. Guo, H. J. Wang, S. J. Poon and G. J. Shiflet, "Ductile Titanium-Based Glassy Alloy Ingots," *Applied Physics Letters*, Vol. 86, No. 9, 2005, Article ID: 091907.

[10] Y. J. Huang, J. Shen, J. F. Sun and X. B. Yu, "A New Ti-Zr-Hf-Cu-Ni-Si-Sn Bulk Amorphous Alloy with High Glass-Forming Ability," *Journal of Alloys and Compounds*, Vol. 427, No. 1-2, 2007, pp. 171-175.

[11] X. F. Wu, Z. Y. Suo, Y. Si, L. K. Meng and K. Q. Qiu, "Bulk Metallic Glass Formation in a Ternary Ti-Cu-Ni Alloy System," *Journal of Alloys and Compounds*, Vol. 452, No. 2, 2008, pp. 268-272.

[12] H. M. Fu, H. F. Zhang, H. Wang and Z. Q. Hu, "Cu-Based Bulk Amorphous Alloy with Larger Glass-Forming Ability and Supercooled Liquid Region," *Journal of Alloys and Compounds*, Vol. 458, No. 1-2, 2008, pp. 390-393.

[13] Yu V. Milman, B. A. Galanov and S. I. Chugunova, "Plasticity Characteristic Obtained Through Hardness Measurement," *Acta Metallurgica et Materialia*, Vol. 41, No. 9, 1993, p. 2523.

19

Biological Template Based on *ent*-Kaurane Diterpenoid Glycosides for the Synthesis of Inorganic Porous Materials

Ángela B. Sifontes[1*], Mirla Rodriguez[1], David Freire[1], Wendy Rondón[1], Ligia Llovera[1], Edgar Cañizales[2], Franklin J. Méndez[1], Andrea Monaco[1], Yraida Díaz[1]
[1]Centro de Química, Instituto Venezolano de Investigaciones Científicas, Caracas, Venezuela
[2]PDVSA, Intevep, Los Teques, Venezuela

ABSTRACT

Recent studies on the preparation of porous nano-materials revealed that the use of kaurane diterpenoids molecules from steviol as biological template favors the obtaining of metallic oxides with tubular morphology as nanorods or nanofibers. In this sense, the present contribution shows an analysis in order to understand how these glycosides of kaurane diterpenoids control the nucleation and growth of inorganic materials favoring the obtaining of these morphologies. For this purpose, it was necessary to carry out studies of the leaf aqueous extract of *Stevia rebaudiana* by HRTEM, FTIR and ^{1}H-NMR.

Keywords: Diterpene Glycosides; *Stevia rebaudiana*; Template

1. Introduction

Diterpenoids constitute a class of natural compounds among the richest in terms of chemical structure and highly diverse properties [1-3]. Diterpenoids can be subdivided into acyclic (phytanes), bicyclic (labdanes, clerodanes), tricyclic (pimaranes, abietanes, cassanes, rosanes, vouacapanes, podocarpanes), tetracyclic (trachylobanes, kauranes, aphidicolanes, stemodanes, stemaranes, beyeranes, atisan macrocy-clicditerpernes (taxanes, cembranes, daphnanes, tiglianes, ingenanes) and mixed compounds, in accordance with the number and the cyclization patterns displayed by their skeletal [1,2]. Natural compounds of this class have been found in many plants, fungi, sea organisms and insects [1]. In particular, the kauranes represent an important group of tetracyclic diterpenes and their structures are constituted by a perhydrophenantrene unit (*a*, *b* and *c* rings) fused with a cyclopentane unit (*d* ring) formed by a bridge of two carbons between C-8 and C-13 (**Figure 1**) [2]. Several criteria are used for the nomenclature of the kaurane diterpenes, the most frequent being the inversion of the conventional description of stereochemistry when the name is preceded by the prefix "*ent*-" [1,4-6].

*Corresponding author.

Recently, glycosides of the *kaurane diterpene* known as steviol (*ent*-13-hydroxykaur-16-en-19-oic acid) have been employed as biological templates in the preparation of porous nano-materials [4,7]. In this sense, the steviol is a constituent from *Stevia rebaudiana,* a plant native to certain regions of South America (Paraguay and Brazil). The major constituents in the leaves of *S. rebaudiana* are the potently sweet diterpenoid glycosides stevioside, rebaudiosides A and D and dulcoside A (**Figure 2**). The steviol glycoside preparations could be obtained by hot-water extraction from the leaves of the plant, followed by further concentration and purification [4,7]. **Figure 2** shows the main diterpene glycosides present in the *S. rebaudiana* aqueous extract.

Our recent studies on the preparation of porous nanomaterials revealed that the use of diterpene glycoside molecules from steviol as biological template favors the obtaining of metallic oxides with tubular morphology as nanorods [7] or nanofibers [4]. We are interested in understanding how these glycosides of kaurane diterpenoids control the nucleation and growth of inorganic materials, which favor the obtaining of various characteristic morphologies. It is for this reason that the employing of this *ent*-kaurane diterpenoids as biological templates for the preparation of porous nanomaterials is discussed on this communication.

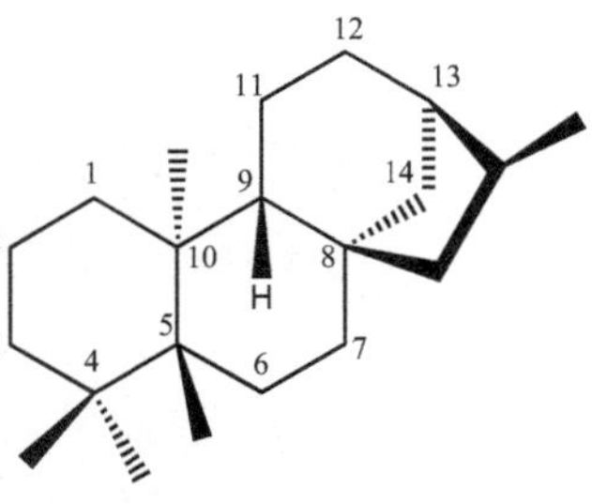

Figure 1. Carbon-skeleton of an *ent*-kaurane diterpene.

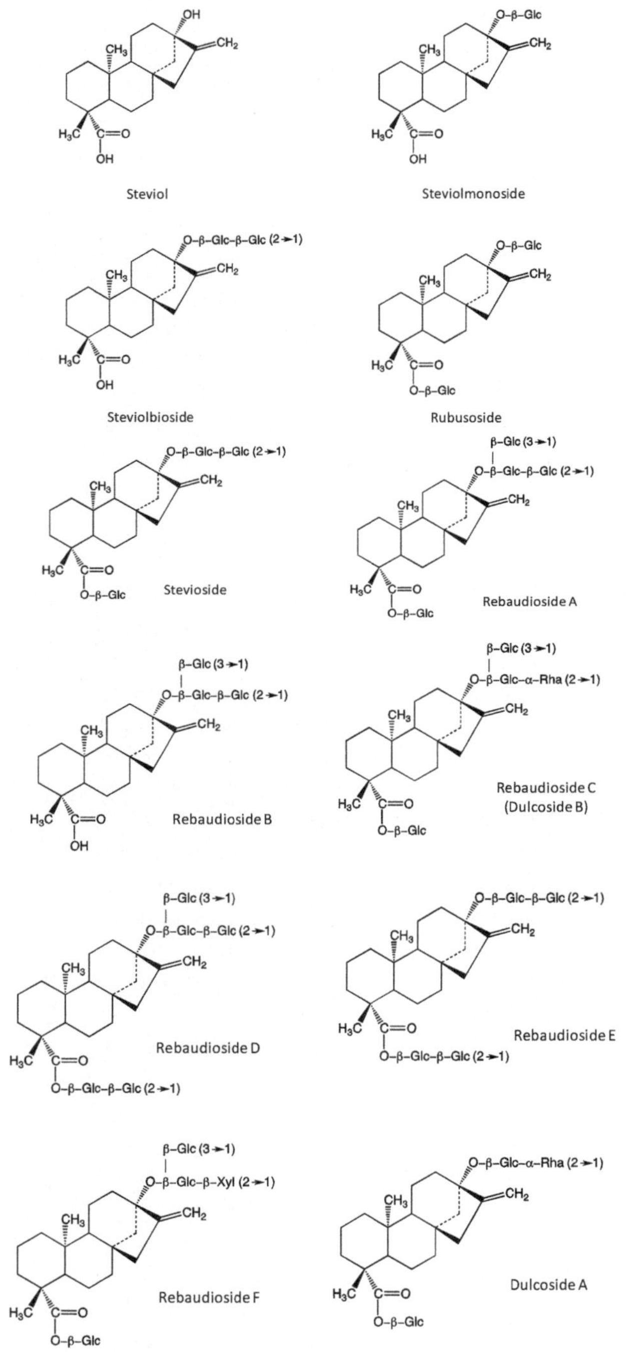

Figure 2. Diterpene glycosides from *S. rebaudiana* aqueous extract.

2. Experimental

Preparation of *Stevia rebaudiana* Leaf Extract

Portions of 1.2 g of leaves of *S. reb audiana* were extracted with 50 mL of hot water (65°C) for 3 h, as described by Nishiyama *et al.* [5]. The crude extract was filtered through a Whatman qualitative filter paper No. 1. Finally the aqueous extract was drying at 60°C.

Characterization of the extract was carried out by Fourier transform infrared (FTIR) recorded with a spectrophotometer Magna-IR 560 from Nicolet. The evaluation by transmission electron microscopy (HRTEM) was performed on a JEOL JEM-2100 microscope with LaB_6 filament (accelerating voltage of 200 kV). The ^{1}H-NMR was recorded on Bruker Advance 500 Mhz spectrometer.

3. Results and Discussion

Recently, derivatives of isosteviol ((4,8,13)-13-methyl-16-oxo-17-norkaurane-18-carboxylic acid I), a diterpene of the kaurane series isolated from the hydrolysis products of glycosides present in *S. reb audiana* plant, have been of particular interest because of the supramolecular structures that can form. The isosteviol has a rigid tetracyclic hydrocarbon skeleton and two polar reactive substituents (carboxyl and carbonyl groups at the C4 and C16 atoms respectively) arranged on the one side of the tetracyclic core (**Figure 3**). This attributed properties to be an ideal starting compound for constructing complex molecular structures capable of non-covalent binding of small and medium-size molecules (**Figure 4**) [6]. Previously, it has been shown that the isosteviol could form an individual molecular complex with aniline in crystal and solutions [6]. Alfonsov *et al*. demonstrated that the supramolecular structure of this complex looks like a double chiral helix cross-linked by aniline molecules (**Figure 4**). Additionally, it was concluded that the chiral spirals are united in bunches due to H-bonding between isosteviol molecules which belong to neighboring spirals. The formation of this complex was attributed to H-bonding between amine group of aniline and carboxyl group of isosteviol [8].

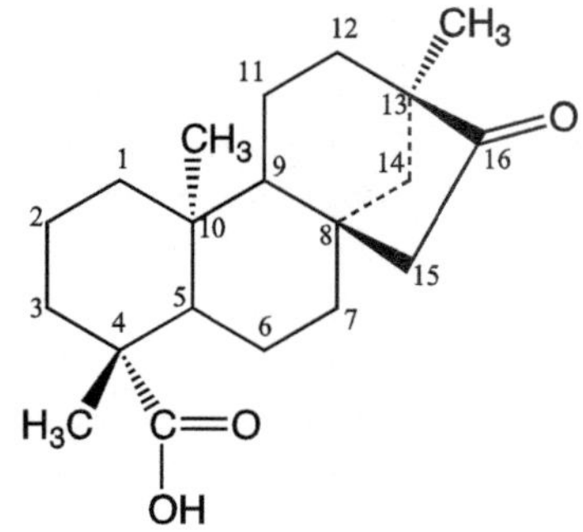

Figure 3. Chemical structure of isosteviol.

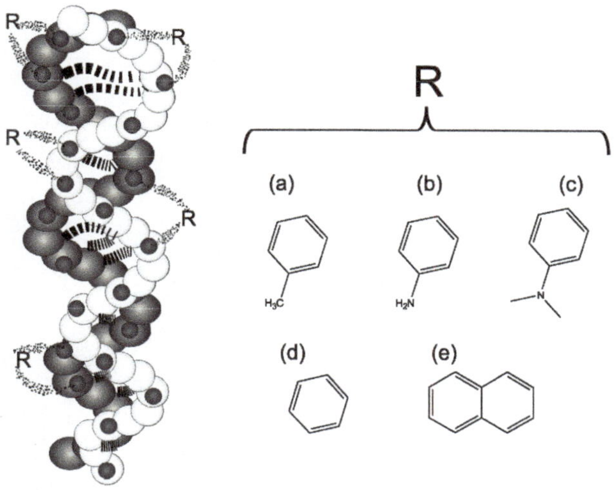

Figure 4. Supramolecular structures look like a double chiral helix cross-linked by aromatic molecules, e.g. (a) toluene; (b) aniline; (c) *N,N*-dimethylaniline; (d) benzene and (e) naphthalene.

The formation of these molecular complexes of isosteviol with other aromatic molecules also has been identified [6,8]. Surprisingly, it is observed that isosteviol can bind molecules, which have no possibility to H-bonding at all (benzene, naphthalene, toluene, xylenes, styrene, ethylbenzene, dimethylanilin). The formed complexes have been reported as isostructural in the space group $P4_32_12$ [9] (**Figure 5**).

According to this, it was established carrying out an HRTEM study of the *S. reba udiana* dried extract, employed in the synthesis of nanosized metal oxides (**Figure 6**). HRTEM observations revealed that the glycosides of kaurane diterpenoids crystallize in cubic or tetragonal structures with diameters from 200 to 500 nm. Recent studies based on the isosteviol derivatives and their analogs have attributed the formation of this crystal packing and supramolecular structures to the localization of the molecular regions with hydrophilic and hydrophobic properties (**Figures** 7 and **8**). The morphology of the supramolecular structures formed depends on the ratio of volumes of hydrophilic and hydrophobic molecular fragments (**Figure 8**) [10]. To explain this phenomenon, the theory of microphase separation was used. This theory was developed previously for block-copolymer and liquid crystal systems [10]. In the HRTEM image can be seem nanostructures with different morphologies, which can be attributed to the different fragments both hydrophilic and hydrophobic of the glycosides present in the aqueous extract of the *S. rebaudiana*. The change of this ratio in the crystal also results in the modification of the morphology type of the structure, analogous to the modification of the supramolecular structure occurring in diblock-copolymers [10,11].

On the other hand, FTIR spectroscopic study of extract

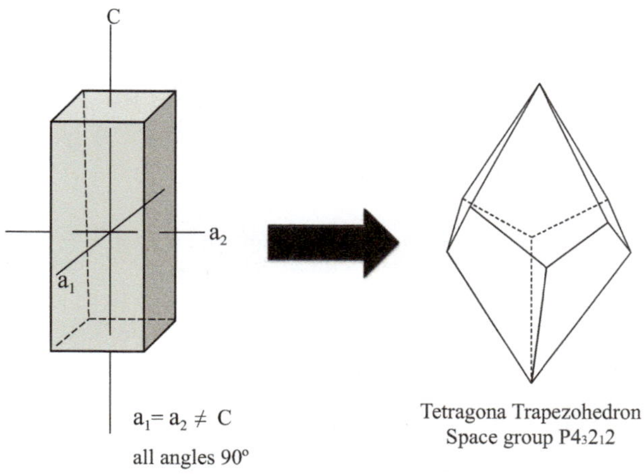

Figure 5. Crystal systems found in the formation of supramolecular structures of *S. rebaudiana.*

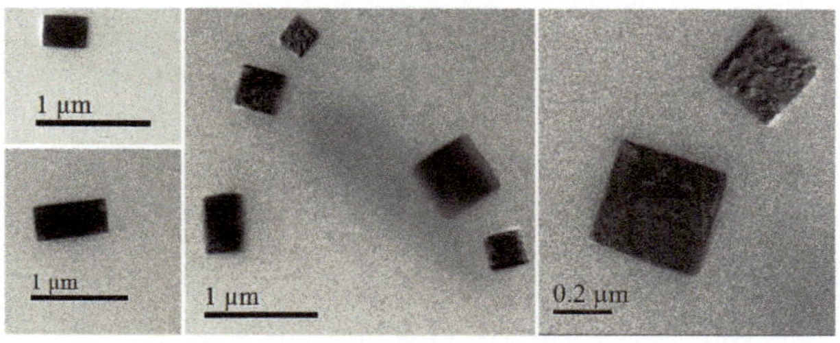

Figure 6. HRTEM image of the *S. rebaudiana* dried extract.

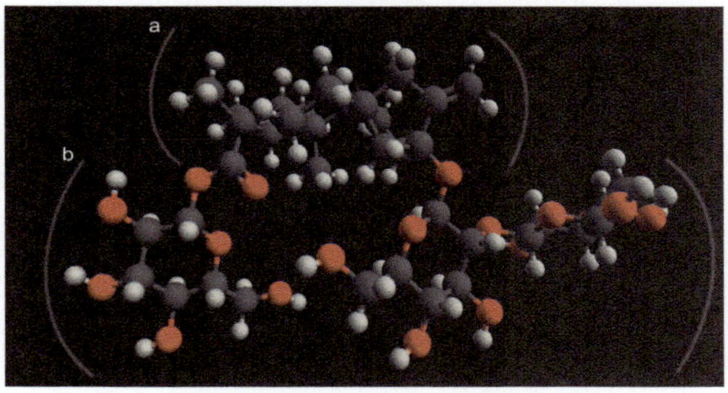

Figure 7. The distribution of the hydrophobic (a) and hydrophilic (b) fragments in the stevioside molecule.

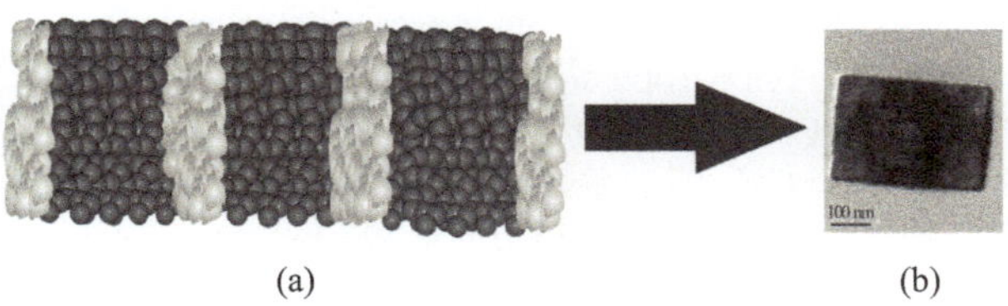

Figure 8. Model of (a) possible distribution of the hydrophilic (dark) and hydrophobic (light) fragments in a glycoside kaurane diterpenoids monocrystaland (b) HRTEM image of the glycoside of kaurane diterpenoids monocrystal from *S. rebaudiana.*

aqueous has been used to compare the changes in the functional groups with regard to the diterpenes isolated from *S. rebaudiana* glycosides (**Table 1**) [12].

Table 1. Fourier transform infrared (FTIR) spectroscopic study of extract aqueous and the diterpenes isolated from *S. rebaudiana* glycosides.

	OH Stretching (cm^{-1})	Alkane CH Stretching (cm^{-1})	C=O Stretching (cm^{-1})	C=C Stretching (cm^{-1})	C-O-C Stretching (cm^{-1})	C=C Bending (cm^{-1})
Steviol* Glycosides	3319 - 3379	2920 - 2938	1720 - 1735	1641 - 1662	1054 - 1070	870 - 894
Extract aqueous		2929.44	-	1608	1075.61	817.54

*FTIR spectral analysis of diterpene glycosides isolated from *Stevia rebaudiana*.

FTIR measurements of the stevia leaf extract aqueous are depicted in **Figure 9**. Different types of vibration may contribute to the broad band observed between 490 cm^{-1} and 817 cm^{-1}: bending vibrations of acetylenic C-H deformation, aliphatic CH_3 deformation, olefinic *cis*-CH=CH, as well as the out-of plane bending vibrations of the alcoholic bonds. The bands in 1385 - 1400 cm^{-1} range can be attributed to aliphatic $-CH_2$ and $-CH_3$ groups or bending modes of O-H bonds.

The absorption band at 1730 cm^{-1} associated with C=O bond stretching that frequently appear in the FTIR spectrum of the glycosides diterpenes "isolated" was not observed in the aqueous leaf extract from *S. rebaudiana.* This seems to indicate that the carbonyl groups participated in the assembly of the polyhedral structures, previously observed in HRTEM images, through interactions with H bound to neighboring spirals of *ent*-kaurane diterpenoids. This imply that the preparation of inorganic structures (metal oxides) using the leaf aqueous extract of *S. rebaudiana* as biological template cannot be carried out through the formation of a metal-carboxylate complex as in the case of other procedures of synthesis, which uses metal alkoxides and oligosaccharides (in acid medium). Another important point to mention is that the organic agent can only act controlling the hydrolysis of metal alkoxide if it contains in its structure active hydrogen atoms, hydroxyl groups, carboxyl, amide acid, imides, sulfonic acids, phenolic or ketoenolics groups [13]. This gave us a broad view of the applications that may have some vegetal extracts as biological templates in the synthesis of inorganic porous materials.

^{1}H-NMR spectrum of Stevia leaf extract is depicted in **Figure 10**. The strong signal observed at $\delta = 4.69$ ppm is due to DHO in D_2O. Signal at $\delta = 0.74$ ppm could be due to aliphatic CH_3-C groups, whereas the signals appearing between $\delta = 1.0$ ppm and $\delta = 1.297$ ppm are related to aliphatic C-CH_2-C and alkoxy CH_3-C-O groups, respectively [14]. It should be noted that any signal appears in the $\delta = 10 - 12$ ppm range in which signals due to carboxylic acids are expected. This is in accordance with was mentioned above in the FTIR study.

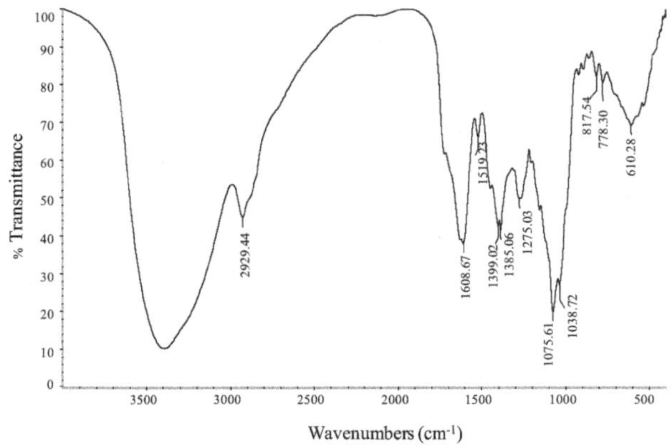

Figure 9. FTIR spectra of the stevia aqueous leaf extract.

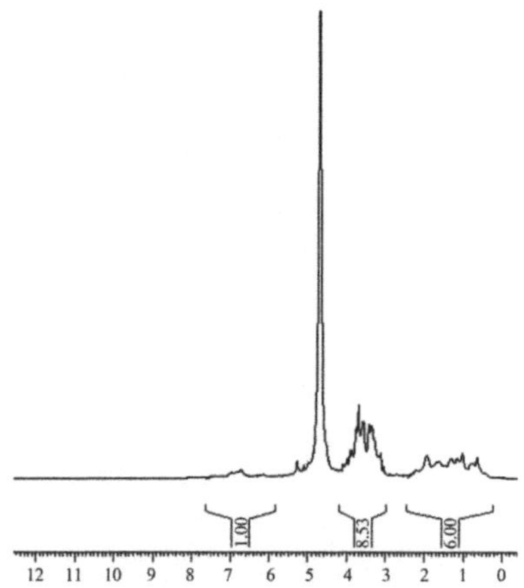

Figure 10. ^{1}H-NMR spectrum of stevia aqueous leaf extract dissolved in D_2O.

3.1. Preparation of Porous Inorganic Materials Using Aqueous Extract of *S. rebaudiana*

Recently, we have described the synthesis of porous materials using *S. reb audiana* aqueous leaf extract [4,7]. The materials prepared by these method exhibit properties potentially useful for many applications in nanotechnology, in particular: high porosity, narrow pore size distributions and particle sizes ranging from 5 - 50 nm. It is important to mention the case of the TiO_2 prepared at room temperature in an aqueous medium, using titanium isopropoxide as precursor and the *S. rebaud iana* leaf extract as template [4]. The anatase phase of the synthesized TiO_2 has a high surface area (252 $m^2 \cdot g^{-1}$), diameter of 25 - 60 nm, length of 100 - 150 nm and morphology of nanofibers (**Figure 11**). The observed morphology was similar to the morphology of alumina nanorods reported in our previous publication [7], which was also prepared employing the metal alkoxide. It is of interest to note that the prepared titania nanofibers are composed of many aggregates of nanocrystals that are added providing the resulting morphology (**Figure 11**). On the other hand, these aggregates are grouped together forming structures that correspond with the supramolecular arrangements, tetragonal and cubic, formed by the kaurane diterpenoids (**Figure 12**).

A proposal of the design of materials using the *ent-kaurane diterpenoids* as biological templates is shown in **Figure 13**. As you can see, the first step of the synthesis is the preparation of aqueous extract, which contains the biological template. Then, you must form a hybrid material together with the inorganic precursor. The formed structures could attach inorganic particles building tubular units (nanorods or nanotubes). Subsequently heat treatment will allow the removal of the biological template to finally obtain the metal oxide with tubular morphology.

3.2. *ent*-Kaurane Diterpenoids from Different Plant Extracts

The sweet *ent-kaurane di terpene gl ycosides*, which are considered in this contribution as potential biological templates, can be obtained from various plant extracts commonly used in the traditional herbal medicine. Among these is important to mention some of them. For example, phytochemical studies of *the Wedelia trilo bata* showed that the plants of this genus are a rich source of *ent-kaurane di terpenoids*. The *ent*-kauranes recently isolated are shown in **Figure 14** [15].

The study of the isolation and characterization of *cussoracoside C ent-kaurane di terpenoids*, from *Cussonia racemosa*, represents a starting point for the investigation of other species such as *Cussonia v antsilana* from the same genus. From the leaves of the plant have been isolated a large amount of cussoracoside C (**Figure 15**) together with four *ent*-kaurane diterpenoid glycosides, named *cussovantosides* A-D, respectively [16].

In addition, it has been also reported the *ent*-kaur-16-en-18-oic acid, isolated from *Croton antisyp hiliticus* root extract, showed in **Figure 16** [17].

Finally, it is highlight the purification of the extract of dried seeds of *Pharbitis n il Choisy* (*Convolvulaceae*), which allowed obtaining six new *ent*-kaurane diterpene glycosides, pharbosides A-F [18]. There are some exam ples in **Figure 17**.

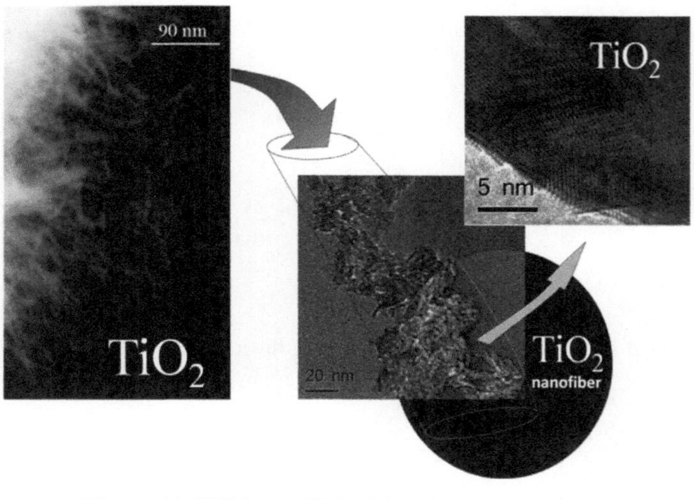

Figure 11. TEM and HRTEM of titania nanofibers.

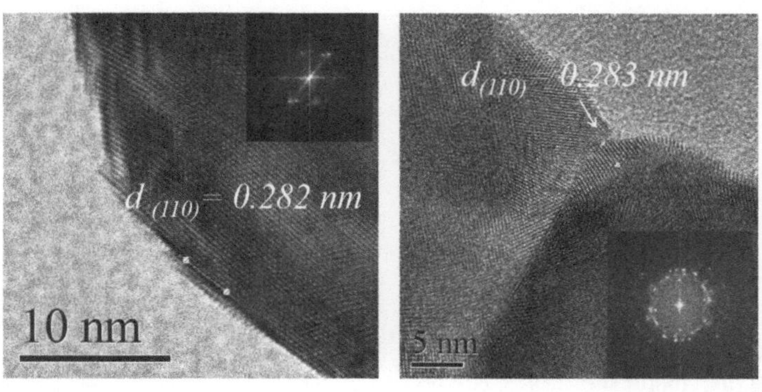

Figure 12. HRTEM image of the TiO_2 prepared at room temperature in an aqueous medium, using titanium isopropoxide as precursor and the *S. rebaudiana* leaf extract as template.

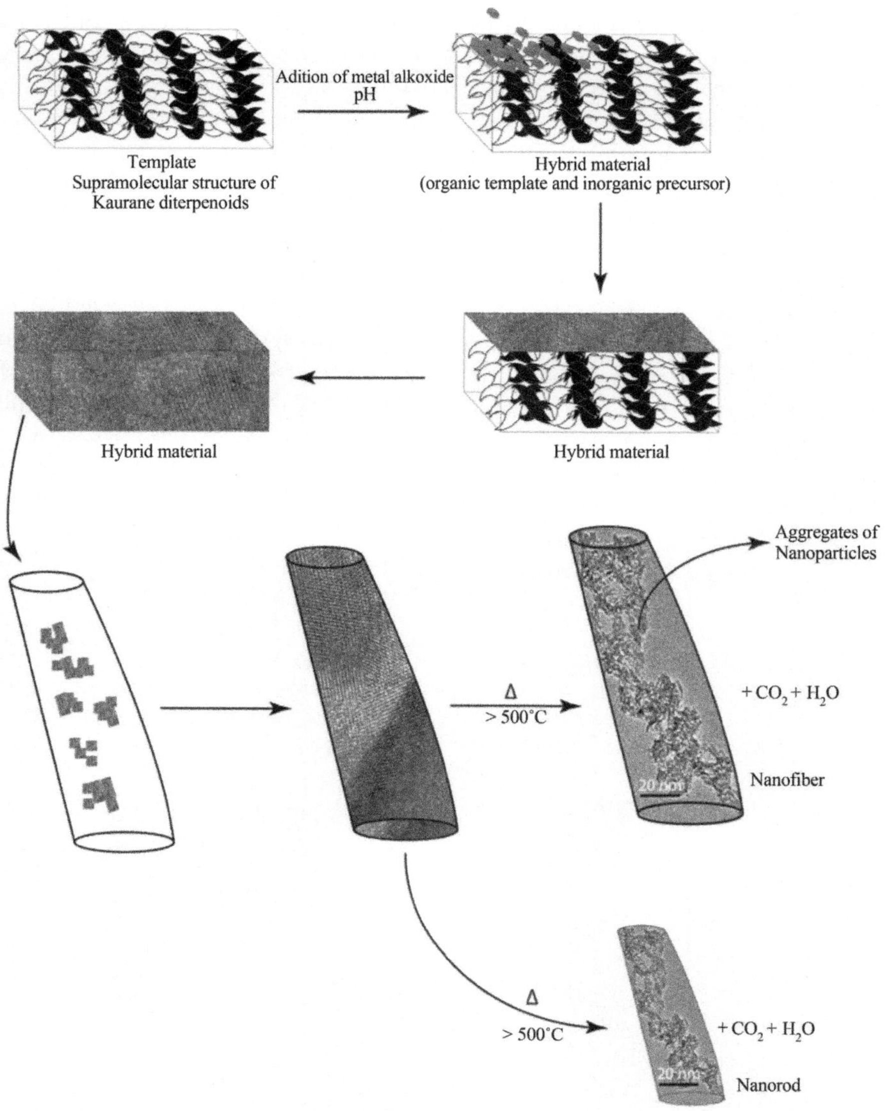

Figure 13. Scheme proposed for the design of materials using the kaurane diterpenoids as biological templates.

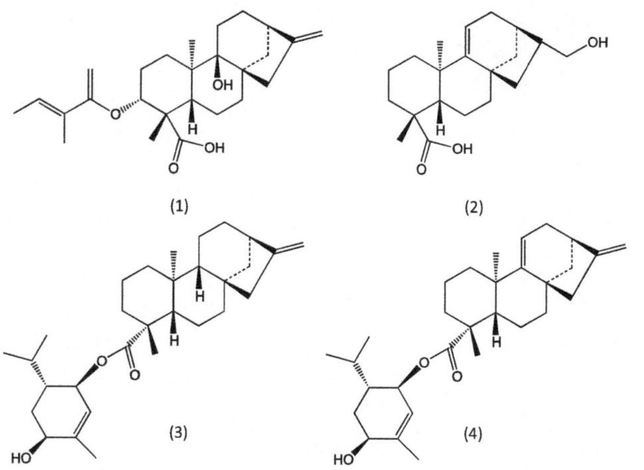

Figure 14. Molecular structure of *ent*-kauranes isolated of *Wedelia trilobata*: 3α-tigloyloxypterokaurene L3 (1), *ent*-17-hydroxykaura-9(11), 15-dien-19-oic acid (2) and wedelobatins A and B (3,4).

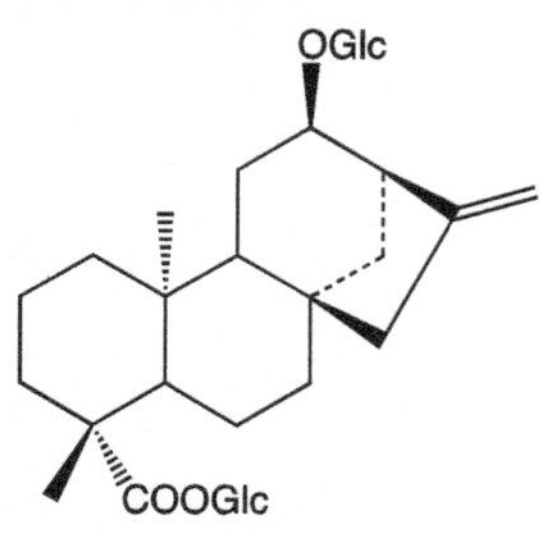

Figure 15. Molecular structure of *cussoracoside C.*

4. Conclusion

The *ent-kaurane di terpenoids* molecules are authentic biological templates with potential applications for the preparation of metal oxides. The HRTEM studies demonstrated that the glycosides of kaurane diterpenoids from *S. reba udiana* form cubic or tetragonal crystal structures. The formation of metal oxide with tubular

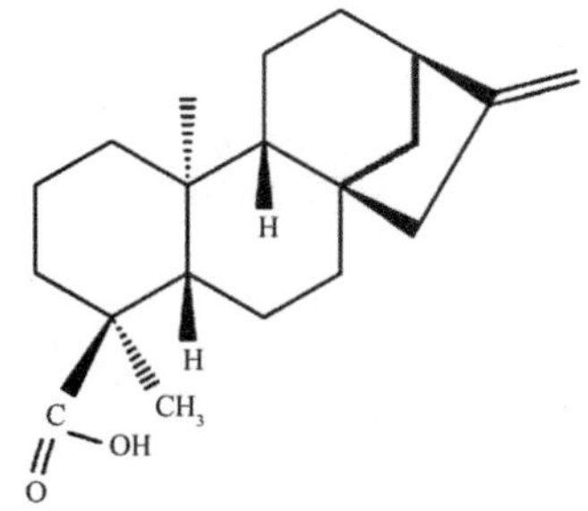

Figure 16. Molecular structure of *ent*-kaur-16-en-18-oic acid.

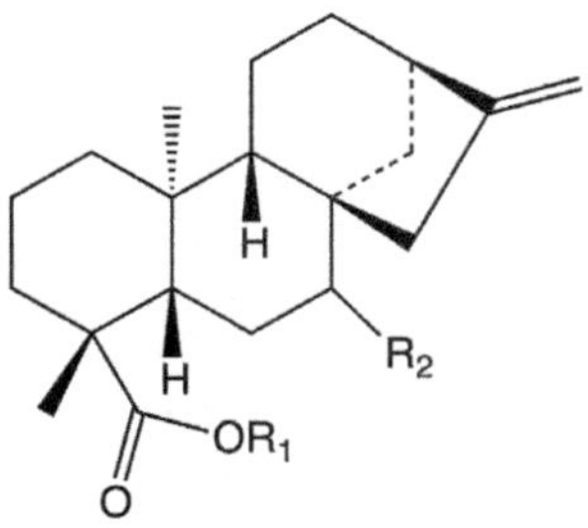

	A	B
R_1	Glc	Glc
R_2	*β-OH*	*α-OH*

Figure 17. Molecular structure of pharbosides A and B.

morphology occurs as a result of the building of hybrid units of *ent-kaurane diterpenoids* molecules and the inorganic precursor, which is subjected to heat treatment at high temperature.

5. Acknowledgements

The authors are grateful to the Instituto Venezolano de Investigaciones Científicas (IVIC) project 1077 for financial support and Lic. Liz Cubillán for FTIR analysis.

REFERENCES

[1] E. Kataev, R. N. Khaybullin, R. R. Sharipova and I. Y. Strobykina, "ent-Kaurane Diterpenoids and Glycosides: Isolation, Properties, and Chemical Transformations," *Review Journal of Chemistry*, Vol. 1, No. 2, 2011, pp. 93-160.

[2] P. Garcia, A. De Oliveira and R. Batista, "Occurrence, Biological Activities and Synthesis of Kaurane Diterpenes and Their Glycosides," *Molecules*, Vol. 12, No. 3, 2007, pp. 455-483.

[3] V. S. Prakash Chaturvedula, M. Upreti and I. Prakash, "Diterpene Glycosides from *Stevia rebaud iana*," *Molecules*, Vol. 16, No. 5, 2011, pp. 3552-3562.

[4] M. Rodríguez, Á. B. Sifontes, F. J. Méndez, E. Cañizales, A. Mónaco, M. Tosta and J. L. Brito, "Synthesis and Characterization of Anatase TiO_2 Nanofibers Using *Stevia rebaudiana* Leaf Aqueous Extract," In: A. Gayathri, Ed., *Recent Research Developments in Material Sciences*, Research Signpost, Kerala, 2013, pp. 45-58.

[5] P. Nishiyama, M. Alvarez and L. G. E. Vieira, "Quantitative Analysis of Stevioside in the Leaves of *Stevia r ebaudiana* by Near Infrared Reflectance Spectroscopy," *Journal of the S cience of Food and Agricu lture*, Vol. 59, No. 3, 1992, pp. 277-281.

[6] V. A. Al'fonsov, O. V. Andreeva, G. A. Bakaleinik, D. V. Beskrovnyi, A. T. Gubaidullin, V. E. Kataev, G. I. Kovylyaeva, A. I. Konovalov, M. G. Korochkina, S. K. Latypov, I. A. Litvinov, R. Z. Musin and I. Y. Strobykina, "Chemistry and Structure of Diterpene Compounds of the Kaurane Series: VI. Isosteviol Esters," *Russian Journal of General Chemistry*, Vol. 73, No. 7, 2003, pp. 1119-1129.

[7] M. Rodríguez, Á. B. Sifontes, F. J. Méndez, Y. Díaz, E. Cañizales and J. L. Brito, "Template Synthesis and Characterization of Mesoporous γ-Al_2O_3 Hollow Nanorods Using *Stevia reb audiana* Leaf Aqueous Extract," *Ceramics International*, Vol. 39, No. 4, 2013, pp. 4499-4506.

[8] V. A. Alfonsov, V. E. Kataev, O. V. Andreeva, D. V. Beskrovniy, G. I. Kovyljaeva, G. A. Bakaleynik, I. Y. Strobykina, I. A. Litvinov and A. I. Konovalov, "The Reception Activity of Isosteviol Isolated from the Plant *Stevia rebaudia na* Bertoni," *Molecular Design and Synthesis of Supra molecular Arch itectures*, Kazan, 27-31 August, 2002, p. 54.

[9] D. Beskrovnyy, A. Gubaidullin and I. Litvinov, "Crystal Structure of Isosteviol and Its Derivatives," *XX IUCr Congress*, Florence, 23-31 August 2005, p. C283.

[10] A. T. Gubaidullin, V. A. Mamedov, I. A. Litvinov, H. Ye and S. Tsuboi, "Synthesis and Comparative Analysis of Molecular and Supramolecular Structures of 4,8-Disubstituted 1,5-Dichloro-2,6-dioxotricyclo[5.1.0.0]octanes," *Monatshefte für Chemie*, Vol. 134, No. 9, 2003, pp. 1229-1240.

[11] G. Wanka, H. Hoffmann and W. Ulbricht, "Phase Diagrams and Aggregation Behavior of Poly(oxyethylene)-Poly(oxypropylene)-Poly(oxyethylene) Triblock Copolymers in Aqueous Solutions," *Macromolecules*, Vol. 27, No. 15, 1994, pp. 4145-4159.

[12] V. S. P. Chaturvedula, "IR Spectral Analysis of Diterpene Glycosides Isolated from *Stevia rebaudiana*," *Food and Nutrition Sciences*, Vol. 03, No. 10, 2012, pp. 1467-1471.

[13] B. K. Davison, "Organic Aluminium Compounds," US Patent No. 3184490, 1965.

[14] R. M. Silverstein, F. X. Webster and D. J. Kiemle, "Spectrometric Identification of Organic Compounds," 7th Edition, John Wiley & Sons, 2005.

[15] B.-J. Ma, C.-N. Wen, Y. Gao, F.-C. Ren, F. Wang and J.-K. Liu, "ent-Kaurane Diterpenoids from the Plant *Wedelia triloba ta*," *Natural Products and Bioprospecting*, Vol. 3, No. 3, 2013, pp. 107-111.

[16] L. Harinantenaina, R. Kasai and K. Yamasaki, "ent-Kau-

rane Diterpenoid Glycosides from a *Malagasy end emic* Plant, *Cussonia vantsilana*," *Phytochemistry*, Vol. 61, No. 4, 2002, pp. 367-372.

[17] S. Pereira, S. Taleb-Contini, J. Coppede, P. Pereira, B. Bertoni, S. Franca and A. M. Pereira, "An ent-Kaurane-Type Diterpene in *Croton anti syphiliticus* Mart," *Molecules*, Vol. 17, No. 8, 2012, pp. 8851-8858.

[18] K. H. Kim, S. U. Choi and K. R. Lee, "Diterpene Glycosides from the Seeds of *Pharbitis nil*," *Journal of Natural Products*, Vol. 72, No. 6, 2009, pp. 1121-1127.

20

Heavy Crude Oil Upgrading: Jazmin Crude

Jesús Alirio Carrillo*, Laura Milena Corredor
Ecopetrol S. A: Refining and Petrochemical, Instituto Colombiano del Petróleo, Bucaramanga, Colombia

ABSTRACT

The Jazmin crude oil is located at the heart of Middle Magdalena in Colombia. It is heavy and sour crude oil with 43 wt.% of vacuum bottoms. It cannot be processed at the conventional refinery without being mixed with other lighter crudes, and should be upgraded to produce synthetic crude with higher concentration of distillates and lower acidity and carbon content. In this paper eight upgrading alternatives are presented. The alternatives include the processing of the crude, reduced crude and vacuum bottoms of the Jazmin crude oil using the following technologies: Distillation, solvent deasphalting, visbreaking, Delayed Coking, and Hydrotreating. The experiments were conducted at pilot scale, and there were used standard analysis techniques such as ASTM. In this study it was found that Jazmine crude oil and its heavy components produce high distillate yields when they were processed with thermal conversion processes. In addition those processes reduce the products acidity. Within the analyzed scheme the one corresponding to the visbreaking of the crude oil and the Delayed Coking of the vacuum bottoms from the visbreaking is perhaps the most attractive, giving 5.9 wt.% of gas, 78.2 wt.% of distillates and 15.9 wt.% of coke.

Keywords: Jazmin Crude; Upgrading; Visbreaking; Deasphalting; Coking; Thermal Conversion; Heavy Crude; Acidity

1. Introduction

The depletion in reserves of light and medium crude oils has focused interest in heavy oils [1,2]. Furthermore, there is currently high demand for 20° - 25° API quality crude oil because the production of Maya and other similar crude oils has been declining in recent years. Exports of Maya crude oil have decreased by about 1 million of barrels per day over the past seven years [3].

Crude oils located in the Middle Magdalena region of Colombia, such as Teca, Nare and Jazmín, are typically viscous and therefore require diluents for their transport by pipeline. They are considered as heavy crude oils. These crude oils do not have high concentrations of coke promoters, so they are sometimes mixed with a pool of crudes suitable for submission to a conventional refining process. However, they are characterized by high acidity (NN > 3.9 mg KOH/g of oil), which is a problem for transportation and refining. An important aspect of the implementation of the thermal processes to acid crudes is that they eliminates or substantially reduces its acidity [4, 5].

Heavy crude oils like Jazmin are cheaper and have lower profits per barrel than conventional oil. However, their refining margin could be improved if they are properly handled. For such a purpose, the conventional processes are based on carbon rejection, hydrogen addition or processes with both technologies [6-12]. Carbon reject technologies includes processes such as visbreaking, thermal cracking and coking while hydrogen addition technologies are classified depending on the type of the reactor used.

In the group of carbon reject technologies, delayed coking and fluid-bed coking continue to be the major process route for producing distillates from vacuum bottoms [13]. Delayed coking (DC) is a thermal process which converts vacuum residue or other residue feedstocks into gas, light products and petroleum coke. DC distillates have a large proportion of olefins which under conditions of temperature, residence time and pollutants tend to polymerize forming gums and degrade their quality. To control these reactions, the DC liquid effluents should undergo hydrotreating processes which stabilize them by the saturation of their double bonds. The coke production ranges from 18 to 30 wt.% of the residual oil, depending on the composition of the feedstock and the operating conditions.

Fluid-bed coking is a continuous process where petroleum coke is combusted to provide process heat, avoid-

*Corresponding author.

ing the used of fuel or natural gas as used in DC. With fluid-bed coking results three are lower coke yields, higher liquid and gas production than with DC.

Visbreaking (VBR) is probably the lowest cost conversion technology. It is a thermal cracking process at low temperatures to reduce the viscosity of the feedstock and produces gases, naphtha and fuel oil and in some cases gas oils which can be used as feedstock for Fluid Catalytic Cracking (FCC). Usually the VBR feedstock is constituted by vacuum bottoms with CCR of 20 wt.% but in some cases it consists of pitch with CCR of 40 wt.%.

Solvent deasphalting process (SDP) is still having high importance in the world [14-16]. SDP is a physical separation process by means of a solvent. Its products are deasphalted oil (DAO) and pitch or deasphalting bottoms (DAB) rich in aromatics with high concentration of impurities such as metals (Ni, V), sulfur, asphaltenes and Conradson carbon. The DAO can either be used for the production of lube oil and paraffinic waxes or as feedstock for the FCC process or for the hydrocracking. The DAB is used either to prepare asphalt or as feed to the visbreaking process. Pitches like DAB had been generally expected to have processing problems in the delayed Coker as unstable feeds due to the concentrated asphaltenes and double bonds.

Foster Wheeler had patented the integration of SDA and DC and claimed that there was a synergistic benefit from combining these two technologies that produced a lower coke yield. The technology is marketed by Foster Wheeler as the ASCOTSM process [16]. The effort to improve these processes continues as it is shown in the works of Bjoror O. for the precipitation of asphaltenes from Athabasca bitumen using α,α,α-trifluorotoluene [17].

2. State of the Art

Heavy crude oils are the alternative to supply refineries in the world. At present heavy crudes of the Orinoco and those from Canadian tar sands undergo upgrading processes to produce synthetic crude oils, which are already commercialized [18]. The main problem for upgrading heavy crude oils is their high concentration of carbon within the vacuum bottoms, however, the heavy crude oil from the Middle Magdalena despite being viscous have low levels of coke promoters (Vacuum bottoms with CCR 17.9 wt.%, and insoluble in n-C_7 4.6 wt.%), which made them suitable for either a process of Delayed Coking (DC) or hydrotreatment (HDT). Thermal processes break down the molecules that cause acidity [17], so the distillates of a Delayed Coking technology should present low values of NN, compared with the same of the original oil.

For crude oil vacuum bottoms handling, the processing facilities installed at the conventional refinery allow the production of asphalts for the pavement of the roads as well as the production of fuel oil, diluting the vacuum bottoms with appropriate solvents or by vacuum bottoms visbreaking. However, the fuel oil consumption in the world is going down and the refinery should be located near shore to facilitate it shipping.

For the upgrading of these crudes different technologies are combined such as fractionation, SDP, thermal cracking (DC and VBR), and HDT mainly. As a result of the application of these technologies, relatively light crudes with low levels of vacuum bottoms, high concentrations of middle distillates and stable diesel, and low sulfur concentrations, low NN, and low nitrogen concentration are obtained. These characteristics allow their use as part of the pool of crudes to a refinery and thus increase middle distillates yields such as diesel and Jet.

3. Experimental

For the upgrading of Jazmin crude oil, operational schemes that combine commercial technologies, using as raw material either crude oil or reduced crude or vacuum bottoms were studied. These cuts were obtained by Jazmin crude oil ASTM distillation. The distillation yields and the properties of the Jazmin crude oil are given in the **Table 1**, column 2. The yields of refined cuts are: atmospheric distillates 19.6 wt.%, and vacuum distillates 35.1 wt.%, so the amount of vacuum bottoms are 45.3 wt.%.

The runs were carried out at pilot plants designed and built in ECOPETROL-ICP [19-22]. The analyses were performed following ASTM methods in the laboratories of ECOPETROL-ICP, certified by ISO 9001.

The operating conditions used in the pilot plants were:

1) For Delayed Coking: load 2000 g, temperature 510°C, pressure 10 psig and run time of 2 h.

2) For Visbreaking: temperature 480°C and residence time 60 seconds.

3) For Solvent Deasphalting: Solvent/Feed ratio was 6/1 vol/vol, the bottom temperature 60°C and the top temperature 100°C. The solvent used for deasphalting was nC_4.

The analyzed operational schemes for the upgrading of the Jazmin crude oil were:

1) Processing of the crude:

a) Visbreaking (VBR) of the crude and Delayed Coking (DC) of the Visbreaked vacuum bottoms.

b) nC_4 Crude Deasphalting and DC of its Demetallized oil (DMO), and

c) nC_4 Crude Deasphalting and DC of the bottoms from Deasphalting (DAB).

2) Processing of the reduced crude:

a) nC_4 Solvent deasphalting of the reduced crude and DC of the bottoms from deasphalting, and

b) Delayed Coking of the reduced crude.

Table 1. Alternatives for the upgrading of the crude Jazmín.

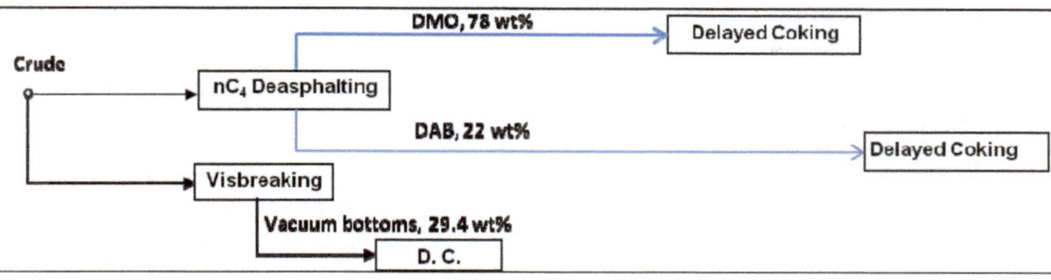

1	2	3	4	5	6	7	8	9	10
		Visbreaking			**Deasphalting and DC of the DMO**			**Deasphalting and DC of the DAB**	
Atmospheric Distillation	**Crude**	**Visbreaking**	**DC of the visbreaking bottoms**	**VB of the crude, and DC of the bottoms from the VB**	**Crude Deasphalting**	**DC of the DMO**	**Deasphalting and DC of the DMO**	**DC of the DAB**	**Deasphalting and DC of the DAB**
Gases		**2.14**	**12.7**	**5.9**		**10.5**	8.2	**6.6**	1.5
Naphthas		12.3	18.9	14.3		19.0	12.2	16.8	2.2
Jet	2.54	1.7	8.3	2.6		11.0	7.0	8.2	1.1
Diesel	7.18	24.9	12.5	26.3		18.0	11.5	14.3	1.9
AGO	9.92	11.3	11.8	12.6		17.0	10.9	14.5	1.9
Vacuum distillation									
LGO	12.18	12.4	13.3	13.8		16.0	10.2	16.6	2.2
MGO	12.85		13.3	1.4		12.0	7.7	15.3	2.0
HGO	10.03	4.8	16.5	6.6		5.0	3.2	11.1	1.5
Vacuum bottoms	**45.3**	29.4	5.4	0.6		2.0	1.3	3.2	0.4
DMO					**78**				78
DAB					**22**		22.0		
DC Distillates			**36.8**			**82.1**		**60.5**	
Coke		1.1	**50.5**	**15.9**		**7.4**	5.8	**32.9**	7.2
Total	**100**	**100**	**100.0**	**100.0**	**100.0**	**100.0**	**100.0**	**100.0**	**100.0**

DC—Delayed Coking; VB—Visbreaking; DMO—Demetallized Oil; DAB—Deasphalter bottoms.

3) Processing of the vacuum bottoms:

a) Delayed coking of the vacuum bottoms.

b) nC_4 Deasphalting of the vacuum bottoms, and DC of the deasphalting bottoms.

c) Visbreaking of the vacuum bottoms.

4. Results and Analysis

4.1. Processing of Crude Jazmín

For the upgrading of the Jazmín crude the schemes presented in the **Table 1** were considered:

4.1.1. Visbreaking of the Crude and Delayed Coking of the Visbreaked Vacuum Bottoms

In this scheme, the crude Jazmín is subjected to the visbreaking process (column 3 of **Table 1**), and the visbreaked vacuum bottoms corresponding to 29.4 wt.% of the crude oil are subjected to DC.

In the column 2 of **Table 1** is shown the composition of Jazmín crude oil; column 3 shows the yields of visbreaked crude. It is clear that the achieved conversion in the visbreaking process of the crude is high as the vacuum bottoms are reduced from 45.3 wt.% in the virgin crude to 29.4 wt.% in the visbreaked crude. That means a conversion of 35.1 wt.%. Moreover, vacuum distillates are reduced from 35.1 wt.% to 17.2 wt.%. That means a conversion of 51 wt.%. As a result of this conversion atmospheric distillates are increased from 19.6 wt.% to 50.2 wt.%.

In the fourth column **Table 1** in bold are given the

yields corresponding to the DC of the visbreaked vacuum bottoms, and in normal is given the simulated distillation of the DC distillates. Although the feedstock to the DC is a visbreaked product, the coke production is relatively low and corresponds to 50.5 wt.%.

In the fifth column of **Table 1** is given the composition of the synthetic crude corresponding to the studied scheme. The yield of diesel is quiet high, 26.3 wt.%, of which most is produced in the visbreaking. In addition, it was found that the vacuum bottoms from the visbreaked Jazmín crude have a relatively high conversion capacity in the DC process. The yield of coke regarding the visbreaked vacuum bottoms is 50.5 wt.% and regarding the synthetic crude of the scheme is 15.9 wt.% (columns 4 and 5).

This is a scheme that reduces the amount of diluents for the transport and it requires low investment in upgrading technology, because DC processes only 29.4 wt.% of the crude oil. Both distillates (from the Visbreaking and from DC) are unstable and required treatment with hydrogen.

As reference for comparison in the first column of the **Table 2** are given the properties of the vacuum bottoms from the virgin crude. The promoters of coke increases significantly in the visbreaked vacuum bottoms as follows: CCR goes from 17.9 wt.% to 34.9 wt.% (an increase of 95 wt.%), and asphaltenes from the SARA analysis goes from 14.5 wt.% to 19.4 wt.% (**Table 2**). For this reason, the coke yield in the process of visbreaked vacuum bottoms DC is high, and reaches 50.5 wt.%, a value that agrees with the rule of thumb on expected coke (1.5 × CCR = 52.4% m).

Another important aspect of this scheme is the acidity reduction as it is shown in the case of the vacuum bottoms. In the crude oil vacuum bottoms NN was 3.93 mg KOH/g of oil, and in the visbreaked vacuum bottoms was 0.8 mg KOH/g of oil. That result confirms that thermal processes reduce the acidity of the feedstock.

4.1.2. nC_4 Crude Deasphalting and Delayed Coking of the Demetallized Oil (DMO)

The yields of nC_4 crude oil Deasphalting are 78.0 wt.% of DMO and 22.0 wt.% of bottoms (DAB, **Table 1**, column 6). Within the same table, in the column 7 in bold the DC yields are given (gas: 10.5 wt.%, distillates: 82.1 wt.%, and coke: 7.4 wt.%), which totalize 100%. The others results correspond to the simulated distillation of the DC distillates, which also correspond to 100%. Because the promoters of coke have been removed from the feed in the previous process of Deasphalting, coke production in the last process is only 7.4 wt.%. The CCR in the feedstock to the DC process is 4.4 wt.% as can be seen from **Table 2**, so the amount of coke is 1.6 times de amount of CCR in the feedstock. The sum of Jet and Diesel is 23.8 wt.% and the yield of naphtha is 15.6 wt.%. Within the distillates, the yield of atmospheric distillates is 65 wt.% and the yield of vacuum distillates is 33.0 wt.%.

In the consolidate of the synthetic crude is shown a coke production of 5.8 wt.% regarding to the original crude (column 8 of **Table 1**), which is a rather low value for the upgrading schemes, however, it should be taking in account the production of 22.0 wt.% of bottoms from deasphalting (DAB), with which the bottoms yield gives a total of 27.8 wt.% respect to the crude oil. DC distillates should be processed in HDT for their stabilization.

This is a scheme of high production of naphtha and high-capacity of the upgrading.

4.1.3. nC_4 Crude Oil Deasphalting and Delayed Coking of the Bottoms from the Deasphalting

The yields from the DC of the deasphalting bottoms are presented in the column 9 of **Table 1** and the composition of the consolidated synthetic crude is given in the column 10. Coke production in the DC of the deasphalting bottoms is 32.9 wt.%, which is a rather low value considering that the DC feed correspond to the bottoms from the Deasphalting, however, according to the rule of thumb, the expected coke yield is 1.5 * CCR in the feedstock, It means 30 wt.%.

In the consolidated scheme the coke yield is 7.2 wt.% and the main product is the DMO with a yield of 78 wt.% (column 10).

This is the scheme that produces the less quantity of solids (coke or bottoms from the Deasphalting process). It is also a low investment process with an average conversion, providing feed to the FCC process. DC distillates require hydrotreating.

4.2. Processing of the Reduced Jazmín Crude 370°C+

The reduced crude is 80.4 wt.% of the whole crude. The analyzed schemes are (**Table 3**):

1) nC_4 Solvent deasphalting of the reduced crude and DC of the bottoms from Deasphalting, and

2) Delayed Coking of the reduced crude.

The product quality of the studied schemes is presented in the **Table 4**.

4.2.1. nC_4 Solvent Deasphalting of the Reduced Crude and DC of the Deasphalter Bottoms

In the column 2 of **Table 3** in bold are given the yields of the Deasphalting process, and the others results correspond to the DMO composition. In the column 3 the composition of the bottoms from the Deasphalting process is given. In the column 4 the yields of the DC of the deasphalter bottoms are given. In the column 5 the composition of the synthetic crude is given.

Regarding reduced crude, the DMO yield is 69.5 wt.%

Table 2. Quality of the products from the upgrading of the crude Jazmín.

	Vacuum Bottoms	Visbreaked Vacuum Bottoms	DC of VBR vacuum bottoms	DMO	DAB	DC of DMO	DC of DAB
Density at 5°C, g/cm^3	1.0262	1.0745	0.9395	0.9684	1.049	0.92	0.925
API	6.3	0.1	19		3.3		21.3
CCR, wt.%	17.9	34.9	4.9	4.4	20	2.2	1.97
Sulphur, wt.%	1.7	2.1	1.6	1.5		1.3	1.646
Ashes, wt.%	-	0.458					
Ca, ppm	2666.0	1128		54.4	1109		
Ni, ppm	130.7	213		17.379	214.88		
V, ppm	167.27	243		15.684	280.7		
Na, ppm	38.5	29					
Ni + V, ppm	298.0	456		33	496		
Refractive index				1.5184			
i-nC_7, wt.%	4.6	25.7			16.25		
i-nC_5, wt.%	11.3	32.3			36.07		
Viscosity-1, cP	30800 @100°C	225000 @120°C		33			
Viscosity-2, cP	4950 @120°C	69500 @130°C		82 @80°C			
Basic N, wt.%	0.284	0.4	0.2	0.124	0.389	0.11	0.18
Total N, ppm	7451	1.2	6130	3679			5302
N.N, mg KOH/g	3.93	0.8					
SARA analysis, wt.%							
S	8.9	5.3					
A	55.2	45.3					
R	21.4	30.0					
A	14.5	19.4					

VBR—Visbreaked; DC—Delayed Coking; DMO—Demetallized Oil; DAB—Deasphalter bottoms.

and the deasphalting bottoms yields are 30.5 wt.%. Within the DMO, 37 wt.% are vacuum bottoms. This result corresponds to low concentration of CCR within the feedstock (11.2 wt.%). Within de DAB, the main component correspond to the vacuum bottoms (75 wt.%, column 3 of **Table 3**).

The Delayed Coking of the DAB produces 31.7 wt.% of coke, which is a low value considering that the feed is deasphalter bottoms and this is due to the low concentration of coke promoters within the feed stock (CCR = 26.3 wt.%, **Table 4**) and to the fact that many of the components of the CCR are soluble in cyclohexane [23]. According to the rule of thumb, the expected coke is 39.45 wt.%.

In the consolidated scheme (synthetic crude oil—column 5 of **Table 3**), coke production is only 7.8 wt.%.

The main processes in this scheme are physical separations, and the conversion is very low and corresponds to the DC process. The upgrading costs in this scheme are low. This is because in the DC is processed only 21.2 wt.% of the feedstock. The vacuum bottoms are reduced from 45.3 wt.% to 21.5 wt.%. That is a substantial reduction. DC distillates must be subjected to hydrotreating. The metals concentration within the DMO is normal for the FCC feedstock or for its Hydrocracking.

4.2.2. Delayed Coking of Reduced Crude

In the column 6 of **Table 3** in bold are given the delayed Coking yields of the reduced crude, and the other results correspond to the composition of the DC distillated.

In this alternative the production of coke is very low (11.7 wt.% with respect to the virgin crude—**Table 3**,

Table 3. Alternatives for the upgrading of the Jazmín reduced crude oil.

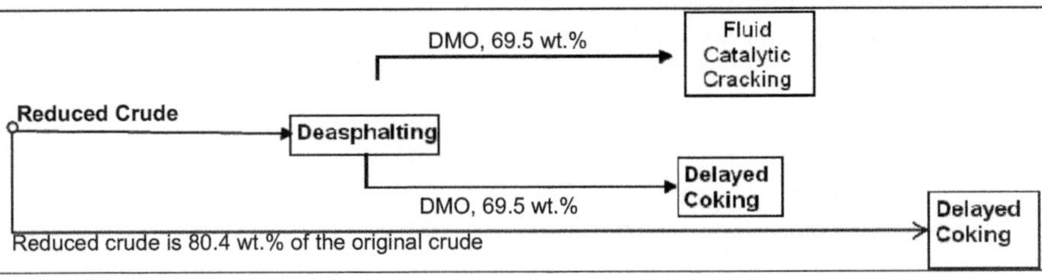

	1	2	3	4	5	6	7
		Deasphalting of the reduced crude and DC of the bottoms from the Deasphalting				**DC of the reduced crude**	
		Deasphalting and DMO composition	**Bottoms from Deasphalting**	**Delayed Coking of the DAB**	**Sintetic Crude from the scheme**	**DC of the reduced crude**	**Synthetic crude from the scheme**
Atmospheric distillation	**Crude**						
Gases				9.2	2.3	**12.1**	9.7
Naphtahs				10.6	2.6	20.0	11.7
Jet	2.5			4.1	3.6	8.0	7.3
Diesel	7.2			7.1	8.9	11.0	13.7
AGO	9.9	9.0	3.0	8.3	17.0	14.0	18.2
Vacuum Destilation							
LGO	**12.2**	16.0	4.0	8.3	11.0	18.0	10.6
MGO	**12.9**	26.0	9.0	11.2	17.3	16.0	9.4
HGO	**10.0**	12.0	9.0	5.9	8.2	13.0	7.7
Vacuum bottoms	**45.3**	37.0	75.0	3.5	21.5	0.0	0.0
DC Distillates						**73.4**	
DMO		**69.5**					
Bottoms from Deasphalting		**30.5**					
Coke				31.7	7.8	**14.5**	11.7
Total	**100.0**	**100.0**	**100.0**	**100.0**	**100.0**	**100.0**	**100.0**

DC—Delayed Coking; DMO—demetalized Oil; DAB—Bottoms from Deasphalting.

column 7), and the conversion is high, especially to middle distillates (Jet + Diesel), which goes from 9.7 wt.% in the original oil to 21.0 wt.% in the synthetic crude. The vacuum bottoms are totally converted. According to the rule of thumb, the expected coke was 1.5 * CCR within the feedstock, it means 16.8 wt.%, and the explanation for this behavior consist in the fact that many components of the CCR are soluble in cyclohexane [23].

This alternative produces high yields of atmospheric distillates; it is costly because it involves processing all reduced crude, which accounts 80.4 wt.% of the whole crude. Distillates from DC must be subjected to hydrotreating.

When comparing the quality of the studied schemes products (**Table 4**), we observed expected behaviors. The Ni + V concentration within the DMO is 31.2 ppm, so we should consider it hydrotreating to adequate feed to catalytic cracking process.

4.3. Processing of the Vacuum Bottoms from the Jazmin Crude Oil

The analyzed schemes were (**Table 5**):

1) Delayed coking of the vacuum bottoms.

2) nC_4 Deasphalting of the vacuum bottoms, and DC of the deasphalter bottoms.

3) Visbreaking of the vacuum bottoms.

In the column 3 of **Table 5** in bold the main products of DC are shown: gas, distillates and coke, and the other results correspond to the simulated distillation of distil-

Table 4. Quality of the products from the upgrading of reduced crude.

	Reduced crude (RC)	DMO	DAB	DC of DAB	DC of RC
Density @ 15°C, g/cm^3	1.0086	0.9817	1.0608	0.9314	0.936
API	8.7		1.8	20.3	19.6
CCR, wt.%	11.2	4.7	26.3	2.99	3.6
Sulphur, wt.%	2.25	1.6	2.3	1.533	1.5
Ca, ppm	345	41.5	1541		
Ni, ppm	79	17.0	195.78		
V, ppm	93	14.2	253.25		
Na, ppm	69				
Ni + V, ppm		31.2	449.0		
Refractive Index		1.5269			
insolubles in nC_7, wt.%	2.6		11.8		
insoluubles in nC_5, wt.%	6.2		23		
Viscosity-1, cP	577 a 120°C	92			
Viscosity-2, cP	79200 a 135°C	270			
Basic N, wt.%	0.218	0.1	0.4	0.2	0.17
N Total, ppm	6387	3828		5414	5570
N. N, mg KOH/g	7.4				
Pour point, °C	42				
SARA analysis, wt%				DMO-Demetallized Oil	
S	18.5	34.9		DAB-Bottoms from	
A	54	51.1		deasphalting	
R	25.1	14		DC-Delayed Coking	
A	2.4			RC-Reduced crude	

lates from DC. Column 4 shows the synthetic crude from this scheme before hydrotreating and column 5 the same products after Hydrotreating. Columns 6 to 9 present data from the scheme of vacuum bottoms deasphalting and DC of deasphalter bottoms. Column 9 shows the synthetic crude from this scheme.

Column 10 presents the data corresponding to the Visbreaking of the vacuum bottoms and column 11 shows the synthetic crude regarding to this scheme.

The composition of all synthetic crude is given considering not only the products of the given scheme of processing, but also include the components of oil which do not fall in the transformations given.

4.3.1. Delayed Coking of the Vacuum Bottoms (Table 5)

Coke production in relation to vacuum bottoms is of 22.2 wt.% (column 3), which is a fairly low value for this type of feed. This is due to the low concentration of coke precursors. The insolubles in n-C_7 are completely soluble in cyclohexane [23]. The CCR in the vacuum bottoms is 17.9 wt.%, and the insolubles in n-C_7 are 4.6 wt.% (**Table 6**). The typical values of CCR in the vacuum bottoms are above 25 wt.% (for the Castilla crude oil they are 37 wt.%).

In the consolidated for crude oil, the coke production is only 10.1 wt.%, a value similar to the production of coke in the consolidated scheme of DC of reduced crude. In the synthetic crude the production of naphtha is 6.2 wt.% and the yield of middle distillates (Jet and Diesel) is increased from 9.7 wt.% to 16.8 wt.% (column 5). Hydrogen treatments applied to DC distillates do not significantly alter their composition (columns 4 and 5 of **Table 6**).

This is a scheme of high conversion where only vacuum bottoms are processed; thereby the size of the up-

Table 5. Vacuum bottoms upgrading alternatives.

1	2	3	4	5	6	7	8	9	10	11
		Delayed Coking (DC) of the vacuum bottoms			**Deasphalting of vacuum bottoms and DC of DAB**				**Visbreaking (VR) of the vacuum bottoms**	
Atmospheric Distillation	**Crude oil**	**DC**	**Synthetic crude**	**Synthetic crude + HDT**	**Deas-phalting**	**DAB**	**DC of DAB**	**Consolidated**	**VR**	**Synthetic Crude**
Gases		**9.9**	4.5	4.5			**9.5**	2.8	2.0	0.9
Naphtha		20.0	6.2	5.5			19.0	3.1	5.7	2.6
Jet	2.5	9.0	5.3	5.3	2.5		8.0	3.9	0.9	2.9
Diesel	7.2	12.0	10.9	11.5	7.2		13.0	9.3	3.7	8.9
AGO	9.9	12.0	13.6	13.9	9.9		11.0	11.7	4.7	12.1
Vacuum distillation										
LGO	12.2	13.0	16.2	16.8	12.2		14.0	14.5	3.5	13.8
MGO	12.9	14.0	17.2	17.5	12.9	2.0	13.0	15.0	12.2	18.4
HGO	10.0	20.0	16.2	15.0	10.0	10.0	14.0	12.3	15.7	17.1
Vacuum bottoms	**45.3**					88.0	8.0	1.3	49.7	22.5
DMO					**36.0**			16.3		
DAB					**64.0**					
DC distillates		**67.9**					**57.0**			
Coke		**22.2**	10.1	10.1			**33.5**	9.7	1.8	0.8
Total	**100**	**100**	**100.0**	**100.0**	**100.0**	**100.0**	**100.0**	**100.0**	**100.0**	**100.0**

DC—Delayed Coking; DAB—Bottmos from Deasphalting; VR—Visbreaking; HDT—Hydrotreating.

grader is reduced with respect to the processing of reduced crude.

4.3.2. nC_4 Deasphalting of the Vacuum Bottoms and DC of the Bottoms from the Deasphalter (Table 5)

The DMO yield with respect to the vacuum bottoms is 36.0 wt.% and the yield of deasphalter bottoms is 64.0 wt.%. This low yield of DMO is an indicative of the high aromaticity of this type of vacuum bottoms, which is confirmed by the SARA analysis (**Table 6**), where the sum of aromatics + Resins + asphaltenes is 91.1 wt.%. This type of crude should not be processed in a low conversion scheme because the fuel oil production is quiet high. The deasphalter bottoms yield with respect to the crude oil is 29.0 wt.%, a very high value.

Within the DC of the DAB are produced 9.5 wt.% of gases, 57.0 wt.% of distillates and 33.5 wt.% of coke (**Table 5**, Column 8), and within the distillates 51.0 wt.% correspond to atmospheric distillates and 41.0 wt.% to the vacuum distillates.

In the synthetic crude produced in this scheme (column 9), is observed that this scheme produces very low quantity of residues, only 9.7 wt.%, but the DMO concentrates very high amount of metals (282 ppm, **Table 6**) and this is neither good for the FCC feedstock not for the Hydrocracking.

4.3.3. Visbreaking of the Vacuum Bottoms (Table 5)

The Visbreaking of the vacuum bottoms produces a semi-synthetic crude oil in which the concentration of the cut corresponding to the vacuum bottoms is significantly reduced. It changes from 45.3 wt.% in the original crude oil to 22.5 wt.% in the semi-synthetic crude oil. On the other hand, the gas oils are increased from 45.0 wt.% in the raw material to 61.3 wt.% in the semi-synthetic crude (columns 2 and 11). The yield of coke is normal for this type of process.

The Visbreaking bottoms in the **Table 6** correspond to liquid effluents from the Visbreaking process, excluding naphtha.

This scheme is an ideal arrangement for a low conver-

Table 6. Properties of the products from the vacuum bottoms upgrading.

	Vacuum Bottoms	DC of vacuum bottoms	DMO	DAB	Visbreaking bottoms	Vacuum bottoms from VBR
Density @ 15˚C, g/cm^3	1.0262	0.9227	1.0284	1.0314	1.0459	1.0745
API	6.3	21.8	6	5.6	3.7	0.1
CCR, wt.%	17.9	4	18.3	25.8	25.9	34.9
Sulphur, wt.%	1.73	1.64	2.0	2.1	2.1	2.1
Ca, ppm	2666.0		852.5	337	982	1128
Ni, ppm	130.7		124.7	111.7	153	212.7
V, ppm	167.3		157.8	158.6	200	242.8
Na, ppm	38.5		33.2	43.5	32	29.4
Ni + V, ppm	298		282	270	354	455
i-nC_7, wt.%	4.63			9.16	17.3	25.7
i-nC_5, wt%	11.27			21.43	20.9	32.3
Viscosity-1, cP	30800 a 100˚C		322000 @80˚C	4780 @120˚C		0
Viscosity-2, cP	4950 a 120˚C		30100 @100˚C	1880 @140˚C		225000
Básico N, wt.%	0.284	0.186	0.284	0.312	0.373	0.441
N Total, ppm	7451	6681				1.21
N.N, mg KOH/g	3.93				2.88	0.8
Pour point, ˚C					51	126
Flash point, ˚C					173	323
SARA analysis, wt.%						
S	8.9				15.1	5.3
A	55.2				45.3	45.3
R	21.4				22.7	30.0
A	14.5				12.4	19.4

DC—Delayed Coking; DMO—Demetallized oil; VBR—Visbreaking.

sion refinery.

5. Cost of Technologies for Evaluated Schemes

The calculations were performed with the values presented in **Table 7** and on the basis of 50.000 BOPD.

In the **Table 8** are given the values of the different technological alternatives for the processing 50 KBOPD of crude Jazmín. The lower cost alternative is the Visbreaking of the Vacuum, bottoms and Hydrotreating of the distillates obtained from the VBR, followed by the alternatives of Visbreaking of the crude and DC of the DAB.

6. Conclusions

1) Jazmin crude oil and its heavy components are subjected to thermal processes with excellent results due to the low concentration of coke promoters. Coke production in relation to vacuum bottoms is of 22.2 wt.%, which is a fairly low value for this type of feed. This is due to the low concentration of coke precursors. The insolubles in n-C_7 are completely soluble in cyclohexane.

2) The scheme corresponding to the visbreaking of the crude oil and the DC of the vacuum bottoms from the visbreaking is perhaps the most attractive for its application because it has low capital cost and solves from the beginning the problem of crude acidity.

3) The best scheme to use depends on the needs of each refinery and its installed facilities.

4) Of the studied schemes, the less expensive because of the used technology is the Visbreaking of the vacuum bottoms, and Hydrotreating of the liquid effluents to stabilize all future products.

Table 7. Technology costs.

	Value of the technology	Taken value
	US$/Bl	US$/Bl
Atmospheric Distillation		1000
Atmospheric and Vacuum Distillation	750 - 2200	2000
Deasphalting	1850 - 8000	4000
Delayed Coking	5800 - 12,000	8000
Hidrotreating	3000 - 3500	3000
Visbreaking	1800 - 3500	3000
Crude for the process, bl/day		50,000

Table 8. Investment cost of the proposed schemes.

	Economic impact of the studied upgrading alternatives of the Jazmin crude	MMUS$
1	Visbreaking (VR) of the vacuum bottoms and HDT of the distillates from VR	236
2	Visbreaking of the crude and Delayed Coking of the Visbreaked vacuum bottoms	275
3	Deasphalting of the crude 199˚C+ and DC of the bottoms obtained in the deasphalting.	319
4	DC of the Vacuum bottoms and HDT of DC distillates	327
5	n-C4 Deasphalting of the vacuum bottoms; DC of the DAB, and HDT of the distillates from DC	331
6	n-C4 Deasphalting of the reduced crude and DC of the Deasphalted bottoms	334
7	Delayed Coking of reduced crude	478
8	Deasphalting of crude 199˚C+ and DC of its DMO	608

5) Thermal upgrading schemes not only increase distillates yield and reject carbon, but also reduce the acidity of the products.

REFERENCES

[1] http://www.pacificrubiales.com/operations/colombia/production.html

[2] http://www.pacificrubiales.com/operations/colombia/production.html

[3] J. Plotkin, "Exploiting Unconventional Heavy Oils," PTQ-Q1, 2007, pp. 65-67.

[4] M. Aldescu, "High-Quality Products from Sour Heavy Crudes," PTQ Q3, 2012, pp. 115-120.

[5] J. A. Carrillo, L. G. Garzón, L. M. Corredor and H. J. Picon, "Visbreaking of Jazmin Crude Oil: Transport Alternatives," *Preprints of Papers-American Chemical Society, Division of Fuel Chemistry*, Vol. 52, No. 2, 2007, pp.

[6] M. G. Bienstock, *et al.*, Patent USA 6086751, 2000.

[7] M. Aldescu, "Challenges of Heavy Crude Processing," Sour/Heavy, 2012, pp. 127-132. www.ptq.com

[8] L. Castañeda, J. Muñoz and J. Ancheyta, "Combined Process Schemes for Upgrading of Heavy Petroleum," *Fuel*, Vol. 100, 2012, pp. 110-127.

[9] R. Martínez-Palou, *et al.*, "Transportation of Heavy and Extra-Heavy Crude Oil by Pipeline: A Review," *Journal of Petroleum Science and Engineering*, Vol. 75, No. 3-4, 2001, pp. 274-282.

[10] M. Rana, *et al.*, "A Review of Recent Advances on Process Technologies for Upgrading of Heavy Oils and Residua," *Fuel*, Vol. 86, No. 9, 2007, pp. 1216-1231.

[11] J. Le Page, S. Chatila and M. Davidson, "Residue and Heavy Oil Processing," Technip, Paris, 1992.

[12] M. Gray, "Upgrading Petroleum Residues and Heavy Oils," Marcel-Dekker, 1994.

[13] J. Speight, "Chapter 5: Thermal Cracking The Refinery of the Future, and Hydrotreating (HDT)," 2011, pp. 147-180.

[14] R. Radmanesh, E. Chan and M. Gray, "Modeling of Mass Transfer and Thermal Cracking during the Coking of Athabasca Residues," *Chemical Engineering Science*, Vol. 63, No. 6, pp. 1683-1691.

[15] J. Gearhart and L. Garwin, "Resid-Extraction Process Offers Flexibility," *Oil & Gas Journal*, Vol. 74, No. 24, 1976, pp. 63-66.

[16] R. Newcomer and R. Soltau, "Heavy Oil Extraction ups FCC Feed at First Three-Stage Grass Roots ROSE Unit, in Kansas," *Oil & Gas Journal*, Vol. 80, No. 26, 1982, pp. 108-110.

[17] J. S. Plotkin, "Exploiting Unconventional Heavy Oils," PTQ Q1, 2007, pp. 65-67.

[18] J. Elliott and M. McGrath, "Cost Effective Conversion of Heavy Residues," March, 2009, pp. 51-58.

[19] "Liquid-Liquid in Continuous Extraction Pilot Plant," Gaceta 448 del 9 del Julio, 1997, p. 216.

[20] Patent Colombian, "Visbreaking Pilot Plant," Gaceta 444 del 7 de Mayo, 1997, p. 200.

[21] J. A. Carrillo, J. Cáceres, G. Vela and H. Bueno, "Solventes de Desasfaltado," *CT&F*, Vol. 1, No. 2, 1996, pp. 67-76.

[22] J. A. Carrillo and F. Pantoja, "Evaluación de Fondos de vacío en la Viscorreducción," Ingeniería Química, Mayo. 1996, pp. 97-103.

[23] J. A. Carrillo, H. J. Picón, L. G. Garzón and L. M. Corredor, "Delayed Coking: Castilla and Jazmín Crude oils," *ACS National Meeting, Fuel Chemistry*, Boston, 23 August 2007.

21

Modeling a General Equation for Pool Boiling Heat Transfer

Mohammed Salah Hameed[1*], Abdul Rahman Khan[2], A. A. Mahdi[3]
[1]Chemical Engineering Department, Higher Colleges of Technology, Abu Dhabi, United Arab Emirates
[2]Department of Chemical Engineering, Kuwait University, Kuwait City, Kuwait
[3]Mechanical Engineering Department, University of Technology, Baghdad, Iraq

ABSTRACT

It is recognized that the nucleate pool boiling data available in literature are mainly related to four known correlations, each differs from the other by a varying magnitude of constant coefficients, depending on restrictive experimental conditions. The present work is concerned in developing an empirically generalized correlation, which covers the entire range of nucleate boiling with a minimum possible deviation from experimental data. The least squares multiple regression technique is used to evaluate the best coefficient value used in the correlations. An empirical correlation that fits a broader scope of available data has been developed by a non-linear solution technique leading to the following equation: $Nu_B = R_3\left(\frac{\Delta P}{\rho_L}\right)^{-0.65}(Pe_B)^{0.82}\left(\frac{\rho_V}{\rho_L}\right)^{1.1}(\mathrm{Pr})^{R_4}$ where the coefficients R_1 and R_3 both represent the effect of surface-liquid combination. They are assessed independently for the used surface material and liquid.

Keywords: Pool Boiling; Nucleate Boiling; Linear and Non-Linear Technique Methods; Heat Transfer

1. Introduction

Boiling is a complex process and an intensive work is needed for its understanding. Within the last decades, several nucleate boiling models were formulated and could be grouped in two main categories: a) Bubble Agitation Models and b) Macro/Micro Layer Evaporation Models.

The bubble agitation models are based on the principle of agitating the liquid, but they carry away little heat. The heat transfer is considered within the turbulent forced convection. The obtained empirical pool boiling heat transfer models employ dimensionless groups based on both fluid and solid properties while the main constant in the model is found to depend on the geometry of the heater. The models found in literature are useful within the range of database used in developing their derivation.

Bubble agitation mechanism together with Helmholtz-instability mechanism can be used either to explain the heat transfer at the low heat flux regime or to explain CHF (Critical Heat Flux). They cannot account for the continuity of the pool-boiling curve. On the other hand, the macro/micro layer evaporation reproduces the pool-boiling curve from the nuclear boiling to transition boiling. The macro/micro layer models play an important role in high heat flux region. The liquid layer includes the micro layer underneath the bubble and the macro layer on the base of coalescence and dries out periodically [1]. Heramura & Katto [1] assume the liquid-vapor interface is stationary and the entire surface heat flux contributes to macro layer evaporation.

Several numerical models were proposed based on the macro layer theory among that of Maruyama *et al*. [2]. Zhao *et al*. [3] put forward a model for transient pool boiling heat transfer. The model employed is too high heating rate to be realized in practical experiments for a horizontal surface. He *et al*. [4] concluded that the macro layer model is more suitable for the high heat flux regime. Dhir [5] confirmed that numerical simulations are not a substitute for detailed experiments. The experimental results are needed to validate the simulations. Numerical simulations provide additional insights into the boiling phenomena.

[*]Corresponding author.

Within the late decade, many researchers worked on viewing the pool boiling in microgravity (in the absent of buoyancy) to understand the lower limit of forced convection. Several workers are Lee [6], Herman [7], Wan & Zhao [8], and Kubota *et al.* [9].

Ji *et al.* [10] enhanced the pool boiling heat transfer in microgravity by using porous coating heating surface at atmospheric pressure and slightly moderate superheats.

Other researchers [11,12] enhanced the pool boiling by using nanofluid (water mixed with extremely small amount of nanosized particles). They concluded the enhancement of the thermal conductivity and convection heat transfer capability of the suspended particles of nanometer in size for many volume fractions of nanofluids.

The results of workers [6-12] can be used as a guidance in formulating proper equations that can be used in design. The aim of the present work is to use bubble agitation models to obtain a generalized empirical correlation that gives the best possible representation of collected data. Pool of data is collected from literature for various liquids effects with different plain test surfaces. For this purpose, linear and non-linear programming techniques were used in the evaluation of the proposed correlation.

2. Theoretical Analysis of Bubble Agitation Models

The primary requirement for nucleation to occur or for a nucleus to subsist in a liquid is that the liquid should be superheated. There are two types of nuclei. One type is formed in a pure liquid; it can be either a high energy molecular group resulting from thermal fluctuations of liquid molecules, or a cavity resulting from a local pressure reduction such as that occurs in accelerated flow. The other type, formed on a foreign object can be either a cavity on the heating wall or suspended foreign material with a non-wetted surface.

Rohsenow [13] assumes that the movement of bubbles at the instant of breaking away from the heating surface is of prime importance and obtained Equation (1) for heat transfer in the region of nucleation pool boiling.

$$\frac{C_L\left(T_s - T_{sat}\right)}{H_{fg}} = C_{sf}\left(\frac{q}{\mu_L H_{fg}}\sqrt{\frac{\sigma g_c}{g\left(\rho_L - \rho_V\right)}}\right)^{0.33} \Pr^r \quad (1)$$

The recommended variation of r is within 0.8 to 2.0. Evaluation of C_{sf} from experimental results of many workers [14] prove to be a parameter which does not pick out only the nucleation ability of heating surface but contains the effect of physical properties of liquid.

Rohsenow [15] proposed the surface factor C_{sf} to prescribe the condition of heating surface in nucleate boiling. Various investigators [14] utilized this factor in their determination of empirical expressions. The surface factor is defined by Equation (1), generally known as Rohsenow empirical correlation.

Forster and Zuber [16] indicated that small bubbles grow rapidly and large ones slowly, but the degree of agitation in the surrounding liquid due to bubble growth remains the same. They derived the following empirical correlation:

$$\frac{q C_L \rho_L \sqrt{\pi \alpha_L}}{K_L H_{fg} \rho_V}\left(\frac{2\sigma}{\Delta P}\right)^{1/2}\left(\frac{\rho_L}{\Delta P}\right)^{1/4} = 0.0015\,\mathrm{Re}_b^{0.62}\,\Pr_L^{1/3} \quad (2)$$

Equation (2) predicts the same heat transfer coefficient for a liquid boiling on any hot surface (for all heterogeneous cases only) or boiling in bulk (for all homogeneous cases only). Rohsenow's Equation (1) was developed and applied to the heterogeneous case only.

Forster and Greif [17] suggested a different approach by considering that the mechanism of high heat transfer rate, during nucleate boiling, is mainly due to the liquid-vapor exchange. They obtained a dimensional empirical correlation, for the pool boiling heat flux q in water at 100 - 4763 kN/m^2, as shown in Equation (3).

$$q = (4.3)\left(10^{-5}\right)\frac{K_L T_{sat} \Delta P^2}{\sigma^{1/2}\left(H_{fg}\rho_V\right)^{3/2}}\left[\left(C_L T_{sat}\alpha^{1/2}\right)^{1/4}\left(\frac{\rho_L}{\rho_V}\right)^{5/8}\left(\frac{\mu_L C_L}{K_L}\right)^{1/3}\right] \quad (3)$$

This correlation is not as widely verified as that of Rohsenow.

Gupta and Varshney [18] obtained experimental data for boiling heat transfer, using distilled water, benzene and toluene as liquids over a heated horizontal cylinder made of stainless steel. Their data was correlated by the following dimensionless empirical correlation:

$$Nu_B = 1.39\left(Pe_B\right)^{0.7}\left(\frac{\rho_V}{\rho_L}\right)^{0.21}(\Pr)^{-0.21} \quad (4)$$

where Nu_B and Pe_B are the Nusselt and Peclet number of boiling respectively. Or it can be written as:

$$\frac{h}{K_L}\sqrt{\frac{\sigma g_c}{\left(\rho_L - \rho_V\right)g}} = 1.39\left(\frac{q \rho_L C_L}{\rho_V H_{fg} K_L}\sqrt{\frac{\sigma g_c}{\left(\rho_L - \rho_V\right)g}}\right)^{0.7}\left(\frac{\rho_V}{\rho_L}\right)^{0.21}\left(\frac{\mu_L C_L}{K_L}\right)^{-0.21} \quad (5)$$

In order to derive a general correlation based on bubble agitation phenomena to be more versatile than the correlations existing in literature, a search was made through published work in literature and found that the following four well known empirical correlations referred to in most publications:

a) Rohsenow correlation (Equation(1))
b) Forster and Zuber correlation (Equation (2))
c) Forster and Greif correlation (Equation (3))
d) Gupta and Varshney correlation (Equation (5)).

For the sake of analysis, experimental data collected from many literatures [19-24] tabulated as heat flux (q), surface temperature (T_s), heat transfer coefficient (h), and coefficient h^* (equal to $h/q^{0.7}$). Moreover, physical and thermodynamic properties collected at the reported experimental conditions from literature [25-29] to be used in the analysis. The properties include, liquid thermal conductivity (K_L), liquid heat capacity (C_L), density of vapor (ρ_V), surface tension of liquid (σ), saturation temperature (T_{sat}), density of liquid (ρ_L), latent heat of vaporization (H_{fg}), and viscosity of liquid (μ_L).

The above stated correlations are of dimensionless form with the exception of the Forster and Greif correlation (Equation (4)). These equations can be represented by general equation as shown in **Appendix**. Many modifications to linear correlations have been tried to minimize the sum of squares of errors and to conclude some general correlations.

3. Results and Discussion

Boiling heat transfer studied earlier indicated that several variables are important in nucleate boiling such as pressure, fluid properties, surface condition, boiling temperature, kind and relative amount of impurities. The practical data showed that changes in magnitude of these properties and conditions could significantly affect pool boiling heat transfer.

A graphical analyses for 56 sets of literature data was used in studying the effect of heat flux, (q), and operating pressure (P), on boiling heat transfer coefficient, (h). **Figures 1-3** show the variation of heat transfer coefficient with heat transfer flux, (q). The lines in the figures are the best-fit lines of the reported data. **Figure 1** plotted for various liquids at different operating pressure and test surface. **Figure 2** corresponds to various liquid-surface combinations at constant atmospheric pressure. **Figure 3** reflects the behavior of various metal surfaces and operating pressures for the same liquid.

All the data can be represented by the empirical, Equation (6), with an average percentage error ranging from 0.012 to 11.8

$$h = h^* q^{0.7} \tag{6}$$

The proportionality constant h^* is proved to be a function of pressure and liquid-surface combination. Cichelli

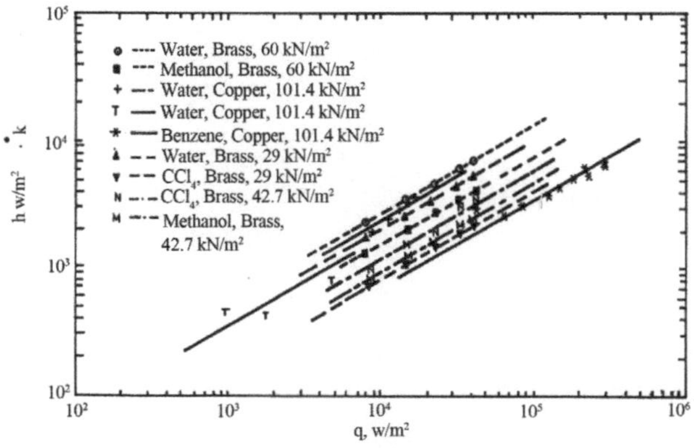

Figure 1. Heat transfer coefficient (h) versus heat flux (q) at various pressures.

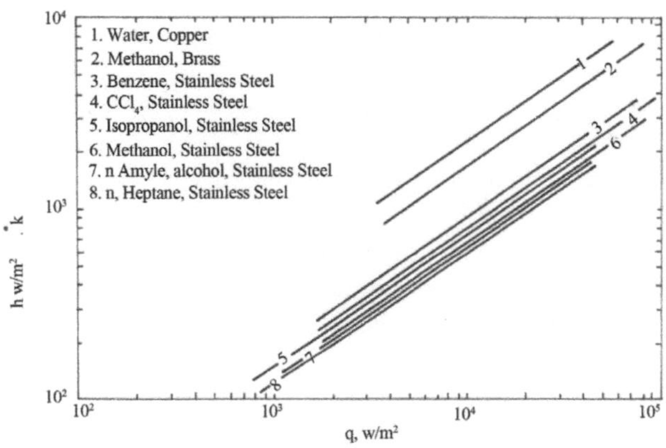

Figure 2. Heat transfer coefficient (h) versus heat flux (q) at atmospheric pressure.

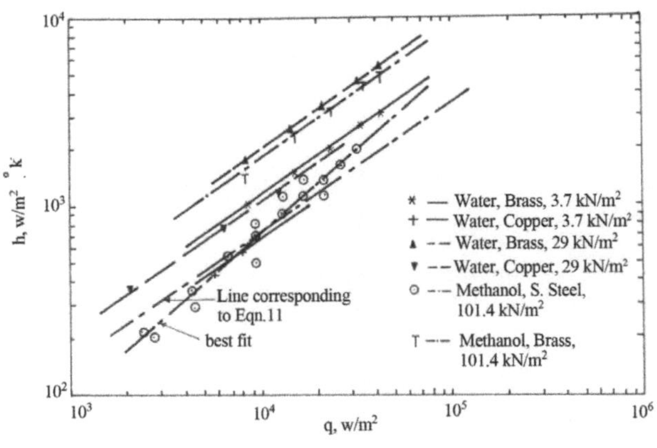

Figure 3. Heat transfer coefficient (h) versus heat flux (q) for various metal surfaces.

and Bonilla [19] confirmed that the coefficient of heat transfer increases with absolute operating pressure in the nucleate boiling zone. They reported the following correlation:

$$h = 5400(P)^{0.44} \tag{7}$$

Figure 4 shows the variation of the proportionality constant (h^*) as function of pressure (P) for a definite liquid-surface combination. Equation (8) represents the relationship between h^* and P, that was obtained from best data fit of **Figure 4**.

$$h^* \alpha P^{0.25} \qquad (8)$$

The overall dependence of (h) on operating pressure (P) and heat flux (q) for different liquid-surface combinations is shown in **Figure 5**.

3.1. Linear Programming Analysis of Empirical Correlations

Equation (A.4) in **Appendix** used to test the validity of the published correlations. It is used to formulate a correlation that shows the best fit of the experimental data. All the data cited in the literature from [19-24] classified as eight liquids (Water, Benzene, Methanol, Carbon Tetrachloride, n-Butanol, Isopropanol, n-Amyl Alcohol, and n-Heptane) and four surfaces (Brass, Copper, Nichrome, and Stainless Steel) at different operating pressures grouped in 56 data sets.

The applicability of the four empirical correlations, Equations (1)-(3) and (5), in representing the data was test by using linear-programming; that by fixing some of

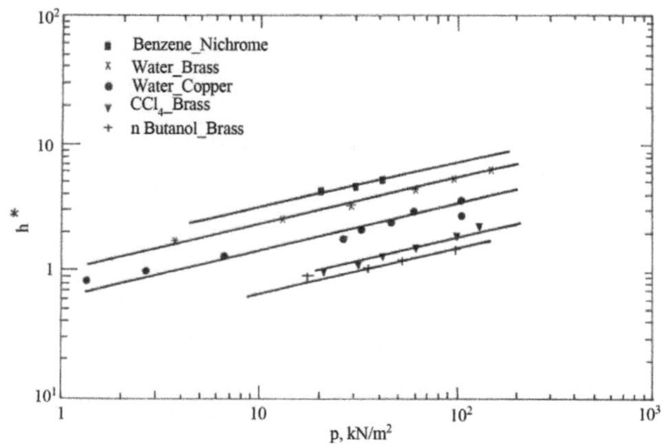

Figure 4. Proportionality constant (h^*) versus operating pressure (P).

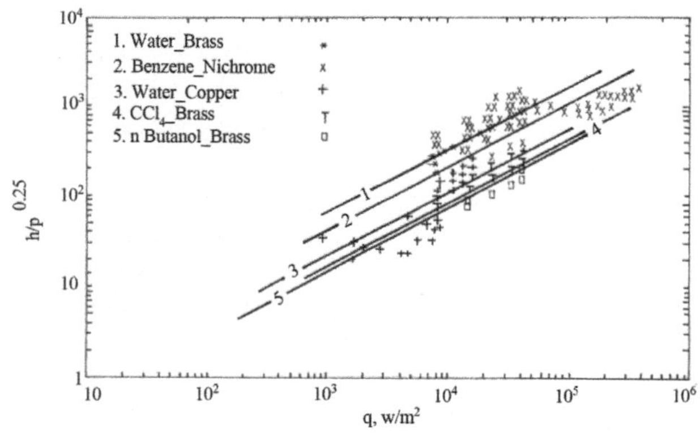

Figure 5. Heat transfer coefficient (h) dependence on operating pressure and heat flux (q) .

the coefficients and evaluating the others and the average percentage error is used as test criteria for comparison purposes.

3.1.1. Rohsenow's Correlation

Equation (9) is a general expression for Rohsenow's correlation while the exact expression stated as in Equation (1).

$$\frac{C_L\left(T_s - T_{sat}\right)}{H_{fg}} = C_{sf}\left(\frac{q}{\mu_L H_{fg}}\sqrt{\frac{\sigma}{\rho_L - \rho_V}}\right)^{R_1}\left(\frac{C_L \mu_L}{K_L}\right)^{R_4} \qquad (9)$$

The coefficient C_{sf} reported, in the literature, to vary with each liquid-surface combination and it is independent of pressure [22]. The validity of Equation (9) was tested for the entire collected data by applying the least-squares method.

In the initial analysis of data, the pressure was assumed constant and the obtained results showed inconsistency in the calculated values of constants for various experimental conditions tested. The inconsistency in values of constants is most likely due to pressure effect, which was not considered as variable during the initial analysis of data. On the next try, a pressure parameter was introduced in an attempt to narrow the variation in the values of constants for different systems and to conclude general correlation. Four different expressions of pressure parameter cited from literature [16], and stated in Equation (10), was used and expected to have an effect on heat transfer in the region of pool boiling nucleation:

$$R_3\left(\frac{\Delta P}{\rho_L}\right)^{R_2}, R_3\left(\frac{\Delta P}{\rho_V}\right)^{R_2}, R_3\left(\frac{\Delta P}{\sigma}\right)^{R_2}, \text{ and } R_3\left(\frac{\Delta P}{\rho_L}\right)^{R_2}\left(\frac{\Delta P}{\sigma}\right)^{R_5} \qquad (10)$$

Each of these pressure expressions selected to replace the coefficient C_{sf} in Equation (9) and the obtained results were compared and checked. The analysis found that the pressure parameter $R_3\left(\frac{\Delta P}{\rho_L}\right)^{R_2}$ gives the lowest minimum percentage error between the others forms and selected to replace C_{sf} and modify Equation (9) to the form showing in Equation (11).

$$\frac{C_L\left(T_s - T_{sat}\right)}{H_{fg}} = R_3\left(\frac{\Delta P}{\rho_L}\right)^{R_2}\left(\frac{q}{\mu_L H_{fg}}\sqrt{\frac{\sigma}{\rho_L - \rho_V}}\right)^{R_1}\left(\frac{C_L \cdot \mu_L}{K_L}\right)^{R_4} \qquad (11)$$

To make the present work more general, the analysis was repeated for various liquids at certain test surface and different operating pressures, that by calculating all the coefficients (R_1, R_2, R_3, and R_4) of Equation (11) for each set of data. The values of coefficients were selected, by the help of Equation (A.4), on the basis of using Equation (11) with the lowest average percentage error.

3.1.2. Gupta and Varshney Correlation

Equation (12) is a general expression for Gupta and Varshney empirical correlation, while Equation (5) represent the exact form.

$$\frac{h}{K_L}\sqrt{\frac{\sigma g_c}{(\rho_L-\rho_V)g}}=R_3\left(\frac{q\rho_L C_L}{\rho_V H_{fg} K_L}\sqrt{\frac{\sigma g_c}{(\rho_L-\rho_V)g}}\right)^{R_1}\left(\frac{\rho_V}{\rho_L}\right)^{R_6}\left(\frac{\mu_L C_L}{K_L}\right)^{R_4} \tag{12}$$

The pressure term is not included in the correlation of Gupta and Varshney as was the case of Rohsenow's empirical correlation, Equation (1). Equation (12) differs from Rohensow's Equation (9) by including the density ratio term and classifying other dimensionless term in well-known groups. A similar data analysis used for Equation (12) as it was with the case of Rohsenow's Correlation. By substituting the various pressure forms of Equation (10) in place of coefficient R_3 in Equation (12), it is found that pressure expression $R_3\left(\Delta P/\rho_L\right)^{R_2}$ gave the minimum average percentage error. This term is then chosen as the best fit expression for pressure and used in Equation (12) to obtain Equation (13).

$$\frac{h}{K_L}\sqrt{\frac{\sigma g_c}{(\rho_L-\rho_V)g}}=R_3\left(\frac{\Delta P}{\rho_L}\right)^{R_2}\left(\frac{q\rho_L C_L}{\rho_V H_{fg} K_L}\sqrt{\frac{\sigma g_c}{(\rho_L-\rho_V)g}}\right)^{R_1}\left(\frac{\rho_V}{\rho_L}\right)^{R_6}\left(\frac{\mu_L C_L}{K_L}\right)^{R_4} \tag{13}$$

In order to make the present work more general, the analysis repeated for various liquids at specific test surface and different operating pressures then follow the same procedure as in case of Rohsenow's correlation. The analysis concluded that the modified Equation (13) provides a better data fit than that of Gupta and Varshney Equation (5).

3.1.3. Forster and Zuber Correlation

Equation (14) is a general expression for Forster and Zuber empirical correlation, while the exact form is given in Equation (2).

$$\frac{qC_L\rho_L}{K_L H_{fg}\rho_V}\sqrt{\frac{\pi K_L}{\rho_L C_L}}=R_3\left(\frac{2\sigma}{\Delta P}\right)^{R_5}\left(\frac{\rho_L}{\Delta P}\right)^{R_2}\left[\frac{\rho_L}{\mu_L}\left(\frac{C_L\rho_L(T_s-T_{sat})}{H_{fg}\rho_V}\sqrt{\frac{\pi K_L}{\rho_L C_L}}\right)^2\right]^{R_1}\left(\frac{\mu_L C_L}{K_L}\right)^{R_4} \tag{14}$$

The pressure term is taken care of in Forster and Zubers correlation, Equation (2). By applying a similar procedure as in previous cases, it was found that the validity of the above equation is restricted to specific experimental data near critical temperature difference. By checking the above results for Forster and Zuber empirical correlation, Equation (14), it is found that the overall average percentage errors is very high in predicting the published data under consideration.

3.1.4. Forster and Greif Empirical Correlation

Equation (15) is a general expression for Forster and Greif dimensional empirical correlation, while the exact form is given in Equation (3).

$$h=R_1(T_s-T_{sat})^{R_2}(\sigma)^{R_3}(\rho_L)^{R_4}(\Delta P)^{R_5}(K_L)^{R_6}(H_{fg})^{R_7}(\rho_V)^{R_8}(C_L)^{R_9}(\mu_L)^{R_{10}} \tag{15}$$

Using the same sets of data analyzed previously gave a higher average percentage error as compared with the dimensionless empirical correlations as shown in **Table 1**.

The variation of some thermodynamic properties as function of pressure and temperature is not reported in literature and these properties were considered constant during the calculations, which may be the cause of the large average percentage error. Moreover, this empirical correlation dealt with fluid side effect of the problem and ignored the surface side effect on the nucleate boiling behavior. This might have added more error to the validity of this correlation.

The linear programming analysis recommended the use of the modified correlations of Rohsenow and Gupta & Varshney as they give closer prediction to the experimental data than in case of using Forster & Zuber and Forster & Greif modified correlations as showing in **Table 1**. Hence, the last two correlations excluded from any further analysis.

Table 1. Linear programming results for Equation (11) to Equation (15).

No.	No. of Sets	Material of Test Surface	Equation (11)		Equation (13)		Equation (14)		Equation (15)	
			APE	CC	APE	CC	APE	CC	APE	CC
1	6	Nichrome	8.9	0.995	2.1	0.935	18.5	0.968	35.0	0.845
2	15	Copper	14.1	0.965	22.4	0.995	82.3	0.903	134.8	0.882
3	22	Brass	21.1	0.807	25.1	0.933	80.7	0.740	65.3	−0.04
4	13	Stainless Steel	15.6	0.858	16.2	0.553	78.4	0.586	74.0	0.228

Where APE = Average Percentage Errors; CC = Correlation Coefficient.

3.2. Non-Linear Programming Analysis of Empirical Correlations

The data were re-analyzed by non-linear programming methods by using the modified correlations of Rohsenow, Equation (11), and Gupta & Varshney Equation (13) in an attempt to improve the correlations for lower average percentage error.

By assume that: $\left[\frac{q}{\mu_L H_{fg}}\sqrt{\frac{\sigma}{(\rho_L-\rho_V)}}\right]$ in Equation (4) is equal to X and $\left[\frac{q\rho_L C_L}{\rho_V H_{fg} K_L}\sqrt{\frac{\sigma g_c}{(\rho_L-\rho_V)g}}\right]$ in Equation (13) is equal to Y, then use either of the following equations:

Binominal Expression:

$$\left[R_3 (X \text{ or } Y)^{R_1} = \left(R_7 + R_8 (X \text{ or } Y)\right)^{R_9}\right] \quad (16)$$

or non-linear Expression:

$$\left[R_3 (X \text{ or } Y)^{R_1} = R_7 + R_8 (X \text{ or } Y)^{R_9}\right] \quad (17)$$

to replace the $R_3 X^{R_1}$ in Equation (11) or $R_3 Y^{R_1}$ in Equation (13) respectively. It is found from data fittings that a better representation can be obtained by using the binominal expression, Equation (16), with $\left(R_7 + R_8 (X)\right)^{R_9}$ in place of $R_3 X^{R_1}$ in Equation (11) and using the binominal expression, Equation (16), with $\left(R_7 + R_8 (Y)\right)^{R_9}$ in place of $R_3 Y^{R_1}$ in Equation (13).

Various forms of expressions for the constant R_7 was tried for both Rohsenow and Gupta & Varshney modified correlations. It was found that keeping R_7 as a constant and independent of other parameters yielded a better data representation. It was also concluded that the average percentage error would not be improved by using the non linear equations instead of the linear equation as showing in **Table 2**.

General Empirical Correlation

By using the best data fit for Equation (11) for different surfaces (Nichrome, Copper, Brass, and Stainless Steel), the variation of the powers of pressure expression term and Prandtl number in the equation were found to be approximately equal to 0.08 and 1.0 respectively leading to the following generalized Equation (18).

$$\frac{C_L (T_s - T_{sat})}{H_{fg}} = R_3 \left(\frac{\Delta P}{\rho_L}\right)^{0.08} \left[\frac{q}{\mu_L H_{fg}}\sqrt{\frac{\sigma g_c}{(\rho_L-\rho_V)g}}\right]^{R_1} \left(\frac{\mu_L C_L}{K_L}\right)^{1.0} \quad (18)$$

The coefficients R_1 and R_3 represent the effect of surface-liquid combination. They are assessed independently for each surface by the least-squares linear regression method and the results are stated in **Table 3**.

A similar analysis tried for Equation (13) and concluded that the powers of pressure term, Peclet number (Pe_B), and density ratio term (ρ_V/ρ_L) were relatively independent of surface—liquid combination as compared with the coefficient R_3 and the power of Prandtl number, R_4. The best form of Equation (13) was tested for different data sets and concluded Equation (19).

$$\frac{h}{K_L}\sqrt{\frac{\sigma g_c}{(\rho_L-\rho_V)}} = R_3 \left(\frac{\Delta P}{\rho_L}\right)^{-0.65} \left[\frac{q\rho_L C_L}{\rho_V H_{fg} K_L}\sqrt{\frac{\sigma g_c}{(\rho_L-\rho_V)g}}\right]^{0.82} \left(\frac{\rho_V}{\rho_L}\right)^{1.1} \left(\frac{\mu_L C_L}{K_L}\right)^{R_4} \quad (19)$$

A similar way was followed for Equation (19), to that of Equation (18), in finding the power R_4 and the coefficient R_3 and their best values are given in **Table 3**. The result of analysis of Equation (18) and Equation (19) listed in **Table 3** suggested the use of Equation (19) in preference to Equation (18).

The applicability of Equaiton (19) was examined for different surfaces as showing in **Figures 6-9**. The equation found to fit well for all the data with the exception of Brass. The deviation in the results for Brass is due to

Table 2. Comparison of linear, binominal, and non linear expressions versions of Equations (11) and (13).

No. of Sets	Material of Test Surface	Equation (11)						Equation (13)					
		Linear		Binomial		Non Linear		Linear		Binomial		Non Linear	
		APE	CC	APE	CC	APE	CC	APE	CC	APE	CC	APE	CC
6	Nichrome	8.9	0.995	13.2	0.994	15.0	0.996	2.1	0.935	2.6	0.925	2.5	0.995
15	Copper	14.1	0.965	29.3	0.955	17.6	0.980	22.4	0.995	31.7	0.999	32.9	0.991
22	Brass	21.1	0.807	59.5	0.875	98.0	0.576	25.1	0.933	45.9	0.968	45.7	0.967
13	Stainless Steel	15.6	0.858	17.0	0.665	25.9	0.796	16.2	0.553	18.7	0.989	20.2	0.987

Table 3. Linear programming results for Equations (18) and (19).

No.	No. of Sets	Test Surface Material	Equation (18) R_3	R_1	Average Percentage Error	Correlation Coefficient	Equation (19) R_3	R_4	Average Percentage Error	Correlation Coefficient
1	6	Nichrome	0.0019	0.684	15.5	0.976	19145.12	−1.300	6.4	0.954
2	15	Copper	0.0080	0.256	25.4	0.891	5438.43	−0.758	15.3	0.978
3	22	Brass	0.0045	0.228	21.1	0.907	4383.20	−0.132	36.0	0.771
4	13	Stainless Steel	0.0093	0.300	27.6	0.705	4944.28	−0.470	27.3	0.893

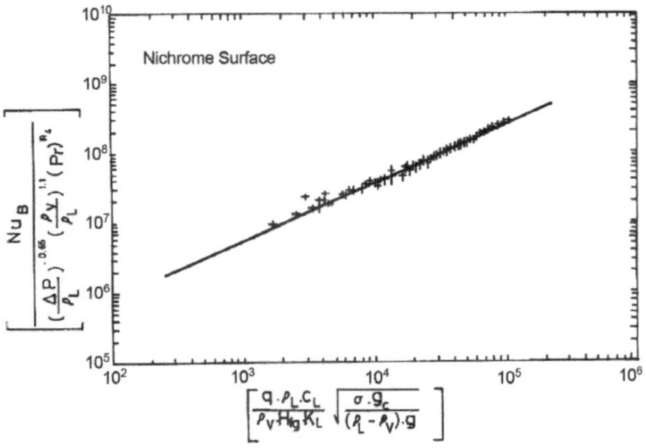

Figure 6. Experimental data predictions using Equation (19) for Nichrome surface.

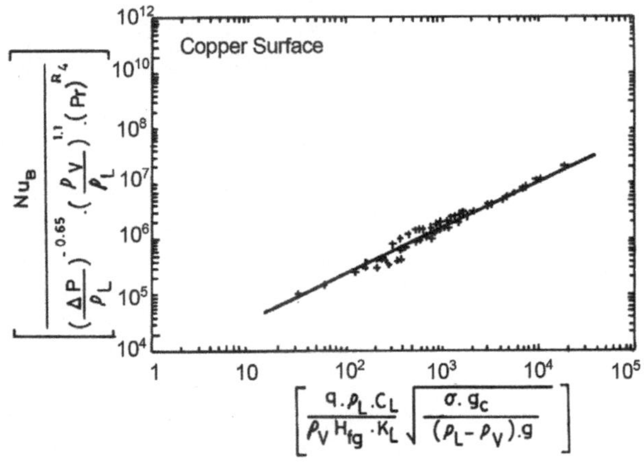

Figure 7. Experimental data predictions using Equation (19) for Copper surface.

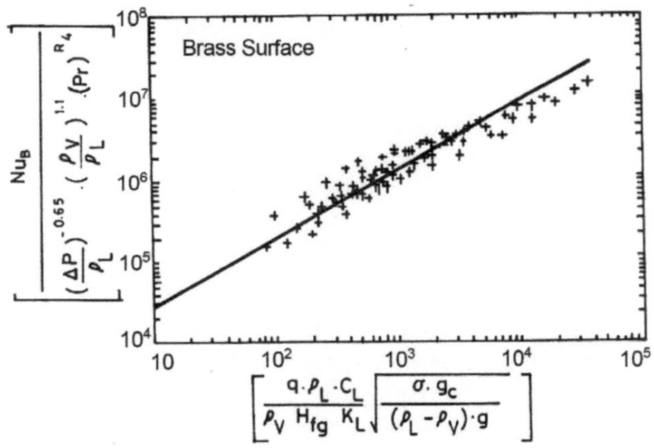

Figure 8. Experimental data predictions using Equation (19) for Brass surface.

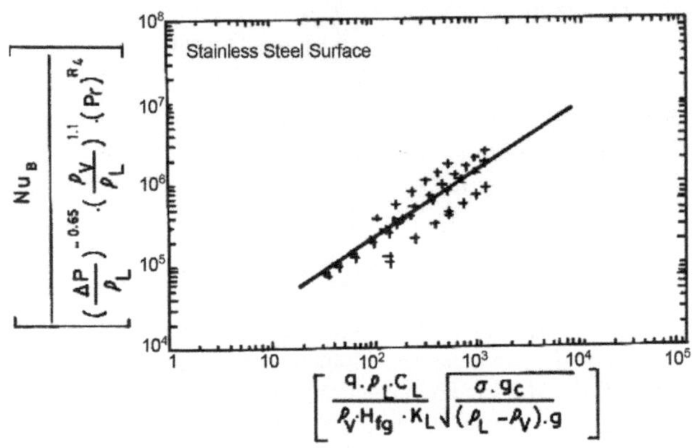

Figure 9. Experimental data prediction using Equation (19) for Stainless Steel surface.

limited available data at very low pressure. Equation (20) represents the dimensionless form of Equation (19).

$$Nu_B = R_3\left(\frac{\Delta P}{\rho_L}\right)^{-0.65}\left(Pe_B\right)^{0.82}\left(\frac{\rho_V}{\rho_L}\right)^{1.1}\left(\mathrm{Pr}\right)^{R_4} \quad (20)$$

The analysis concluded that Equation (20) is valid for the entire available data and represent a more generalized correlation than the correlations found in literature.

4. Conclusions

A graphical analysis concluded that the empirical Equation (6) is showing the effect of heat flux (q) and operating pressure (P) on the boiling heat transfer coefficient (h).

$$h = h^* q^{0.7} \quad (6)$$

where h^* is a function of pressure and for different liquid-surface combinations, it is found to vary with the pressure as follows:

$$h^* \alpha P^{0.25} \quad (8)$$

56 sets of literature data were tested on each of the four known correlations, Rohsenow, Forster & Zuber, Forster & Greif, and Gupta & Varshney, by using the linear and non-linear programming solution. The concluded results show that any of these correlations does not fit the entire data satisfactory. To improve their predictions, the correlations were modified, including additional parameters in an attempt to close up the deviation in the values of calculated parameters. The modified correlations of Rohsenow and Gupta & Varshney responded better to the applied modification than that of Foster & Zuber and Foster & Grief and they were considered for further analysis.

The least squares multiple regression technique [30,31] is used to evaluate the best possible values of the constant coefficients in the correlation. The cumulative error squares were minimized by using an ordinary optimum seeking technique. Linear, binominal & non-linear correlations were tested in concluding the final correlation.

The use of non-linear solution technique did not improve correlations 11 and 13 that were concluded by the linear technique and hence Equation (20) gives the best representation of the entire tested data.

REFERENCES

[1] Y. Haramura and Y. Katto, "A New Hydrodynamic Model of Critical Heat Flux, Applicable Widely to Both Pool and Forced Convection Boiling on Submerged Bodies in Saturated Liquids," *International Journa l of Heat and Mass Transfer*, Vol. 26, No. 2, 1983, pp. 389-399.

[2] S. Maruyama, M. Shoji and S. Shimizu, "A Numerical Simulation of Transition Boiling, Heat Transfer," *Proceedings of the 2nd JSM E-KSME Thermal Engineering Conference*, Kitakyushu, 19-21 October 1992, pp. 345-348.

[3] Y. H. Zhao, T. Masuoka and T. Tsuruta, "Theoretical Studies on Transient Pool Boiling Based on Microlayer Model (Mechanism of Transition from Nonboiling Regime to Film Boiling)," *Trans. JSME* (*B*), Vol. 63, No. 607, 1997, pp. 218-223.

[4] Y. He' M. Shoji and S. Maruyama; "Numerical Study of High Heat Flux Pool Boiling Heat Transfer," *International Journal o f Heat and Mass Transfer* , Vol. 44, No. 12, 2001, pp. 2357-2373.

[5] V. K. Dhir; "Numerical Simulation of Pool-Boiling Heat Transfer," *AICHEJ*, Vol. 47, No. 4, 2001, p. 813.

[6] H. S. Lee, "Mechanisms of Steady-State Nucleate Pool Boiling in Microgravity," *Annals of th e New York Academy of Sciences*, Vol. 974, 2002, pp. 447-462.

[7] H. Herman, "Some Parameter Boundaries Governing Microgravity Pool Boling Modes," *Annals o f the New York Academy of Science*, Vol. 644, 2006, pp. 629-649.

[8] S. X. Wan and J. Zhao, "Pool Boiling in Microgravity: Recent Results and Perspectives for the Project DEPA-SJ10," *Microgravity Scien ce a nd Technolog y*, Vol. 20, No. 3-4, 2008, pp. 219-224.

[9] C. Kubota, *et al* ., "Experiment on nucleate Pool Boiling in Microgravity by using Transparent Heating Surface-Analysis of Surface Heat Transfer Coefficients," *Journal of Physics*: *Conference Series*, Vol. 327, 2011, Article ID: 012040.

[10] X. Ji, J. Xu, Z. Zhao and W. Yang, "Pool Boiling Heat Transfer on Uniform and Non-uniform Porous Coating Surfaces," *Experimental Thermal and Fluid Science*, Vol. 48, 2013, pp. 198-212.

[11] S. H. You, J. H. Kim and K. H. Kim, "Effect of Nano Particle on Critical Heat Flux of Water in Pool Boiling Heat Transfer," *Applied Physics Letters*, Vol. 83, No. 16, 2003, pp. 3374-3376.

[12] S.K. Das, N. Putra and W. Roetzel, "Pool Boiling Characteristics of Nano-Fluid," *International Journal of Hea t and Mass Transfer*, Vol. 46, No. 5, 2003, pp. 851-862.

[13] W. M. Rohsenow and J. A. Clark, "Heat Transfer and Fluid Mechanics," Inst. Stanford University Press, 1951, p. 193.

[14] K. Nishikawa and Y. Fujita, *Memoirs of th e Faculty of Engineering, Kyushu University*, Vol. 36, No. 2, 1976, p. 136.

[15] W. M. Rohsenow, *Trans. ASME*, Vol. 74, 1952, p. 966.

[16] H. K. Forster and N. Zuber, "Dynamics of Vapor Bubbles and Boiling Heat Transfer," *AIChE Journal*, Vol. 1, No. 4, 1955, p. 531.

[17] H. K. Forster and R. A. Grief, *Journal of Heat Transfer*,

Vol. 81, No. 1, 1959.

[18] S. C. Gupta and B. S. Varshney, *International Journal of Heat and Mass Transfer*, Vol. 23, 1980, p. 219.

[19] M. T. Cichelli and C. F. Bonilla, *Trans. AICHE*, Vol. 41, 1945, p. 755.

[20] D. S. Cryder and A. C. Finalborgo, *Trans. AICHE*, Vol. 33, 1937, p. 346.

[21] C. V. Sternling and L. J. Tichacek, "Heat Transfer Coefficients for Boiling Mixtures: Experimental Data for Binary Mixtures of Large Relative Volatility," *Chemical Engineering Science*, Vol. 16, No. 3-4, 1961, pp. 297-337.

[22] I. A. Rabin, R. T. Beaubuef and G. E. Commerford, *Chemical Eng ineering Progress Symposium S eries*, Vol. 61, No. 57, 1965, p. 249.

[23] G. S. Yadav and O. P. Chawla, *Mechanical Engineering Division*, Vol. 60, Part ME. 6, 1980, p. 229.

[24] A. A. Mahdi, "Pool Boiling of Saturated Liquids," M.Sc. Thesis, University of Technology, Baghdad, 1983.

[25] E. A. Farber and E. L. Scorah, *Trans. ASME*, Vol. 70, 1948, p. 369.

[26] ESDU Engineering Sciences Data Unit, Chemical Engineering Series, Physical Data, Vol. 2, 1978.

[27] ESDU Engineering Science Data Unit, Chemical Data, Vol. 4, 1978.

[28] E. W. Washburn, "International Critical tables of Numerical Data, Physics, Chemistry and Technology," McGraw-Hill Book Company Inc., New York, 1926.

[29] R. H. Perry and C. H. Chilton, "Chemical Engineers Handbook," 4th Edition, McGraw-Hill Kogakusha, Ltd., 1963.

[30] C. Brice, H. A. Luther and O. W. James, "Applied Numerical Methods," John Wiley and Sons, Inc., Hoboken, 1969, p. 272.

[31] P. R. Adby and M. A. H. Demoster, "Introduction to Optimisation Methods," Chapman and Hall, 1974.

Appendix

Sum of Squares of Errors

Equations (1) to (4) can be represented by a general equation:

$$y_i = \left(x_{i1}^{a1}\right)\left(x_{i2}^{a2}\right)\cdots\left(x_{ik}^{ak}\right)$$

where $i = 1,2,3,\cdots,N$

$$y_i = \prod_{j=1}^{k} x_{ij}^{aj} \quad \text{(A.1)}$$

This equation represents a general form of those correlations and simplified by taking logarithms of both sides.

$$\ln y_i = a_1 \ln x_{i1} + a_2 \ln x_{i2} + a_3 \ln x_{i3} + \cdots + a_k \ln x_{ik}$$

or

$$\ln y_i = \sum_{j=1}^{k} a_j \ln\left(x_{ij}\right) \quad \text{(A.2)}$$

For N number of data readings there will be N number of linear equations, while for the determination of k coefficients only k equations are required. A least squares multiple regression technique [30,31] was used to evaluate the best possible coefficient from raw data readings. The cumulative error squares minimized by an ordinary optimum seeking technique [31] resulting into k number of equations to provide k number of coefficients for the entire data. These equations mathematically represented in the form:

$$E_i = \ln y_i - \sum_{j=1}^{k} a_j \ln\left(x_{ij}\right) \quad \text{(A.3)}$$

The sum of squares of errors is expressed as:

$$E_s = \sum_{i=1}^{N}\left(\ln y_i - \sum_{j=1}^{k} a_j\left(x_{ij}\right)\right)^2 \quad \text{(A.4)}$$

Nomenclature

C_L—Heat capacity of liquid, J/kg·°C
C_{sf}—Surface factor
F_μ—Nucleation factor
g—Acceleration of gravity, m/s^2
g_c—Conversion ratio, kg m/kg·s^2
h—Heat transfer coefficient W/m^2·°C
h^*—Proportionality constant
H_{fg}—Latent heat of vaporation, J/kg
K_L—Thermal conductivity, W/m °C
Nu_B—Nusselt number for boiling = $\left(h/K_L\right)\left(\sigma g_c/\left(\rho_L - \rho_V\right)g\right)^{0.5}$
P—Operating pressure (kN/m^2)
Pe_B—Peolet number for boiling = $\left(q/\rho_V H_{fg}\alpha_L\right)\left(\sigma g_c/\left(\rho_L - \rho_V\right)g\right)^{0.5}$
Pr—Prandtl number
q—Heat Flux, W/m^{2}s
$R_1, R_2, R_3, \cdots, R_9$ —coefficients
Re—Reynolds number
Re$_b$—Reynolds number for bubbles =

$$\frac{\rho_L}{\mu_L}\left(\frac{C_L\rho_L\Delta T\sqrt{\pi\alpha_L}}{H_{fg}\rho_V}\right)^2$$

Sr—Superheat ratio = $\dfrac{C_L\left(T_s - T_{sat}\right)}{H_{fg}}$

T_s—Test surface temperature, °C
T_{sat}—Saturation temperature, °C

Greek Letters

α_L Thermal diffusivity, m^2/s, $K_L/\left(r_L C_L\right)$
ΔP Pressure difference corresponds to (T_s–T_{sat}), kN/m^2
μ_L Viscosity of liquid, kg/m·s
ρ_V Density of vapor, kg/m^3
ρ_L Density of liquid, kg/m^3
σ Surface tension, kg/s^2

Subscripts

b Refers to bubble property
B Refers to boiling condition
L Refers to liquid condition
sf Refers to surface factor
s Refers to surface condition
sat Refers to saturation condition
v Refers to vapor condition

Permissions

The contributors of this book come from diverse backgrounds, making this book a truly international effort. This book will bring forth new frontiers with its revolutionizing research information and detailed analysis of the nascent developments around the world.

We would like to thank all the contributing authors for lending their expertise to make the book truly unique. They have played a crucial role in the development of this book. Without their invaluable contributions this book wouldn't have been possible. They have made vital efforts to compile up to date information on the varied aspects of this subject to make this book a valuable addition to the collection of many professionals and students.

This book was conceptualized with the vision of imparting up-to-date information and advanced data in this field. To ensure the same, a matchless editorial board was set up. Every individual on the board went through rigorous rounds of assessment to prove their worth. After which they invested a large part of their time researching and compiling the most relevant data for our readers. Conferences and sessions were held from time to time between the editorial board and the contributing authors to present the data in the most comprehensible form. The editorial team has worked tirelessly to provide valuable and valid information to help people across the globe.

Every chapter published in this book has been scrutinized by our experts. Their significance has been extensively debated. The topics covered herein carry significant findings which will fuel the growth of the discipline. They may even be implemented as practical applications or may be referred to as a beginning point for another development.

The editorial board has been involved in producing this book since its inception. They have spent rigorous hours researching and exploring the diverse topics which have resulted in the successful publishing of this book. They have passed on their knowledge of decades through this book. To expedite this challenging task, the publisher supported the team at every step. A small team of assistant editors was also appointed to further simplify the editing procedure and attain best results for the readers.

Our editorial team has been hand-picked from every corner of the world. Their multi-ethnicity adds dynamic inputs to the discussions which result in innovative outcomes. These outcomes are then further discussed with the researchers and contributors who give their valuable feedback and opinion regarding the same. The feedback is then collaborated with the researches and they are edited in a comprehensive manner to aid the understanding of the subject.

Apart from the editorial board, the designing team has also invested a significant amount of their time in understanding the subject and creating the most relevant covers. They scrutinized every image to scout for the most suitable representation of the subject and create an appropriate cover for the book.

The publishing team has been involved in this book since its early stages. They were actively engaged in every process, be it collecting the data, connecting with the contributors or procuring relevant information. The team has been an ardent support to the editorial, designing and production team. Their endless efforts to recruit the best for this project, has resulted in the accomplishment of this book. They are a veteran in the field of academics and their pool of knowledge is as vast as their experience in printing. Their expertise and guidance has proved useful at every step. Their uncompromising quality standards have made this book an exceptional effort. Their encouragement from time to time has been an inspiration for everyone.

The publisher and the editorial board hope that this book will prove to be a valuable piece of knowledge for researchers, students, practitioners and scholars across the globe.

List of Contributors

Zongkuan Liu, Lei Zhang, Jian Jiang, Cheng Bian, Zichao Zhang and Zifeng Gao
Chemical Engineering Department, School of Chemical Engineering, Xi'an Jiaotong University, Xi'an, China

Hamza M. Habib
Department of Mathematics, College of Science, Salman Bin Abdualziz University, Al-Kharj, KSA

G. Ibrahim and Osama Noman Saleh
Department of Basic Engineering Science, Faculty of Engineering, Menoufia University, Shebin El-Kom, Egypt

Abdellah Eid, Ouarda Benlounes and Smain Hocine
Laboratoire de Chimie Appliquée et de Genie Chimique, Universite Mouloud Mammeri, Tizi-Ouzou, Algeria

Hikmat S. Hilal
Laboratory of Semiconductor and Solar Energy Research, Chemistry Department, An-Najah N. University, Nablus, Palestine

Chérifa Rabia
Laboratoire de Chimie du Gaz Naturel, Institut de Chimie USTHB, Alger, Algeria

Atheel Hassan Alwash
School of Chemical Engineering, Universiti Sains Malaysia, Penang, Malaysia
Department of Chemistry, College of Science, Al-Nahrain University, Baghdad, Iraq

Ahmad Zuhairi Abdullah
School of Chemical Engineering, Universiti Sains Malaysia, Penang, Malaysia

Norli Ismail
School of Industrial Technology, Universiti Sains Malaysia, Penang, Malaysia

Amadou Kouyate
Laboratoire de Chimie-Physique, UFR-SSMT, Université de Cocody, Abidjan, Côte d'Ivoire

Faculté des Sciences et Techniques, Campus de Saint-Jérôme, Aix-Marseille Université, Marseille, France

Ange Privat Ahoussou
Laboratoire de Chimie-Physique, UFR-SSMT, Université de Cocody, Abidjan, Côte d'Ivoire

Jacques Rogez and Pierre Benigni
Faculté des Sciences et Techniques, Campus de Saint-Jérôme, Aix-Marseille Université, Marseille, France

José R. Parga
Institute Technology of Saltillo, Saltillo, Mexico

Guillermo Tiburcio Munive, Jesús L. Valenzuela and Víctor V. Vazquez
Departament of Chemistry and Metallurgy, University of Sonora, Hermosillo, Mexico

Gregorio González Zamarripa
Faculty of Metallurgy, University of Coahuila, Monclova Coah, México

Lekan Taofeek Popoola
Department of Petroleum and Chemical Engineering, Afe Babalola University, Ado-Ekiti, Nigeria

Gutti Babagana
Department of Chemical Engineering, University of Maiduguri, Maiduguri, Nigeria

Alfred Akpoveta Susu
Department of Chemical Engineering, University of Lagos, Lagos, Nigeria

Sandy Eap and Alice Ferrand
INSERM, Institut National de la Santé et de la Recherche Médicale, "Osteoarticular and Dental Regenerative Nanomedicine" Team, Strasbourg, France

Valérie Machi, Olivier Huck and Nadia Benkirane-Jessel
INSERM, Institut National de la Santé et de la Recherche Médicale, "Osteoarticular and Dental Regenerative Nanomedicine" Team, Strasbourg, France
Université de Strasbourg, Faculté de Chirurgie Dentaire, Strasbourg, France

Vincent Ball
Université de Strasbourg, Faculté de Chirurgie Dentaire, Strasbourg, France

Reza Rezazadeh
Shahid Hasheminejad Gas Processing Company (S.G.P.C), Khangiran, Iran

Sima Rezvantalab
Department of Chemical Engineering, Urmia University of Technology, Urmia, Iran

Morgana de Vasconcellos Araújo and Severino Rodrigues de Farias Neto
Department of Chemical Engineering, Center of Science and Technology, Federal University of Campina Grande (UFCG), Campina Grande, Brazil

Antonio Gilson Barbosa de Lima
Department of Mechanical Engineering, Center of Science and Technology, Federal University of Campina Grande (UFCG), Campina Grande, Brazil

Tomomi Uchiyama
EcoTopia Science Institute, Nagoya University, Nagoya, Japan

Sou Kusamichi
Bridgestone Cycle, Ageo, Japan

Dong-Heui Kwak and Mi-Sug Kim
Department of Environmental and Chemical Engineering, Seonam University, Namwon, Republic of Korea

Jae-Seung Kim
Department of Environmental Engineering, Chonbuk National University, Jeonju, Republic of Korea

Young-Youl Oh
Yoyo Korea Agricultural Association, Jeongehb, Republic of Korea

Soon-Ok Noh, Byung-Ok So, Su-Young Jung, Su-Jin Jung and Soo-Wan Chae
Clinical Trial Center for Functional Foods, Chonbuk National University Hospital, Jeonju, Republic of Korea
Department of Medical Nutrition Therapy, Chonbuk National University, Medical School, Jeonju, Republic of Korea

Huaxiang Lin, Luyao Ding, Weihua Deng, Xuxu Wang and Jinlin Long
Research Institute of Photocatalysis, Fujian Provincial Key Laboratory of Photocatalysis, State Key Laboratory Breeding Base, Fuzhou University, Fuzhou, China

Qun Lin
First Affiliated Hospital, Fujian Medical University, Fuzhou, China

Ghazi AlQahtani, Ahmed Alzahabi, Ernee Kozyreff, Ismael R. de Farias Jr. and Mohamed Soliman
Texas Tech University, Lubbock, USA

Antonio José Ferreira Gadelha, Severino Rodrigues de Farias Neto and Ramdayal Swarnakar
Department of Chemical Engineering, Center of Science and Technology, Federal University of Campina Grande, Campina Grande, Brasil

Antonio Gilson Barbosa de Lima
Department of Mechanical Engineering, Center of Science and Technology, Federal University of Campina Grande, Campina Grande, Brasil

Akkarat Wongkaew
Department of Chemical Engineering, Faculty of Engineering, Burapha University, Chonburi, Thailand

Pichet Limsuwan
Department of Physics, Faculty of Science, King Mongkut's University of Technology Thonburi, Bangkok, Thailand

Loubna Nasri
Department of Pharmaceutical Engineering, University Constantine 3, Constantine, Algeria

Salima Bensaad
Department of Chemistry, University Constantine 1, Constantine, Algeria

Zouhir Bensetiti
Unisignal Inc., Brossard, Canada

Nasr-Eddine Chakri
LASEA, Department of Physic, University of Badji Mokhtar, Annaba, Algeria
LASEA, Department of Chemistry, University of Badji Mokhtar, Annaba, Algeria

Badis Bendjemil
LASEA, Department of Chemistry, University of Badji Mokhtar, Annaba, Algeria
University of Guelma, Guelma, Algeria

M. Baricco
Università di Torino, Torino, Italy

Ángela B. Sifontes, Mirla Rodriguez, David Freire, Wendy Rondón, Ligia Llovera, Franklin J. Méndez, Andrea Monaco and Yraida Díaz
Centro de Química, Instituto Venezolano de Investigaciones Científicas, Caracas, Venezuela

Edgar Cañizales
PDVSA, Intevep, Los Teques, Venezuela

Jesús Alirio Carrillo and Laura Milena Corredor
Ecopetrol S. A: Refining and Petrochemical, Instituto Colombiano del Petróleo, Bucaramanga, Colombia

Mohammed Salah Hameed
Chemical Engineering Department, Higher Colleges of Technology, Abu Dhabi, United Arab Emirates

Abdul Rahman Khan
Department of Chemical Engineering, Kuwait University, Kuwait City, Kuwait

A. A. Mahdi
Mechanical Engineering Department, University of Technology, Baghdad, Iraq